Molecular Basis of Chromatographic Separation

Esther Forgács
Tibor Cserháti
Central Research Institute for Chemistry
Hungarian Academy of Sciences
Budapest, Hungary

CRC Press
Boca Raton   New York

Library of Congress Cataloging-in-Publication Data

Catalog record is available from the Library of Congress.

This book contains information obtained from authentic and highly regarded sources. Reprinted material is quoted with permission, and sources are indicated. A wide variety of references are listed. Reasonable efforts have been made to publish reliable data and information, but the author and the publisher cannot assume responsibility for the validity of all materials or for the consequences of their use.

Neither this book nor any part may be reproduced or transmitted in any form or by any means, electronic or mechanical, including photocopying, microfilming, and recording, or by any information storage or retrieval system, without prior permission in writing from the publisher.

CRC Press LLC's consent does not extend to copying for general distribution, for promotion, for creating new works, or for resale. Specific permission must be obtained in writing from CRC Press LLC for such copying.

Direct all inquiries to CRC Press LLC, 2000 Corporate Blvd., N.W., Boca Raton, Florida 33431.

© 1997 by CRC Press LLC

No claim to original U.S. Government works
International Standard Book Number 0-8493-7696-3
Printed in the United States of America 1 2 3 4 5 6 7 8 9 0
Printed on acid-free paper

AUTHORS

Esther Forgács was born in Budapest, Hungary in 1957. After she graduated as a pharmacist from the Semmelweis University of Medicine, Budapest in 1981, she went on to obtain her M.D. degree at the Eötvös Lóránd University in 1992. She was awarded the Ph.D. (Candidate) degree in 1993. She has been employed by the chromatographic group of the Central Research Institute for Chemistry of the Hungarian Academy of Sciences since 1989.

Dr. Forgács is the author or co-author of more than 80 papers dealing with the various aspects of chromatographic separation and data evaluation techniques. Her main fields of interest are the possible applications of new high-performance liquid chromatographic supports and the use of mathematical methods for comparison of the retention characteristics of supports.

Tibor Cserháti was born in Hungary in 1938. He graduated from the Chemical Engineering Faculty of the Technical University of Budapest in 1962 and earned his M.D. degree at the same university in 1968. He obtained his Ph.D. (Candidate) degree in 1973. Tibor Cserháti has been a research fellow at the Dairy Research Institute (1965–1973) and a senior research scientist at the Plant Protection Institute of the Hungarian Academy of Sciences (1975–1988). Since 1989 he has been a scientific counsellor at the Central Research Institute for Chemistry of the Hungarian Academy of Sciences. He was awarded the D.Sc. degree by the Hungarian Academy of Sciences in 1988.

Dr. Cserháti is the author or co-author of more than 320 scientific papers. The main fields of interest of Dr. Cserháti are the various chromatographic techniques such as thin-layer, gas-liquid, and high-performance liquid chromatography, and the application of various multivariate mathematical-statistical methods for the elucidation of the relationship between retention behavior and molecular parameters and chemical structure of solutes.

Preface

Chromatography is one of the most rapidly developing fields of separation sciences. Many excellent books and reviews have been published on the various theoretical and practical problems of chromatography; however, the exact role of molecular structural characteristics on chromatographic retention and separation has not been determined. The profound knowledge of the molecular basis of chromatographic separation processes will facilitate the efficient use of chromatography. Researchers will be able to intelligently select the chromatographic method most suitable to a particular type of separation problem and will be able to take full advantage of the possibilities that chromatography offers.

The objectives of the present monograph are the compilation and concise evaluation of the newest results in this rapidly developing domain of chromatography, brief enumeration of the methods applied, critical discussion of the results, and elucidation of the impact of molecular structure on the optimization of a wide range of chromatographic separations. The book is meant to be sufficient in terms of the needs of the average professional who intends to work in this interesting field. We are confident that the book will serve as a valuable reference for researchers and serious students interested in the topics covered. We hope that the book will have some impact on the development and growth of the study of various aspects of the relationship between retention capacity and molecular structure of various solutes. Elucidation of this relationship will promote the successful application of various chromatographic techniques in biotechnology, environmental protection, forensic sciences, etc.

We are grateful to Zsuzsa Bárányos and Viola Sándor-Orosz for their valuable technical assistance.

Esther Forgács
Tibor Cserháti

Dedication

To our children
András and Gergely Forgács and Éva Cserháti

Contents

Chapter 1
Adsorption Phenomena and Molecular Interactions in Chromatography.................3

  I. Adsorption Phenomena in Gas–Liquid Chromatography 3
 II. Adsorption Phenomena in Liquid Chromatography 5
     A. Competitive Adsorption of the Eluent Components 5
     B. Competitive Adsorption of Analytes 7
     C. Molecular Forces in Chromatography 10
        1. Ion–Ion Interactions..10
        2. Ion–Permanent Dipole Interactions 10
        3. Ion–Induced Dipole Interaction 10
        4. Hydrogen Bonds 11
        5. van der Waals Interactions 11
           a. Permanent Dipole–Permanent Dipole Interactions 11
           b. Permanent Dipole-Induced Dipole Interactions 11
           c. Induced Dipole–Induced Dipole Interactions 11
        6. Hydrophobic Interactions 13

Chapter 2
Gas Chromatography 15

  I. Fundamentals 15
 II. Models of GLC Retention and Separation 17
III. Influence of Adsorption Side Effects on Retention Behavior....................24
 IV. Descriptors of GLC Retention and Their Correlation with Molecular
     Parameters 32
     A. Retention Time 32
     B. Capacity Factor 37
     C. The Kováts Index 47
        1. Isocratic Separation Mode 47
           a. Relationship with Physicochemical Parameters 47
           b. Relationship with Geometrical and Structural
              Characteristics..54
        2. Temperature Programming 59
     D. The Takács Retention Index 68
     E. Other Descriptors 73
  V. Application of Solute Parameters Retention Relationships 79

Chapter 3
Liquid Chromatography..91

  I. Thin-Layer Chromatography..92
     A. Principles of Adsorption and Reversed-Phase
        Thin-Layer Chromatography 92

		B.	Molecular Basis of Separation in Adsorption Thin-Layer Chromatography 94
		C.	Molecular Basis of Separation in Reversed-Phase Thin-layer Chromatography 106
II.	High-Performance Liquid Chromatography 116		
	A.	Fundamentals 116	
		1.	Adsorption High-Performance Liquid Chromatography 116
		2.	Reversed-Phase High-Performance Liquid Chromatography 118
		3.	Retention Determination 118
		4.	Development of Column 120
	B.	Direct (Normal)-Phase High-Performance Liquid Chromatography 120	
		1.	Silica Supports 120
			a. Chemical Methods 121
			b. Isotopic Exchange Methods 121
			c. Spectroscopic Methods 121
		2.	Chemical Properties of Silica 122
			a. Acid–Base Properties 122
			b. Silanol Acidity 122
			c. Water Adsorption and Desorption 123
			d. Water Solubility 123
			e. Particle Size 123
			f. Surface Area 124
			g. Porosity 124
			h. Relationship between Surface Area and Porosity 125
			i. Surface Silanols and Polarity 125
			j. Carbon Content and Bonding Density 126
			k. Chemical and Physical Chromatographic Requirements 127
			l. Application of Normal-Phase Chromatography 127
			m. Silanophilic Interaction 128
			n. Reduction of the Effects of Silanol Groups 129
			o. Ionic Interactions 129
			p. Reversed-Phase Character of Silica 132
	C.	Reversed-Phase Chromatography 132	
		1.	Solvent-Stationary Phase Interactions in Reversed-Phase Liquid Chromatography 133
			a. Molecular Interactions and Retention in Liquid Chromatography 134
			b. Stationary Phase Effects in Reversed-Phase Liquid Chromatography 135
		2.	Alkyl-Bonded Silica Phase 137
			a. Effect of Alkyl Chain Length of Bonded Phase on Selectivity 137
			b. Steric Selectivity of Alkyl-Bonded Stationary Phase 138
		3.	Polymer-Based Packing Materials in Reversed-Phase Liquid Chromatography 140
			a. Selectivity of Polymer-Based Packings 140
			b. Steric Selectivity of Polymer Gels 140
			c. Effect of Mobile Phase on Steric Selectivity 142
			d. The Effect of Biporous Structure on Chromatographic Properties 142

    e. The Effect of Polymer Structure on Polar Group Selectivity 142
  4. Correlation of Retention and Selectivity of Separation in Reversed-Phase High-Performance Liquid Chromatography with Interaction Indices and with Lipophilic and Polar Structural Indices 143
    a. Structural Correlations of Lipophilic and Polar Indices 144
  5. Quantitative Structure–Retention Relationship in Reversed-Phase High-Performance Liquid Chromatography 146
  6. Hydrophobicity Concept in Chromatography 150
  7. Correlations between RPHPLC Retention Parameters and 1-Octanol–Water Partition Coefficients 151
  8. Correlation Analysis in Liquid Chromatography of Metal Chelates 152
  9. Validation of Chromatographic Retention Models in Reversed-Phase High-Performance Liquid Chromatography 155
  10. Physicochemical Meaning of QSRR Equations 155
  11. The S-Index in the Retention Equation in Reversed-Phase High-Performance Liquid Chromatography 158
  12. Functional Group Contributions to the Retention of Analytes in Reversed-Phase High-Performance Liquid Chromatography 164
  13. Application of Various Physicochemical Parameters for the Evaluation of Chromatographic Retention Data in RPHPLC 166
  14. Investigation and Characterization of RPHPLC Phases 177
D. Graphitized Carbon Packing Materials 182
  1. Hydrophobic Properties of Carbon Packing 182
  2. Steric Selectivity of Carbon Packing 182
  3. Retention Strength of Porous Graphitized Carbon Column 183
  4. Eluotropic Strength of Organic Solvents on PGC Column 184
  5. Retention Behavior of the PGC Column 185
  6. Application of PGC Column 186
E. Other Metal Oxide-Based Supports 188
  1. Chemistry of Zirconia 188
  2. Purity of Zirconia 191
  3. Chemical Stability of Zirconia 191
  4. Application of Zirconia in Liquid Chromatography 192
    a. Dynamic Chemical Modification 192
    b. Permanent Chemical Modification 193
    c. Physical Screening 193
  5. Aluminum Oxide-Based Liquid Chromatography Support 195
F. New RPHPLC Stationary Phase Materials 204
  1. Donor–Acceptor Type Bonded Stationary Phase 204
  2. Determination of Hydrophobicity with New RPHPLC Phases 204
  3. β-Cyclodextrin-Coated Silica Column 206

Appendix 211

References 215

Index 235

Introduction

Chromatography is a frequently used analytical technique for the separation and quantitative determination of various compounds such as pharmaceuticals, pesticides and other environmental pollutants, food ingredients and additives, toxic substances, drugs, etc. Chromatography has been developed as a powerful and rapid technique with the unique objective of separating compounds with highly similar molecular characteristics even from complicated matrices. The rapid progress of various chromatographic methods and the increasing range of applications quickly revealed that chromatography is more than a very effective separation method based on empirical or semiempirical principles. Chromatographic techniques have reached the point where the average chromatographer can achieve standard spectacular results using empirical rules and experiences. However, it is more and more obvious that successful separation of various solutes (selection of the best separation technique, stationary and mobile phase composition, etc.) requires a profound knowledge of the effects of molecular parameters of solutes, supports, and mobile phases and their possible interactions on retention. The rational design of an optimal separation process involves the expert application of such knowledge. Over the past few years major advances have been made not only in practice, but also in the theory of chromatography. Many excellent papers have been published on the influence of molecular structure on the retention behavior of solutes using both theoretical and practical approximations. In this study the influence of molecular structure on the retention and separation capacity is discussed and a thorough study of the special field of research is offered.

Chapter 1

ADSORPTION PHENOMENA AND MOLECULAR INTERACTIONS IN CHROMATOGRAPHY

Chromatographic separations have a common principle, the distribution of analytes to be separated between stationary and mobile phases. The mobile phase may be a liquid or a mixture of liquids, a gas, or a supercritical fluid with the characteristics of gas and liquid depending on the chromatographic conditions. A huge variety of stationary phases such as inert or adsorptive solids, liquids bonded by adsorptive forces to solids or to the wall of columns, chemically modified solids, etc. can be used for the separation of analytes. The differences in the capacities of analytes to bind to the stationary phase result in different migration velocities and in the separation of the analytes. Many excellent papers[1-7] and books[8-9] have been published on the various aspects of the theoretical principles governing chromatographic separation.

I. ADSORPTION PHENOMENA IN GAS–LIQUID CHROMATOGRAPHY

The retention in GLC is mainly determined by the partition of the analyte between the mobile and stationary phases. However, it was recognized long ago that various adsorption phenomena can considerably influence the retention. Thus the analyte can adsorb on the surface of the liquid stationary phase, on the interface between the liquid stationary phase and its support, on the surface of the support not covered by the liquid, and on the not entirely inert wall of the GC column.[10,11] The adsorption may modify the retention of the analyte and can increase or decrease the efficiency of the chromatographic separation process. Taking into consideration the various adsorption processes the net retention volume (V_N) can be expressed by

$$V_N = V_L \cdot K_L + A_I \cdot K_I + A_A \cdot K_A + A_S \cdot K_S \tag{1}$$

where A_I, A_A, and A_S are the adsorbing surface areas of the gas–liquid surface, the gas–solid surface, and the liquid–solid surface, respectively; the K_i values are the corresponding equilibrium constants.

The total number of the analyte (N_I) at the gas–liquid interface is defined by

$$N_I = b \cdot C_L + (a - b) \cdot C_G + a \cdot C_I \tag{2}$$

where C_L, C_G, and C_I are the concentrations of the analyte in the liquid and gas phases and on the gas–liquid surface, respectively. The concentration of the analyte at the gas–liquid surface is

$$C_I = (d\sigma/da) \cdot - a/RT \qquad (3)$$

where σ is the surface tension of the liquid phase, and a is the activity coefficient of the analyte in the solution. The K_I value can be calculated by

$$K_I = [e^{(\sigma_1 - \sigma_2) \cdot A_2^0 RT} - 1] \cdot V_1^0 \cdot K_L / A_1^0 \qquad (4)$$

where A_1^0 and A_2^0 are the molar areas of the solvent and analyte, respectively, and σ_1 and σ_2 are the corresponding surface tensions. The adsorption of analytes on the gas–solid surface has been discussed in more detail in References 12 to 16. The adsorption of the analytes on the active centers of the support not covered by the liquid stationary phase can be described by the Langmuir isotherm:

$$\Theta = b \cdot p / (1 + b \cdot p) \qquad (5)$$

where Θ and p are the fraction of the surface covered by the analyte and its pressure, respectively, and b is the adsorption coefficient.

The excess concentration of the adsorbed analyte C_A is equal to

$$C_A = A_2^0 \cdot b \cdot C_G \cdot RT / (1 + b \cdot C_G \cdot RT) \qquad (6)$$

The retention increment ($V_{N,A}$) caused by the adsorption of the analyte on the gas–solid surface can be expressed by

$$V_{N,A} = A_A \cdot \delta C_A / \delta C_G = A_A \cdot b_1 / [A_2^0 (1 + b_1 \cdot C_G)^2] \qquad (7)$$

or in the case of infinite dilution:

$$V_{N,A} = A_A \cdot b_1 / A_2^0 = A_A \cdot K_A \qquad (8)$$

The equations discussed previously are valid only when the adsorption energy is the same for each adsorption center on the support surface, the absorbed analytes do not interact with each other, and they form a monolayer. However, the chromatographic supports generally do not comply with the requirements listed above. Taking into consideration the heterogeneity of the adsorptive centers, the fraction of the surface covered by the analyte is expressed by

$$\Theta_T = b \cdot p / [1 + (b \cdot p)^m]^{1/m} \quad 0 < m \leq 1 \qquad (9)$$

where m is considered as a measure of heterogeneity.[17,18]

$$V_{N,A} = A_A \cdot b_1 / A_2^0 [1 + (b_1 \cdot C_G)^m]^{(1 + 1/m)} \qquad (10)$$

The impact of the adsorption of the analyte on the liquid–solid interface can be described by

$$V_{N,S} = A_S \cdot b_2 / [A^0 (1 + b_2 \cdot C_G)^2] \qquad (11)$$

where $V_{N,S}$ is the retention volume increment caused by the adsorption of the analyte on the liquid–solid interface, and

$$b_2 = b_{0'} \cdot e^{(\epsilon_1 - \epsilon_2)/RT} \tag{12}$$

where ϵ_1 and ϵ_2 are the adsorption energies of the analyte and the carrier gas, respectively. The Gibbs equation is also suitable for the description of liquid–solid adsorption:

$$K_S = [e^{(\sigma_1^S - \sigma_2^S) \cdot A_2^0/RT} - 1] \cdot V_1^0 \cdot K_L / A^0 \tag{13}$$

The liquid–solid interface tensions are related to the solvent strength parameter of Snyder:[19,20]

$$\sigma_r^S - \sigma_i^S = e_i^0 \cdot RT \cdot 4 \cdot 50 \cdot 10^{-5} \tag{14}$$

where r refers to a reference solvent.

II. ADSORPTION PHENOMENA IN LIQUID CHROMATOGRAPHY

Due to the presence of multicomponent eluents, the adsorption processes are more complicated in liquid than in gas chromatography. The various components of the eluent mixture are competitively adsorbed on the surface of the stationary phase. Similar competition can occur between the analytes and the impurities of the sample.

A. Competitive Adsorption of the Eluent Components

In the case of a two-component eluent system the chemical potential (μ) of components 1 and 2 can be described by

$$\mu_{1l} = \mu_{1l}^0 + RT \cdot (1 - r_{12}) \cdot \Phi_{1l} + RT \cdot \ln(\Phi_{1l} \cdot \tau_{1l}) \tag{15}$$

$$\mu_{2l} = \mu_{2l}^0 + RT \cdot (r_{21} - 1) \cdot \Phi_{2l} + RT \cdot \ln(\Phi_{2l} \cdot \tau_{2l}) \tag{16}$$

where μ_{1l}^0 and μ_{2l}^0 are the standard chemical potentials in the surface phase; Φ_{1l} and Φ_{2l} are the volume fractions of eluent components 1 and 2 in the l-th surface layer; τ_{1l} and τ_{2l} are the corresponding activity coefficients; $r_{12} = r_1/r_2$; $r_{21} = r_2/r_1$; and r_1 and r_2 are the number of adsorption sites occupied by eluent components 1 and 2, respectively.

The interaction energy (U_{in}) in the adsorbed phase is composed of three components:

$$U_{in} = U_{in}^{(b)} + U_{in}^{(id)} + U_{in}^{(mix)} \tag{17}$$

$$U_{in}^{(b)} = (\Phi_1 \cdot \epsilon_{11}^{(p)} + \Phi_2 \cdot \epsilon_{22}^{(p)} + 2\alpha_{12} \cdot \Phi_1 \cdot \Phi_2) \cdot N_0 / 2 \tag{18}$$

where N_0 is the number of adsorption sites in a surface layer; $\epsilon_{11}^{(p)}$ and $\epsilon_{22}^{(p)}$ are the interaction energies of the molecular pairs 1–1 and 2–2; α_{12} is the interchange energy:

$$\alpha_{12} = \epsilon_{12}^{(p)} - (\epsilon_{11}^{(p)} + \epsilon_{22}^{(p)})/2 \tag{19}$$

$$U_{in}^{(id)} = (a_p + a_v) \cdot (\Phi_{11} \cdot \epsilon_{11}^{(p)} + \Phi_{21} \cdot \epsilon_{22}^{(p)}) \cdot N_0/2 \sum_{l=2} (\Phi_{11} \cdot \epsilon_{11}^{(p)} + \Phi_{21} \cdot \epsilon_{22}^{(p)}) \cdot z \cdot N_0/2 \quad (20)$$

where a_p and a_v are the numbers of the nearest-neighbors adsorption sites in the same surface layer (p) and above (v), and z is the total number of nearest-neighbors adsorption sites.

$$U_{in}^{(mix)} = \sum_{l=1}^{L-1} N_0 \cdot \alpha_{12} \cdot [a_p \cdot \Phi_{1l} \cdot \Phi_{2l} + a_v \cdot (\Phi_{1l} \cdot \Phi_{2(l+1)} + \Phi_{1(l+1)} \cdot \Phi_{2l})]$$
$$+ N_0 \cdot \alpha_{12} \cdot [a_p \cdot \Phi_{1l} \cdot \Phi_{2l} + a_v \cdot (\Phi_{1l} \cdot \Phi_2 + \Phi_1 \cdot \Phi_{2l} - \Phi_1 \cdot \Phi_2)] \quad (21)$$

The activity coefficients can be calculated from the interaction energies:

$$RT \cdot \ln \tau_{1l} = (\delta U_{in}^{(mix)}/\delta N_{1l})_{N_{2l}, N_{1m}, N_{2m}} \quad (22)$$

$$RT \cdot \ln \tau_{2l} = (\delta U_{in}^{(mix)}/\delta N_{2l})_{N_{1l}, N_{1m}, N_{2m}} \quad (23)$$

where τ_{1l} and τ_{2l} are the activity coefficients of components 1 and 2 in the surface layer.

$$\ln \tau_{1l} = A_{12} \cdot [f_v \cdot \Phi_{2l} \cdot (\Phi_{2(l-1)} - \Phi_{1(l-1)}) + f_p \cdot (\Phi_{2l})^2 + f_v \cdot \Phi_{2(l+1)}] \quad (24)$$

$$\ln \tau_{2l} = r_{21} \cdot A_{12} \cdot [f_v \cdot \Phi_{1l} \cdot (\Phi_{1(l-1)} - \Phi_{2(l-1)}) + f_p \cdot (\Phi_{1l})^2 + f_v \cdot \Phi_{1(l+1)}] \quad (25)$$

where $f_p = a_p/z$ and $f_v = a_v/z$.

The adsorption equation in liquid chromatography is

$$(\Phi_{1l} \cdot \tau_{1l}/\Phi_1 \cdot \tau_1) \cdot (\Phi_2 \cdot \tau_2/\Phi_{2l} \cdot \tau_{2l})^{r12} = K_{12}^{(l)} \quad (26)$$

where $K_{12}^{(l)}$ is the adsorption equilibrium constant in the surface layer:

$$K_{12}^{(l)} = e[(\mu_1^0 - \mu_{1l}^0) - r_{12} \cdot (\mu_2^0 - \mu_{2l}^0)] \quad (27)$$

The difference between the standard chemical potentials in the bulk phase (μ^0) and in the surface layer (μ_l^0) is equal

$$\mu_1^0 - \mu_{1l}^0 = \epsilon_{11} - RT \cdot \ln(q_{1l}/q_1) \quad (28)$$

$$\mu_2^0 - \mu_{2l}^0 = \epsilon_{21} - RT \cdot \ln(q_{2l}/q_2) \quad (29)$$

where ϵ_{11} and ϵ_{21} are the adsorption energies of components 1 and 2 and q_i is their molecular partition functions in the adsorbed layer and in the bulk phase.

Many efforts have been devoted to the practical application of the theoretical considerations discussed previously in modern high-performance liquid chromatography.[21-26]

The equations discussed previously are the results of theoretical considerations. However, in the practice of chromatographic separations the adsorption centers on the support surface are not homologous. The distribution of the energy of adsorption of the adsorption centers have been modeled many times either by the Gaussian or Poisson equations, or by the assumption of the random distribution of adsorption centers.[27]

It has been proved many times that the prediction power of the theoretical distribution functions describing the real situation on the support surface is sometimes unsatisfactory. Many excellent studies deal with the effect of surface inhomogeneity on the retention characteristics of supports both in gas[28] and in liquid chromatography.[29-31]

B. Competitive Adsorption of Analytes

Competition occurs not only among the components of the eluent mixture, but also among the analytes present in the sample to be analyzed. In the case of monolayer adsorption and binary eluent mixture, the partition coefficient K can be expressed by

$$K = x_{s1}/x_s = r_s/K^{(1)}_{1s} \cdot (\Phi_1 + K^{(1)}_{21} \cdot \Phi_2)^{-1} \tag{30}$$

where x_{s1} and x_s are the molar concentrations of s in the surface layer and in the bulk eluent, respectively; r_s is the number of sites occupied by one molecule of s; $K^{(1)}_{1s}$ and $K^{(1)}_{21}$ are the equilibrium constants for the adsorption in the surface layer, and Φ values are the volume fractions.

Capacity factor k is related to the partition coefficient:

$$k = K \cdot V_s/V_m \tag{31}$$

where V_s is the molecular volume of component s in the bulk phase and V_m is the molar volume of the eluent mixture.

When K_1 is the partition coefficient in pure component 1:

$$\log (K_1/K) = \log (\Phi_1 + K^{(1)}_{21} \cdot \Phi_2) \tag{32}$$

$$\log (K_1/K) = a' \cdot \alpha_2 \cdot (\epsilon^0_{12} - \epsilon^0_1) \tag{33}$$

where a' is the adsorbent activity according to Snyder; α_2 is the surface area occupied by one molecule of component 2; ϵ^0_{12} is the solvent strength of a two-component eluent system; and ϵ^0_1 is the eluent strength of component 1.

Under ideal conditions the retention can be described by

$$\epsilon^0_{12} = \epsilon^0_1{}^{a' \cdot \alpha_2 \cdot (\epsilon^0_2 - \epsilon^0_1)} + (a' \cdot \alpha_2)^{-1} \cdot \log (\Phi_1 + \Phi_2 \cdot 10) \tag{34}$$

The nonideality of the chromatographic system has been discussed in more detail in References 32 to 34.

When the surface areas occupied by the analyte and by the components of the eluent the equations describing the retention become very complicated.[35,36]

The partition coefficient K in reversed-phase HPLC can be described by

$$\ln K = \ln K^{(in)} + \ln K^{(comp)} \tag{35}$$

where $\ln K^{(in)}$ is the contribution of the analyte-solvent interaction to the partition, and $\ln K^{(comp)}$ is related to the competitive adsorption of solvents on the support surface.

In the case of ternary mobile phases[37] the activity coefficient of the analyte ($\tau_{s(f)}$) in a given phase f is described by

$$RT \ln \tau_{s(f)} = v_s \cdot (\delta_s - \delta_{(f)})^2 \tag{36}$$

where δ_s and $\delta_{(f)}$ are the solubility parameters for the analyte and for the phase, respectively.

$$\ln K = [(\delta_s - \delta_{(m)})^2 - (\delta_s - \delta_{(s)})^2] \cdot v_s / RT \tag{37}$$

where s and m refer to the surface and the mobile phase, respectively. When

$$\delta_{(m)} = \sum_{i=1}^{3} \Phi_i \cdot \delta_i \quad \text{and} \quad \sum_{i=1}^{3} \Phi_i = 1 \tag{38}$$

the partition coefficient becomes

$$\ln K = \{[\delta_s - \Phi_1 \cdot \delta_1 - \Phi_2 \cdot \delta_2 - (1 - \Phi_1 - \Phi_2) \cdot \delta_3]^2 - (\delta_s - \delta_{(s)})^2\} \cdot v_s / RT \tag{39}$$

In the case of constant $\delta_{(s)}$ value the partition coefficient is

$$\ln K = A_1 \cdot \Phi_1 + B_1 \cdot \Phi_2 + A_2 \cdot \Phi_1^2 + B_2 \cdot \Phi_2^2 + \Phi_1 \cdot \Phi_2 \cdot A_2 \cdot B_2 + C \tag{40}$$

where

$$A_1 = (\delta_s - \delta_3) \cdot (\delta_3 - \delta_1) \cdot 2v_s / RT \tag{41}$$

$$B_1 = (\delta_s - \delta_3) \cdot (\delta_3 - \delta_2) \cdot 2v_s / RT \tag{42}$$

$$A_2 = (\delta_1 - \delta_3)^2 \cdot v_s / RT \tag{43}$$

$$B_2 = (\delta_2 - \delta_3)^2 \cdot v_s / RT \tag{44}$$

$$C = [(\delta_3 - \delta_s)^2 - (\delta_{(s)} - \delta_s)^2 \cdot v_s / RT \tag{45}$$

The change of the standard free energy of the transfer of an analyte molecule from the eluent to the stationary phase (ΔF_s) is given by[38]

$$-\Delta F_s = \Delta F_{c,S} + \Delta F_{m,S} \tag{46}$$

where $\Delta F_{c,S}$ is the free energy change of the creation of a cavity in the eluent to accommodate the analyte molecule, and $\Delta F_{m,S}$ is the energy change of the interaction between the analyte molecule and the eluent.

$$-\Delta F_s = C_m^2 \cdot I_m^2 + C_m \cdot C_s \cdot I_m \cdot I_s \tag{47}$$

where s and m refer to the molecules of the analyte and the eluent; I_i are the interaction indexes; and C_i are correcting coefficients for I_i.

In the case of binary eluent mixture

$$I_m = \Phi_1 \cdot I_1 + \Phi_2 \cdot I_2 \tag{48}$$

$$\ln K = -F_s / RT = c + b_1 \cdot \Phi_2 + b_2 \cdot \Phi_2^2 \tag{49}$$

the constants being equal

$$c = [C_m \cdot I_1 \cdot (C_m \cdot I_1 - C_s \cdot I_s)]/RT \qquad (50)$$

$$b_1 = [C_m \cdot (I_2 - I_1) \cdot (C_s \cdot I_s + 2C_m \cdot I_1)]/RT \qquad (51)$$

$$b_2 = C_m^2 \cdot (I_2 - I_1)^2 / RT \qquad (52)$$

The theory of Horváth et al.[39] supposes the reversible association of analyte S with the apolar ligand L on the support surface

$$S + L \rightleftharpoons SL \qquad (53)$$

and the equilibrium constant K is

$$K = [SL]/[S] \cdot [L] \qquad (54)$$

The free energy of association $\Delta F^0_{assoc.}$ can be expressed by

$$\Delta F^0_{assoc.} = \Delta F_{vdw,assoc.} + (\Delta F_{c,SL} + \Delta F_{m,SL}) - (\Delta F_{c,s} + \Delta F_{m,s})$$
$$- (\Delta F_{c,L} + \Delta F_{m,L}) - RT \cdot \ln RT/P_0 \cdot v_m \qquad (55)$$

The free energy of cavity formation $\Delta F_{c,j}$ of the analyte j can be expressed by

$$\Delta F_{c,j} = k_j^e \cdot \alpha_j^{(v)} \cdot \sigma_m \cdot N \cdot [1 - (1 - k_j^s/k_j^e) \cdot (\delta \ln \sigma_m/\delta \ln T + 2\alpha_j^{(v)} \cdot T/3)] \qquad (56)$$

where $\alpha_j^{(v)}$ is the specific surface area of analyte j; σ_m is the surface tension of the eluent mixture; k_j^e is a correction factor expressing the ratio between the energy required for the cavity formation and the energy for expanding the surface of the eluent by the same area; and k_j^s is the corresponding entropy.

The energy ratio can be calculated by

$$k_j^e = 1 + (k^e - 1) \cdot (v_m/v_j)^{2/3} \qquad (57)$$

where k^e is an energy term related to the heat of vaporization.

The term $\Delta F_{m,j}$ is composed of van der Waals ($\Delta F_{m,j,vdw}$) and electrostatic ($\Delta F_{m,j,es}$) components:

$$\Delta F_{m,j} = \Delta F_{m,j,vdw} + \Delta F_{m,j,es} \quad j = S, L, SL \qquad (58)$$

The term $\Delta F_{m,j,vdw}$ can be calculated from the ionization potentials, refractive indexes, and molecular volumes of the analyte and the eluent, and $\Delta F_{m,j,es}$ can be calculated from the corresponding values of the dielectric constants, dipole moment, and polarizability. Assuming that

$$k_j^s = k_j^e = 1 \; j = S, L, SL \qquad (59)$$

$$\Delta F_{m,SL,vdw} = \Delta F_{m,L,vdw} \qquad (60)$$

$$\mu_{SL} = \mu_S \text{ and } \mu_L = 0 \qquad (61)$$

the partition coefficient can be expressed by

$$\ln K = -\Delta F^0_{assoc.}/RT \tag{62}$$

C. Molecular Forces in Chromatography

The molecular forces observed between various molecules are also involved in the process of chromatographic separation. The forces of interest include coulombic interactions between oppositely charged ions, ion–dipole interactions, hydrogen bonds, and hydrophobic interactions such as van der Waals and London dispersion forces.

1. Ion–Ion Interactions

Two oppositely charged particles (q^+ and q^-) separated by a distance (r) exert an interactive force on each other. The potential energy of interaction ($U_{i\text{-}i}$) is given by

$$U_{i\text{-}i} = q^+ \cdot q^-/4\pi \cdot E_0 \cdot E \cdot r \tag{63}$$

where q^+ and q^- are charges of the positive and negative ions, respectively; E_0 is the permittivity of the free space between the charge ions; E is the dielectric constant.

These type of interactions (frequently called salt-bridge or ion-pair formation) can occur between the positively and negatively charged substructures of the analytes and the stationary phase or eluent additives.

2. Ion–Permanent Dipole Interactions

Ion–permanent dipole interactions occur when one of the molecules is an ion and the other a dipole. (Dipoles are molecules that carry no net charge yet have permanent charge separation due to the nature of the electronic distribution within the molecule itself. Vector addition of the electrical forces derived from its partial charge separation leads to the creation of an electric dipole.) The charge asymmetry in the molecule is given a magnitude and direction by calculating the dipole moment (μ). The dipole moment is calculated by multiplying the separate charge with the distance of separation. The existence of a dipole moment requires an asymmetric molecule. The electric dipole is treated like any other charge, although the orientation of the dipoles ($\cos\Theta$) must be taken into consideration in calculating the ion–permanent dipole interactive force ($U_{i\text{-}pd}$):

$$U_{i\text{-}pd} = q \cdot \mu \cdot \cos\Theta/4\pi \cdot E_0 \cdot E \cdot r^2 \tag{64}$$

where q is the charge of the ion species (positive or negative); μ is the permanent dipole moment of the interacting neutral molecule; $\cos\Theta$ is the orientation of the permanent dipole.

These type of interactions can easily occur in adsorption liquid chromatography between the ionizable analyte and the asymmetric molecules of the mobile phase.

3. Ion–Induced Dipole Interaction

Even molecules that have no net dipole moment can have a transient dipole induced in them when they are brought into an electric field. Electronic interactions of this type are called ion–induced dipole interactions and the potential of the interaction ($U_{i\text{-}id}$) depends on the ability of the neutral molecule to be induced into a dipole, that is, on its polarizability (α):

$$U_{i\text{-}id} = q^2 \cdot a / 8\pi \cdot E_0 \cdot E^2 \cdot r^4 \qquad (65)$$

where α is the polarizability of the neutral molecule.

This interaction requires a molecule without permanent dipole moment. As the majority of analytes and eluent components have a considerable permanent dipole moment the importance of this type of interaction in chromatography is negligible.

4. Hydrogen Bonds

Hydrogen bonds are formed by hydrogen atoms located between two electronegative atoms. When a hydrogen atom is covalently attached to a highly electronegative atom such as an oxygen or a nitrogen, it takes a partial positive charge due to the strong electronegativity of the oxygen or nitrogen atoms. This interaction is very common in chromatography. Hydrogen bonds have been many times established between the polar analytes and the hydrophilic adsorption centers of various supports.

5. van der Waals Interactions

When two molecules are in very close proximity, attractive forces called van der Waals interactions occur, which are cohesive forces that vary with respect to distance as $1/r^6$. The van der Waals interactions are generally subdivided into three types, all derivable from electrostatic considerations. They include permanent dipole–permanent dipole, permanent dipole–induced dipole, and induced dipole–induced dipole interactions.

a. Permanent Dipole–Permanent Dipole Interactions

These interactions occur when molecules with permanent dipole interact. In these cases the orienting forces acting to align the dipoles will be countered by randomizing forces of thermal origin. The calculation of the energy of interaction ($U_{pd\text{-}pd}$) includes the terms temperature and distance, indicating that permanent dipole–permanent dipole interactions are sensitive to both temperature (T) and the distance (r) between the interacting molecules:

$$U_{pd\text{-}pd} = -2\mu_1^2 \cdot \mu_2^2 / 3k \cdot T \cdot (4\pi \cdot E_0)^2 \cdot r^6 \qquad (66)$$

where μ_1 and μ_2 are permanent dipole moments of the interacting neutral molecules 1 and 2.

b. Permanent Dipole–Induced Dipole Interactions

Molecules with permanent dipole moment can induce dipole moment in a neutral molecule in a fashion similar to that discussed above for ion–induced dipole interaction. This interaction depends on the polarizability of the neutral molecule, but it is not sensitive to the thermal randomizing forces. The interaction energy ($U_{pd\text{-}id}$) can be calculated by

$$U_{pd\text{-}id} = -2\mu^2 \cdot \alpha / 16\pi^2 E_0 \cdot r^6 \qquad (67)$$

c. Induced Dipole–Induced Dipole Interactions

Induced dipole moment is a time-averaged value. If a snapshot were taken of any neutral molecule at an instant of time, there would be a variation in the distribution of the electrons in the molecule. At this instant, the "neutral" molecule would have a dipole moment. This instantaneous dipole is capable of inducing an instantaneous dipole in other neutral molecule: thus are born induced dipole–induced dipole forces. These forces are also called London or dispersion forces. Dispersion forces fall off very rapidly with distance but can be quite significant for molecules in close proximity. The deformability of the electron clouds, as reflected by α, is obviously important in these interactions. The rigorous calculation for dispersion forces ($U_{id\text{-}id}$) is quite complicated, but an adequate approximation can be used for the calculation:

$$U_{id\text{-}id} = -f(I) \cdot \alpha_1 \cdot \alpha_2 / 16\pi^2 \cdot E_0 \cdot r^6 \tag{68}$$

where α_1 and α_2 are polarizabilities of the neutral molecules 1 and 2; $f(I)$ is a function of the ionization energies (I) of the two molecules:

$$f(I) = 3I_1 \cdot I_2 / 2(I_1 + I_2) \tag{69}$$

where I_1 and I_2 are ionization energies of the interacting neutral molecules 1 and 2.

When the interaction takes place between two identical molecular species having the same polarizability, Eq. 6 can be written as:

$$U_{id\text{-}id} = -f(I) \cdot a^2 / 16\pi^2 \cdot E_0 \cdot r^6 \tag{70}$$

The overall interactive energy for van der Waals forces (U_{vdW}) can be expressed in terms of electrostatically derived interactions defined by the equations listed above:

$$U_{vdW} = -2\mu_1^2 \cdot \mu_2^2 / 3k \cdot T \cdot (4\pi \cdot E_0)^2 \cdot r^6 - 2\mu^2 \cdot \alpha / 16\pi^2 E_0 \cdot r^6$$

$$-f(I) \cdot \alpha_1 \cdot \alpha_2 / 16\pi^2 \cdot E_0 \cdot r^6 \tag{71}$$

Because the above equation is fairly complicated for practical purposes and sometimes the parameters cannot be defined accurately, the attractive interactional energy (U_{vdW}) represented by the van der Waals forces is simplified in the form:

$$U_{vdW} = -A/r^6 \tag{72}$$

where A is a constant that is different for each molecule.

The attractive forces increases rapidly as molecules get closer until the molecules actually contact each other, at which point the interactional energy goes instantly to infinity. However, before the molecules try to occupy the same space, so-called repulsion forces occur between the electron clouds of the molecules as they approach each other. This electron repulsion can be added to the formula for attractive interaction energy (Lennard–Jones potential):

$$U_{vdW} = -A/r^6 + B/r^{12} \tag{73}$$

where A and B are constants that are different for each molecule.

Many results indicate that the van der Waals interactions also play a considerable role in many fields of chromatographic separations. As the measurement of the individual interacting forces discussed above is extremely difficult for such complicated molecular assemblies as analytes–eluent components–surface of stationary phase, little attention has been devoted to the exact determination of this type of interaction.

6. Hydrophobic Interactions

The importance of hydrophobic interactions has been well recognized. Hydrophobic interaction is an attraction in aqueous medium between apolar or amphipathic molecules by which some of the more hydrophobic sites, upon approaching one other, preferentially interact with each other, frequently by van der Waals interactions, which tend to be attractive between low surface energy materials immersed in high-energy solvents such as water. As discussed above, van der Waals forces include a variety of component forces such as permanent dipole–permanent dipole, permanent dipole–induced dipole, and induced dipole–induced dipole interactions. The extent of involvement of these component forces in the chromatographic separation process is obviously different depending on the character of the interacting molecular species.

Chapter 2

GAS CHROMATOGRAPHY

I. FUNDAMENTALS

The term *gas chromatography* summarizes chromatographic methods, when the mobile phase is gas and the stationary phase is solid or liquid (gas–solid chromatography [GSC] or gas–liquid chromatography [GLC]). As GLC has been proved to be superior to GSC in the solution of the majority of analytical problems only the principles of GLC is discussed in this chapter. The common gas chromatographic equipment consists of a carrier gas system, injector, gas chromatographic column, detector, and data processing unit. The carrier gas is preferably a permanent gas with low or negligible adsorption capacity. The nature of carrier gas may influence the separation characteristics of the GLC systems and can modify the sensitivity of the detection. Use of hydrogen as a carrier gas is hampered by its high flammability and by the strict safety regulations concerning its application. Helium is an ideal choice, but the elevated price of helium considerably reduces its use. Nitrogen is the most frequently used carrier gas in common laboratory practice; it is not expensive and many separations can be successfully carried out with nitrogen carrier gas. Some other gases have also been used for the solution of special separation problems or for the elucidation of theoretical principles. The stability and reproducibility of the carrier gas flow rate is a prerequisite of a successful gas chromatographic analysis. Many different injector types have been developed. The main objective of the injectors is delivery of the vaporized sample to the head of the GC column with a minimum initial bandwidth. The gas chromatographic column has to be thermostated at constant temperature (isocratic separation mode) or according to a predetermined temperature program (temperature gradient). As the column temperature is one of the most decisive parameters in GC analysis its exact regulation is of paramount importance. The stationary liquid phase of GC columns has to comply with the following requirements: low vapor pressure, high chemical stability, and relatively low viscosity at the temperature of analysis; selectivity for the sample components under investigation; good wetting capacity both for the surface of the inert support or for the possibly inert wall of the column. Solid supports wetted by the stationary liquid phase are characterized by low adsorption capacity and by large surface area. Inert supports improve peak symmetry and do not interfere with the partition of analytes between the gas and the liquid stationary phases. Both the absence of adsorption centers on the support surface and the large surface area contribute to the even distribution of the wetting agent, which is a prerequisite of high separation performance. Open tubular capillary columns with deactivated capillary wall avoid the problems arising from the remaining adsorption capacity of solid support. Many different detectors and data handling devices have been developed for the sensitive and selective detection and quantitation of sample components. The newest developments are presented in the prospectuses of the great firms producing gas chromatographs.

The retention volume V_R can be defined as the volume of carrier gas needed to elute an analyte from the GLC column. The gas holdup V_M is equal the volume of carrier gas needed to elute an analyte that does not adsorb on the stationary phase. The adjusted retention volume V'_R is

$$V'_R = V_R - V_M \tag{74}$$

The retention volumes of the equation above are measured at the outlet of the column. As the gas volume depends on the pressure drop in the column, the retention volumes have to be corrected by the factor j:

$$V_R^0 = j \cdot F \cdot t_R \tag{75}$$

where F is the flow rate of the carrier gas, and t_R is the retention time (average time needed for an analyte molecule to travel through the column).

$$V_M^0 = j \cdot F \cdot t_M \tag{76}$$

where t_M is the time for a nonretained analyte molecule to travel through the column.

The net retention volume V_N extensively used in theoretical studies (determination of activity coefficients, adsorption coefficients, etc.) can be expressed as

$$V_N = j \cdot V'_R = V_R^0 - V_M^0 = j \cdot (V_R - V_M) \tag{77}$$

The retention index method introduced by Kováts[40] highly increased the reproducibility and reliability of the determination of the retention characteristics. The basis of the method is the comparison of the retention time of a given analyte with the retention times of normal hydrocarbons eluting before and after the analyte. The retention index I can be calculated by the following equation:

$$I = 100 \cdot c + 100 \cdot (\log V_{Nx} - \log V_{Nc}) / (\log V_{Nc+1} - \log V_{Nc}) \tag{78}$$

where subscript x refers to the analyte, c refers to the number of carbon atoms of the *n*-hydrocarbon eluting before the analyte, and c + 1 refers to the number of carbon atoms on the *n*-hydrocarbon eluting after the analyte. The various problems arising from the use of retention indices in temperature programmed GLC have been discussed in more detail in References 41 and 42.

It is generally accepted that the basic process governing the retention of an analyte in GLC is its partition between the mobile gas phase and the stationary liquid phase. The gas–liquid equilibrium constant K defined this partition:

$$K = V_N / V_L \tag{79}$$

where V_L is the volume of the liquid stationary phase. A more profound theoretical treatment and practical applications of GLC can be found in References 43 to 45.

II. MODELS OF GLC RETENTION AND SEPARATION

Much effort has been devoted for the elucidation of the influence of various physicochemical processes on the GLC retention and separation of various analytes.

The exact characterization of stationary phases may considerably increase the precision of the prediction of the retention of any volatile analyte. The retention depends on the various intermolecular forces between the stationary phase and the analyte.[46] The retention indices (I_P^{RX}) can be predicted from the phase constants by[47]

$$I_P^{RX} = I_{NP}^{RX} + a^{RX} \cdot x_p + b^{RX} \cdot y_p + c^{RX} \cdot z_p + d^{RX} \cdot u_p + e^{RX} \cdot s_p \tag{80}$$

where I_{NP}^{RX} is the retention index of the analyte RX on the nonpolar standard phase; x_p, y_p, z_p, u_p, and s_p are phase constants defined as the retention index differences of benzene, ethanol, methylethyl ketone, nitromethane and pyridine, and a^{RX}, b^{RX}, c^{RX}, d^{RX}, and e^{RX} are characteristic coefficients determined by the multiple regression from the retention indices on several stationary phases. The retention can be determined by[48]

$$\log V_g = c + r \cdot R_2 + s \cdot \pi_2^* + a \cdot \alpha_2^H + b \cdot \beta_2^H + l \cdot \log L^{16} \tag{81}$$

where V_g is the specific retention volume for a series of analytes; R_2 is the modified molar fraction; π_2^* is the dipolarity, α_2^H is the hydrogen bond basicity, β_2^H is the hydrogen bond acidity of the analyte; L^{16} is the Ostwald absorption coefficient on hexadecane at 25°C; and r, s, a, b, and l are the characteristic constants of the phases. It was supposed that r is related to the tendency of the phase to interact with π and n-electron pairs of the analyte, s to the phase dipolarity, a and b to the phase hydrogen bond basicity and acidity, respectively, and l represents a combination of general dispersion forces. As the specific retention volume can be calculated from the retention indices by

$$\log V_g = A + B \cdot I/100 \tag{82}$$

Eq. 82 can be rewritten as

$$B \cdot I/100 \approx (c - A) + r \cdot R_2 + s \cdot \pi_2^* + a \cdot \alpha_2^H + b \cdot \beta_2^H + l \cdot \log L^{16} \tag{83}$$

$$I \approx 100 \cdot (c - A)/B + r \cdot R_2 \cdot 100/B + s \cdot \pi_2^* \cdot 100/B + a \cdot \alpha_2^H \cdot 100/B +$$
$$b \cdot \beta_2^H \cdot 100/B + l \cdot \log L^{16} \cdot 100/B \tag{84}$$

It has been previously established that constants c, l, A, and B are interrelated and can be calculated from each other

$$A = 1.112 \cdot c \quad c = 0.896 \cdot A \quad B = 0.4878 \cdot l \quad l = 2.049 \cdot B$$
$$SD = 0.045 \quad SD = 0.040 \quad SD = 0.005 \quad SD = 0.010 \tag{85}$$

Retention indices can be expressed by:

$$I = -100 \cdot 0.11 \cdot c/0.488 \cdot l + R_2 \cdot 100 \cdot r/0.488 \cdot l + \pi_2^* \cdot 100 \cdot s/0.488 \cdot l +$$
$$\alpha_2^H \cdot 100 \cdot a/0.488 \cdot l + \beta_2^H \cdot 100 \cdot b/0.488 \cdot l + \log L^{16} \cdot 100/0.488 \cdot l \tag{86}$$

The precise determination of the retention ratio (r) of two analytes is an important parameter for the selection of the best stationary phase for the successful separation of two or more analytes. Retention ratios can be calculated by

$$\log r = \log V_g^2 - \log V_g^1 = B \cdot (I^2 - I^1)/100 \tag{87}$$

$$\log r = r \cdot R_D + s \cdot \pi_D^* + a \cdot \alpha_D^H + b \cdot \beta_D^H + l \cdot \log L_D^{16} = B \cdot (I^2 - I^1)/100 \tag{88}$$

where R_D, π_D^*, a_D^H, β_D^H, and $\log L_D^{16}$ are the differences in the solute constants of the two analytes. Significant linear correlations were found between the following phase constants:

$$c = -(1.172 \pm 0.057) \cdot B + (0.321 \pm 0.053) \cdot B \cdot \Delta I^{Butanol}/100$$
$$- (0.877 \pm 0.061) \cdot B \cdot \Delta I^{Dioxan}/100$$
$$n = 72 \quad R = 0.9920 \quad s = 0.057 \quad F = 2929 \tag{89}$$

$$l = (2.100 \pm 0.007) \cdot B$$
$$n = 72 \quad R = 0.9998 \quad s = 0.007 \quad F = 181,593 \tag{90}$$

$$r = (0.556 \pm 0.034) \cdot B + (1.701 \pm 0.061) \cdot B \cdot \Delta I^{Benzene}/100$$
$$- (1.18 \pm 0.049) \cdot B \cdot \Delta I^{Pentanone}/100$$
$$n = 72 \quad R = 0.9794 \quad s = 0.035 \quad F = 1094 \tag{91}$$

$$s = (0.154 \pm 0.042) \cdot B + (1.889 \pm 0.107) \cdot B \cdot \Delta I^{Pentanone}/100$$
$$- (0.314 \pm 0.090) \cdot B \cdot \Delta I^{Dioxan}/100$$
$$n = 72 \quad R = 0.9983 \quad s = 0.043 \quad F = 14,119 \tag{92}$$

$$a = (0.153 \pm 0.051) \cdot B + (2.716 \pm 0.049) \cdot B \cdot \Delta I^{Butanol}/100$$
$$- (0.374 \pm 0.162) \cdot B \cdot \Delta I^{Benzene}/100 - (0.402 \pm 0.137) \cdot B \cdot^{Pentanone}/100$$
$$- (0.903 \pm 0.180) \cdot B \cdot \Delta I^{Dioxan}/100$$
$$n = 72 \quad R = 0.9988 \quad s = 0.048 \quad F = 11,939 \tag{93}$$

The data clearly show that the phase coefficients can be calculated from the retention indices or from the differences of retention indices, that is, the contribution of the various physicochemical parameters of analytes to their retention behavior on stationary phases can be calculated from the retention indices or from the retention index differences. These results also suggest the stationary phase with optimal separation characteristic can be selected with the exact knowledge of the physicochemical parameters of analytes to be separated. The calculated parameters of some analytes are listed in Table 1. However, it has to be borne in mind that the interactions between the various substructures of analytes and stationary phases cannot be described on the molecular level by the calculated physicochemical parameters. The retention of nonpolar analytes can also be described by:

$$(\log V_g) - A = N \cdot B + P \cdot B \cdot \Delta I^{Benzene}/100 \tag{94}$$

where $P \cdot B \cdot \Delta I^{Benzene}/100$ is an additional dispersion term. These parameters were also determined from the equations listed above (Table 2). The data in Table 2 suggest that the

TABLE 1

Standard Deviations and Solute Coefficients from Correlations of $(\log V_g) - c$

Analyte	log L	R	π	a	(s)
Dodecane	5.66	−0.02	0.0	0.03	0.025
Cyclohexane	1.94	0.20	0.14	−0.02	0.029
Benzene	2.79	0.60	0.47	0.00	0.019
Diethyl benzene	4.71	0.61	0.50	−0.02	0.019
Terachloromethane	2.85	0.44	0.32	0.02	0.036
Dioxan	2.79	0.50	0.77	−0.02	0.032
Tetrahydrofurane	2.53	0.29	0.57	−0.03	0.035
Trioxepane	3.15	0.66	0.88	−0.01	0.034
Pentanone-2	2.79	0.09	0.68	0.00	0.017
Valeraldehyde	2.87	0.15	0.63	0.00	0.019
Isobutylacetate	3.18	0.06	0.59	0.00	0.015
Butanol-1	2.61	0.25	0.42	0.37	0.016
2-Methyl pentanol-1	3.53	0.21	0.41	0.35	0.016
3-Methyl pentanol-3	3.19	0.19	0.42	0.21	0.030

From Rohrschneider, L., *Chromatographia*, 38, 1994, 679–688. With permission.

TABLE 2

Standard Deviations (s) and Analyte Coefficients for GC-Polarity Model from Correlations of $(\log V_g) - A$ with Polarity Constants B and $B \cdot \Delta I^{Benzene}$

Analyte	N	P	(s)
Decane	10.00	0.0	0.0
Dodecane	12.00	0.0	0.0
Benzene	6.39	1.03	0.007
Diethyl benzene	10.42	0.98	0.027
Cyclohexane	6.54	0.27	0.035
Terachloromethane	6.34	0.78	0.036
Dioxan	6.52	1.46	0.064
Tetrahydrofurane	6.28	1.05	0.063
Trioxepane	7.41	1.72	0.064
Pentanone-2	6.39	1.21	0.088
Valeraldehyde	6.55	1.17	0.076
Isobutylacetate	7.14	1.03	0.069
Butanol-1	6.28	1.64	0.127
2-Methyl pentanol-1	8.16	1.56	0.120
3-Methyl pentanol-3	7.31	1.26	0.085

From Rohrschneider, L., *Chromatographia*, 38, 1994, 679–688. With permission.

retention behavior of analytes can be adequately predicted also by the use of a slightly different model.

Retention in GLC is a complex process involving partitioning and interfacial adsorption. The process can been described by[49]

$$V_N^* = V_L \cdot K_L + A_{GL} \cdot K_{GL} + A_{LS} \cdot K_{GLS} \tag{95}$$

where V_N^* is the net retention volume per gram of column packing; V_L is the volume of the liquid stationary phase per gram of column packing; K_L is the gas–liquid partition coefficient; A_{GL} is the gas–liquid interfacial area per gram of packing; K_{GL} is the adsorption coefficient at the gas–liquid interface; A_{LS} is the liquid–solid interfacial area per gram of packing; and K_{GLS} is the adsorption coefficient at the liquid–solid interface. The exactness of Eq. 95 has been proved for many different GLC systems.[50-55]

The contribution of the interfacial adsorption to the overall retention has been determined for a wide variety of analytes using ten different stationary phases. The results are compiled in Table 3. It has been concluded from the data that gas–liquid partitioning is the dominant retention mechanism at about 100°C column temperature; the impact of interfacial adsorption on the retention is higher at lower temperature and at lower phase loading. Apolar analytes (i.e., hydrocarbons) have a higher tendency for interfacial adsorption on polar stationary phase. The importance of interfacial adsorption is higher on highly cohesive stationary phases and on liquid organic salts. Orientation, induction and charge-transfer interactions, and hydrogen bond formation are the polar intermolecular interactions involved in the retention mechanism.

A theoretical study tried to find relationship between the peak width in capillary GC under high column pressure drop and the molecular parameters of the carrier gas and

TABLE 3
Contribution of Interfacial Adsorption (%) to the Retention of some Typical Solutes on Ten Stationary Phases at Two Temperatures

	Stationary Phase									
	SQ	OV-105	OV-17	OV-225	QF-1	CW-20M	THPED	QPTS	TCEP	DEGS
					Phase Loading (%)					
Compound	15.5	16.0	16.1	15.4	16.0	20.7	16.5	15.9	16.0	15.7
Temperature 81.2°C										
1	7.0	2.9	2.3	1.9	2.3	23.0	2.6	3.2	47.9	66.3
2	7.8	3.5	2.1	7.0	2.9	11.9	5.0	6.0	8.3	32.3
3	8.1	3.7	2.0	4.6	2.6	10.9	3.1	4.9	10.2	26.5
4	8.6	3.3	1.4	6.7	2.8	7.2	5.8	5.3	3.9	9.5
5	15.5	4.5	2.3	6.0	5.5	8.9	0.7	2.1	5.2	14.7
6	8.1	3.9	2.2	11.1	3.6	6.6	4.7	3.9	12.0	15.2
7	8.1	5.6	2.1	13.7	4.0	6.1	2.3	5.1	14.9	33.1
8	7.8	5.6	2.1	12.2	3.9	7.2	14.7	5.4	16.0	37.1
9	7.5	5.9	2.0	12.8	4.1	—	6.3	—	18.4	43.4
10	11.4	6.0	2.1	27.3	5.4	—	4.4	—	19.2	38.8
11	8.0	5.6	2.1	14.1	4.1	5.0	13.5	5.5	17.2	34.6
12	9.2	4.4	3.5	9.1	1.7	4.9	14.4	5.9	0.2	6.6
Temperature 121.2°C										
1	1.5	5.4	0.6	6.3	1.7	4.5	3.7	4.5	21.7	26.9
2	2.1	4.2	0.6	0.6	6.0	5.9	4.3	3.5	7.1	11.2
3	0.9	6.0	0.4	6.0	1.9	3.7	3.3	2.4	1.9	7.3
4	1.0	5.4	0.8	3.9	2.3	1.8	4.3	2.9	3.3	7.0
5	2.6	4.1	0.6	10.8	2.3	1.9	0.5	3.0	2.0	6.6
6	2.8	5.1	0.8	4.2	3.2	4.7	2.4	1.6	12.5	11.4
7	2.3	5.2	0.8	2.3	1.9	4.0	2.3	2.6	1.2	7.5
8	2.6	5.7	0.6	3.0	1.7	2.9	15.5	2.7	1.2	5.3
9	2.8	5.4	0.2	1.4	1.8	3.5	6.4	2.7	2.0	2.8
10	3.1	5.5	0.6	2.1	1.9	2.7	4.4	1.6	0.6	8.4
11	2.5	5.4	0.5	5.5	1.9	2.1	8.2	2.8	2.8	12.5
12	1.2	4.1	0.9	3.2	4.7	2.3	10.0	3.7	0.6	11.8

Note: 1 = tridecane; 2 = oct-2-yne; 3 = methyloctanoate; 4 = heptan-2-one; 5 = heptan-1-ol; 6 = ethylbenzene; 7 = acetophenone; 8 = benzonitrile; 9 = nitrobenzene; 10 = N-methylaniline; 11 = N,N-dimethylaniline; 12 = dioxane.

From Poole, S. K., Kollie, T. O., and Poole, C. F., *J. Chromatogr. A*, 664, 1994, 229–251. With permission.

the analytes.[56] The variance of the peak (τ^2, in units of time squared) can be correlated with the following parameters:

$$\tau^2 = [\mu \cdot L^2 \cdot G(k)/3 \cdot D_{out} \cdot P_{out}]/[1 + (v_{out,opt}/v_{out})^2] \quad (96)$$

$$G(k) = 1 + 6 \cdot k + 11 \cdot k^2 \quad (97)$$

$$D_\tau = D_{out}/D_{m,out} \quad (98)$$

where μ is the viscosity of the carrier gas, in units of (mass per length) per time; k is the capacity ratio; D_m is the self-diffusivity of a mobile phase, in units of length squared per time; P_{out} is the outlet pressure, in units of (mass per length) per time squared; v_{out} and $v_{out,opt}$ are carrier gas outlet linear velocity and its optimum, respectively, in units of length per time; and D_τ is the reduced diffusivity of the analyte. The peak variance can also be expressed by

$$\tau^2 = [\mu \cdot L^2 \cdot G(k)/3 \cdot D_\tau \cdot D_{m,out} \cdot P_{out}]/[1 + (v_{out,opt}/v_{out})^2] \quad (99)$$

The reduced diffusivity of an analyte is related to the molecular properties of the carrier gas and the analyte:

$$D_\tau = D_{out}/D_{m,out} = [(1/M_m + 1/M_s)^{1/2}/(d_m + d_s)^2] \cdot [(2 \cdot d_m)^2/(2/M_m)^{1/2}] \quad (100)$$

where M_m and M_s are the molecular mass of the carrier gas and the analyte; d_m and d_s are the collision diameters of the mobile phase and the analyte molecules. Because generally $M_s \gg M_m$ Eq. 100 can be simplified to:

$$D_\tau = 2 \cdot (2)^{1/2}/(1 + d_s/d_m)^2 \quad (101)$$

Including the molecular parameters in the equations yields:

$$\mu/3 \cdot D_{out} \cdot P_{out} = \sigma_{m,out}/3.6 \cdot P_{out} = M_m/3 \cdot 6 \cdot RT \quad (102)$$

where σ is the density, in units of mass per length cubed.

$$\tau^2 = (L^2 \cdot M_m \cdot G(k)/3.6 \cdot D_\tau \cdot RT)/[1 + (v_{out,opt}/v_{out})^2]$$

$$= (L^2 \cdot G(k)/1.2 \cdot D_\tau \cdot \mu_m^2)/[1 + (v_{out,opt}/v_{out})^2] \quad (103)$$

When the velocity of the carrier gas is higher than the optimum, the previous equation becomes

$$\tau_{min} = \tau_{min}(D_\tau,k) = L(M_m \cdot G(k)/3.6 \cdot D_\tau \cdot RT)^{1/2} = [(G(k)/1.2 \cdot D_\tau)^{1/2}] \cdot L/\mu_m \quad (104)$$

where τ_{min} is the analyte-specific lower bound for the standard deviation of the peak. The absolute lower bound ($\tau_{min,abs}$) is

$$\tau_{min,abs} = L \cdot (M_m/3.6 \cdot RT)^{1/2} = L/1.2^{1/2} \cdot \mu_m \quad (105)$$

$$\tau_{min} = \tau_{min}(D_\tau,k) = [(G(k)/1.2 \cdot D_\tau)^{1/2}] \cdot \tau_{min,abs} \quad (106)$$

It has been concluded from the previous theoretical considerations that the peak width does not depend on the column diameter, and the absolute lower bound of peak width does not depend on the column diameter, the nature of the analyte, or its retention. The relationship between the retention of polyhalogenated biphenyls (PHBs) and their various molecular parameters has been vigorously discussed.[57] Thus, direct molecular descriptors,[58,59] molecular connectivity indices,[60] and molecular orbital calculations have been successfully used for the prediction of the retention behavior of PHBs on various GC stationary phases. It has been recently established that the retention of PHBs on DB-210-CB capillary columns[61] can be described by using only the number, type, and position of the halogen substituents. The best relationships were

$$I = (1631.6 \pm 9.1) + (230.7 \pm 5.8) \cdot N_{Cl} + (319.6 \pm 6.5) \cdot N_{Br} + (440.4 \pm 31.0) \cdot N_I$$
$$- (169.0 \pm 13.5) \cdot N_{OrCl} - (175.3 \pm 15.3) \cdot N_{OrBr} - (214.0 \pm 51.4) \cdot N_{OrCl}$$
$$n = 53, \quad r^2 = 0.990, \quad s = 42, \quad F = 736 \tag{107}$$

$$I = (1645.3 \pm 10.3) + (227.5 \pm 5.6) \cdot N_{Cl} + (315.1 \pm 6.4) \cdot N_{Br} + (426.7 \pm 30.0) \cdot N_I$$
$$- (20.5 \pm 8.4) \cdot N_{OrF} - (168.8 \pm 12.8) \cdot N_{OrCl} - (177.0 \pm 14.5) \cdot N_{OrBr} - (214.0 \pm 48.8) \cdot N_{OrCl}$$
$$n = 53, \quad r^2 = 0.991, \quad s = 40, \quad F = 701 \tag{108}$$

where N_X is the number of halogens of type X and N_{OrX} is the number of halogens located at the ortho position. The reduced impact of halogens at the ortho position was tentatively explained by the supposition that these substituents are less exterior and hence less exposed, which decrease their capacity to interact with the stationary phase. It has been further established than dispersion forces have a higher impact on the retention of PHBs than the polar forces do.

It has been assumed that the L^{16} value (partition coefficient of the analyte between gas and n-hexadecane) can be related to the groups, structural characteristics, and interactional contributions of the analyte X:

$$\log(L^{16}X) = \sum_i l_i \cdot FG_i + \sum_j m_j \cdot SC_j + \sum_k n_k \cdot IC_k \tag{109}$$

where i, j, and k are the identification numbers of the groups, structural contributions, and interaction contributions, respectively; l, m, and n are the regression coefficients for the contributions of a group, a structural contribution, and an interactional contribution, respectively; FG, SC, and IC are the numbers of groups and structural and interactional contributions, respectively.

Calculations were carried out on two data sets containing 336 monofunctional analytes (set A) and 481 mono- and polyfunctional analytes (set B). The results of the calculations are compiled in Tables 4 to 6. The l_i values differ considerably, indicating the marked impact of various molecular substructures on the retention. The lowest and highest values were observed for fluorine and iodine, respectively. This result may be due to the different polarizability of these halogens. The values of regression coefficient for structural arrangement are significantly lower than those of subgroups. This result suggests that the influence of cyclic arrangement of analytes is less important in the determination of the retention behavior. It has been concluded from the data in Table 6 that the interactional contributions exert a significant impact on the retention of various analytes.

TABLE 4

Regression Coefficients l_i, and Their Standard Deviations s_l and Frequencies x_l of Group i for Data Sets A and B

i	Structure	Set A			Set B		
		l_i	s_l	x_l	l_i	s_l	x_l
1	CH_3-	0.333	0.009	620	0.358	0.009	834
2	$-CH_2-$	0.504	0.002	1256	0.499	0.002	1492
3	$-CH$	0.488	0.016	129	0.437	0.017	193
4	$>C<$	0.456	0.029	30	0.350	0.032	46
5	$H_2C=$	0.198	0.027	13	0.280	0.020	59
6	$-HC=$	0.471	0.022	21	0.450	0.017	77
7	$>C=$	0.567	0.050	4	0.609	0.024	20
8	$HC\equiv$	0.085	0.034	11	0.063[a]	0.046	13
9	$-C\equiv$	0.595	0.035	11	0.558	0.046	13
10	$F-$	−0.114	0.045	4	−0.121	0.060	11
11	$Cl-$	0.894	0.030	9	0.786	0.020	50
12	$Br-$	1.293	0.027	11	1.319	0.036	13
13	$I-$	1.766	0.034	7	1.756	0.046	8
14	$-O-$	0.302	0.026	16	0.346	0.016	114
15	$-CHO$	0.982	0.026	13	1.014	0.030	18
16	$-CO-$	1.098	0.023	22	1.087	0.025	34
17	$HCOO-$	1.078	0.028	11	1.070	0.035	12
18	$-COO-$	1.108	0.019	43	1.105	0.020	65
19	$-COOH$	1.589	0.024	15	1.590	0.032	15
20	$-CN$	1.213	0.029	10	1.209	0.039	10
21	$-NH_2$	0.798	0.032	8	0.791	0.043	8
22	$>NH$	0.746	0.047	4	0.711	0.062	4
23	$-N<$	0.575	0.068	2	0.508	0.089	2
24	$-NO_2$	1.524	0.034	7	1.518	0.046	7
25	$-OH$	0.739	0.015	52	0.762	0.015	101
26	$-SH$	1.312	0.026	13	1.323	0.033	14
27	$-S-$	1.435	0.037	7	1.406	0.048	7

[a] Statistically not significant.

From Havelec, P. and Sevcik, J. G. K., *J. Chromatogr. A*, 677, 1994, 319–329. With permission.

TABLE 5

Regression Coefficients m_j and Their Standard Deviations s_m and Frequencies x_m of Specified Structural Contributions j for Data Sets A and B

j	Structure	Set A			Set B		
		m_j	s_m	x_m	m_j	s_m	x_m
1	3-Ring				0.117[a]	0.064	6
2	4-Ring				0.182	0.092	2
3	5-Ring	0.001[a]	0.037	8	0.060[a]	0.040	12
4	6-Ring	0.040[a]	0.033	14	0.084	0.035	18
5	7-Ring	0.169	0.042	6	0.207	0.052	6
6	8-Ring	0.299	0.090	1	0.337	0.121	1
7	trans	−0.090[a]	0.079	2	−0.027[a]	0.046	9
8	cis	0.074[a]	0.062	3	0.094[a]	0.061	5
9	Oxygen atom in ring				0.051[a]	0.030	22

[a] Statistically not significant.

From Havelec, P. and Sevcik, J. G. K., *J. Chromatogr. A*, 677, 1994, 319–329. With permission.

III. INFLUENCE OF ADSORPTION SIDE EFFECTS ON RETENTION BEHAVIOR

Four models have been proposed for the description of the adsorption side effects on the retention behavior in GLC:[63] capillary tube (Figure 1A), nonporous spheres (Figure 1B) and porous particles (Figure 1C) entirely covered with the liquid film, as well as porous particles with pores filled with the liquid (Figure 1D). To simplify the calculations the uniformity of the film thickness and a one-dimensional diffusion through the thin liquid film were assumed. The differential equations describing the four models have been developed with the following boundary and zero initial conditions: $c(z = 0,t)$ and $c(z = \infty,t)$ as well as $c(z,t = 0) = c_i(z,t = 0)$, respectively. Here c is the concentration of analyte in the gas phase (g/cm^3), t is time (s), and z is a distance coordinate for voids (cm).

TABLE 6

Regression Coefficients n_k and Their Standard Deviations s_n and Frequencies x_n of Interactional Contributions k for Data Set B

k	Structure	Set B		
		n_k	s_n	x_n
1	>CF$_2$	0.152	0.073	4
2	>CCl$_2$	−0.117	0.025	24
3	>CBr$_2$	0.715	0.041	4
4	>CI$_2$	1.602	0.128	1
5	>C(F,Cl)	−0.012[a]	0.043	6
6	=C−O− $\mid$ O $\mid$	−0.083	0.033	31
7	−COCO−	−0.128[a]	0.088	2
8	−COCH$_2$CO−	0.471	0.122	1
9	−COOC$_n$H$_{2n}$−OOC−(n = 1–3)	1.163	0.063	4
10	>OHCOH<	0.146	0.060	5

[a] Statistically not significant.

From Havelec, P. and Sevcik, J. G. K., *J. Chromatogr. A*, 677, 1994, 319–329. With permission.

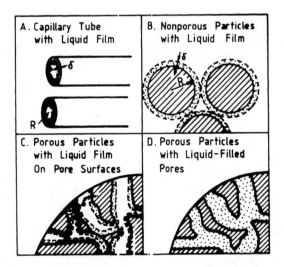

FIGURE 1
Schematic representation of four models for gas-liquid-solid chromatography. R is the radii of the cylindrical tube and the spherical particles; δ is the thickness of the liquid film. (From McCoy, B. J., *Chromatographia*, 36, 1993, 234–240. With permission.)

Model A (Capillary covered with thin, uniform liquid film):
In the voids

$$\delta c/\delta t + v \cdot \delta c/\delta z = D_o \cdot \delta^2 c/\delta z^2 - A_p \cdot k_p \cdot (c - c_s) \quad (110)$$

In the liquid film

$$\delta c_L/\delta t = D_L \cdot \delta^2 c_L/\delta y^2 \quad (111)$$

Boundary conditions

$$A_p \cdot D_L \cdot \delta c_L/\delta y = K_d \cdot \delta c_L/\delta t \quad (y = 0) \quad (112)$$

$$k_p \cdot (c - c_s) = -D_L \cdot \delta c_L/\delta y \quad (y = \delta; c_s = K_H \cdot c_L) \quad (113)$$

where v is the average velocity (cm/s); D_o is the hydrodynamic dispersion coefficient (cm^2/s); A_p is the wetted surface area per volume of sphere or cylinder (1/cm); k_p is the mass-transfer coefficient between gas in voids and liquid (cm/s), c_s is the concentration of analyte in the gas at the liquid surface (g/cm^3); c_L is the concentration of analyte in the liquid (g/cm^3); D_L is the diffusion coefficient of analyte in liquid (cm^2/s); K_d is the adsorption constant for analyte at the liquid–solid interface; K_H Henry's law constant for analyte distributed between gas and liquid (volume of liquid per volume of gas [cm^3/cm^3]).

Model B (nonporous particles covered with thin, uniform liquid film):
In the voids

$$\epsilon \cdot \delta c/\delta t + v_o \cdot \delta c/\delta z = D_o \cdot \delta^2 c/\delta z^2 - (1 - \epsilon) \cdot A_p \cdot k_p \cdot (c - c_s) \quad (114)$$

In the liquid film

$$\delta c_L/\delta t = D_L \cdot \delta^2 c_L/\delta y^2 \quad (115)$$

Boundary conditions

$$A_p \cdot D_L \cdot \delta c_L/\delta y = K_d \cdot \delta c_L/\delta t \quad (y = 0) \quad (116)$$

$$k_p \cdot (c - c_s) = -D_L \cdot \delta c_L/\delta y \quad (y = \delta; c_s = K_H \cdot c_L) \quad (117)$$

where ϵ is interparticle void fraction; v_o is superficial velocity in voids = volume flow rate per cross-sectional area = ϵ_v.

Model C (porous particles covered with thin, uniform liquid film):
In the voids

$$\epsilon \cdot \delta c/\delta t + v_o \cdot \delta c/\delta z = D_o \cdot \delta^2 c/\delta z^2 - (1 - \epsilon) \cdot 3/R \cdot k_p \cdot (c - c_i(R)) \quad (118)$$

In the pores

$$\beta \cdot \delta c_i/\delta t = D_{Gi} \cdot 1/r^2 \cdot \delta(r^2 \cdot \delta c_i/\delta r)/\delta r - A_p \cdot k_G \cdot (c_i - c_s) \quad (119)$$

In the liquid film

$$\delta c_L/\delta t = D_L \cdot \delta^2 c_L/\delta y^2 \quad (120)$$

$$D_{Gi} \cdot \delta c_i/\delta r = k_p \cdot (c - c_i) \quad (r = R) \quad (121)$$

Boundary conditions

$$\delta c_i/\delta r = 0 \quad (122)$$

$$A_p \cdot D_L \cdot \delta c_L/\delta t = K_d \cdot \delta c_L/\delta t \quad (y = 0) \quad (123)$$

$$k_G \cdot (c_i - c_s) = -D_L \cdot \delta c_L/\delta y \quad (y = \delta; c_s = K_H \cdot c_L) \quad (124)$$

where z is a distance coordinate for voids (cm); R is the radius of capillary or spherical particle (cm); c_i is the concentration of the analyte in the pore gas (g/cm³); β is the particle porosity; c is the concentration of analyte in the gas phase (g/cm³); D_{Gi} is the effective intraparticle diffusion coefficient (cm²/s); r is the radial coordinate of a spherical particle (cm); and k_G is the mass-transfer coefficient between pore gas and liquid (cm/s).

Model D (porous particles filled with liquid):
In the voids

$$\epsilon \cdot \delta c/\delta t + v_o \cdot \delta c/\delta/z = D_o \cdot \delta^2 c/\delta z^2 - (1 - \epsilon) \cdot 3/R \cdot k_p \cdot (c - K_H \cdot c_L(R)) \quad (125)$$

In the pores

$$\beta \cdot \delta c_L/\delta t = D_{Li} \cdot 1/r^2 \cdot \delta(r^2 \cdot \delta c_i/\delta r)/\delta r - \delta c_a/\delta t \quad (c_a = K_d \cdot c_L) \quad (126)$$

Boundary conditions

$$\delta c_L/\delta r = 0 \quad (r = 0) \quad (127)$$

$$D_{Li} \cdot \delta c_L/\delta r = k_p \cdot (c - K_H \cdot c_L) \quad (r = R) \quad (128)$$

where c_a is the concentration of adsorbed analyte (g/cm³), and D_{Li} is the effective intraparticle diffusion coefficient (cm²/s).

The first (μ_1) and second (μ_2) temporal moments of the equations listed above were also calculated:

$$\mu_1 = (1 + \Gamma) \cdot z/v \quad (129)$$

$$\mu_2 = (g \cdot D_o/v^2 + \Phi) \cdot 2 \cdot z/v \quad (130)$$

Model A (capillary covered with thin, uniform liquid film):

$$\Gamma = (A_p \cdot \delta + K_d)/K_H \quad (A_p = 2/R) \quad (131)$$

$$g = (1 + \Gamma)^2 \quad (132)$$

$$\Phi = \Gamma^2/A_p \cdot k_p + (A_p^2 \cdot \delta^2/K_H + 3 \cdot \Gamma \cdot K_d) \cdot \delta/3 \cdot D_L \cdot A_p \tag{133}$$

Model B (nonporous particles covered with thin, uniform liquid film):

$$\Gamma = (A_p \cdot \delta + K_d) \cdot (1 - \epsilon)/\epsilon \cdot K_H \quad (\Gamma = (A_p \cdot \delta + K_d)/K_H; A_p = 2/R) \tag{134}$$

$$g = (1 + \Gamma)^2/\epsilon \tag{135}$$

$$\Phi = [\Gamma^2/A_p \cdot k_p + (A_p^2 \cdot \delta^2/K_H + 3 \cdot \Gamma \cdot K_d) \cdot \delta/3 \cdot D_L \cdot A_p] \cdot (1 - \epsilon)/\epsilon \tag{136}$$

Model C (porous particles covered with thin, uniform liquid film):

$$\Gamma = [\beta + (A_p \cdot \delta + K_d)/K_H] \cdot (1 - \epsilon)/\epsilon = (\beta + \Gamma) \cdot (1 - \epsilon)/\epsilon$$
$$(\Gamma = (A_p \cdot \delta + K_d)/K_H) \tag{137}$$

$$g = (1 + \Gamma)^2/\epsilon \tag{138}$$

$$\Phi = [\Gamma^2/A_p \cdot k_p + (A_p^2 \cdot \delta^2 + 3 \cdot \Gamma \cdot K_d) \cdot \delta/3 \cdot D_L \cdot A_p \cdot K_H] \cdot (1 - \epsilon)/\epsilon +$$
$$(\Gamma^2 \cdot R^2/15) \cdot (1/D_{Gi} + 5/R \cdot k_p) \cdot \epsilon/(1 - \epsilon) \tag{139}$$

Model D (porous particles filled with liquid):

$$\Gamma = (\beta + K_d)/K_H \cdot (1 - \epsilon)/\epsilon \tag{140}$$

$$g = (1 + \Gamma)^2/\epsilon \tag{141}$$

$$\Phi = (\beta + k_d)^2 \cdot R^2/15 \cdot K_H \cdot (1/D_{Li} + 5/R \cdot k_p \cdot K_H) \cdot (1 - \epsilon)/\epsilon \tag{142}$$

The equations demonstrate that gas velocity, particle radius, Henry's coefficient, adsorption constant, and the thickness of the liquid film have the highest effect on the retention time whereas the peak shape is mainly determined by the dispersive processes such as intraparticle diffusion, liquid film diffusion, external mass transfer, and hydrodynamic dispersion.

According to the equations the separation efficiency (resolution R_s) for analytes A and B can be calculated by

$$R_s = (\mu_{1B} - \mu_{1A})/[(\mu_{2B})^{1/2} + (\mu_{2A})^{1/2}] \tag{143}$$

It has been concluded that the above equations are suitable not only for the exact description of the retention behavior of various analytes, but also for the comparison of the impact of different equilibrium and transport parameters on the retention characteristics of any GSL chromatographic system.

Not only the adsorptive characteristics of the support, but also the physicochemical properties of coating agent on various supports exert a marked influence on the retention of various analytes.[64] Thus, it has been established that PEG 20M is in two forms on the surface of a silanized support; the characteristics of the inner layer in contact with the surface of the support are different from the outer layer, which is not influenced by the adsorptive forces between the coating agent and the support surface.[65,66] The structure and chromatographic properties of poly(ethylene glycol) 20M layers on an acid-washed, nonsilanized diatomaceous-earth type support (Chromaton N AW) was studied by inverse GC. Calculations were carried out as follows:

$$V_S = (t_{R'} \cdot F_a \cdot j)/(m_{sup}) \cdot (273.16)/(T_F) \cdot (P_0 - P_W)/P_0 \qquad (144)$$

where V_S is the retention volume extrapolated to 1 g support and 0°C temperature; $t_{R'}$ is the adjusted retention time; F_a is the carrier gas flow at the temperature of the soap bubble flowmeter; j is the compressibility correction factor; m_{sup} is the mass of support in the column; T_F is the temperature of the flow meter (K); P_0 is the ambient pressure; and P_W is the partial pressure of water at T_F.

$$V_g^0 = \Delta V_S / \Delta \tau \qquad (145)$$

where V_g^0 is the specific retention volume related to 1 g liquid stationary phase at 0°C temperature, due only to the partition of the analyte between the gas and the liquid stationary phases; and τ is the stationary phase loading (amount of PEG 20M coated on 1 g support).

$$K = V_g^0 \cdot T_c \cdot p / 273.16 \qquad (146)$$

where K is the partition coefficient of the analyte between the liquid and the stationary phases; T_c is the column temperature; and p is the density of the stationary phase at the temperature of the column.

$$V_S = V_g^0 \cdot \tau + A_g \qquad (147)$$

where A_g is the part of the retention volume V_S caused by the adsorption of the analyte on the interface. The parameters V_g^0 and A_g can be determined from the linear dependence of V_S vs. τ.

The stationary phase can be divided in two parts:

$$\tau = \tau_F + \tau_T \qquad (148)$$

where τ_F is the part of the stationary phase that does not melt at its normal melting temperature, and τ_T is the part of the stationary phase not influenced by the solid support.

In each instance the retention volume showed a stepwise change at higher τ values at the melting temperature of PEG 20M. This fact may be due to the higher solubility of analytes in the molten layer of PEG 20M. However, the relationship between the retention volume and column temperature depended considerably on the character of the analyte. The relatively polar ethanol is strongly adsorbed on the surface of the uncoated support. The coating of support with increasing amount of PEG 20M resulted in the decrease of specific retention volume (Figure 2). In contrast to the polar ethanol, the apolar *n*-octane is not adsorbed on the surface of uncoated support; therefore its specific retention volume does not depend considerably on the stationary phase loading ratio (Figure 3). Chloroform is very soluble in PEG 20M and adsorbs negligibly on the support surface. These characteristics account for the high effect of τ on its specific retention volume under the melting point of PEG 20M (Figure 4).

The data discussed above indicate that not only the column temperature and the stationary phase loading ratio, but also the adsorption capacity of analytes on the support as well as their solubility in the stationary phase may markedly influence the separation of any set of analytes. These results were supported by the findings that the partition

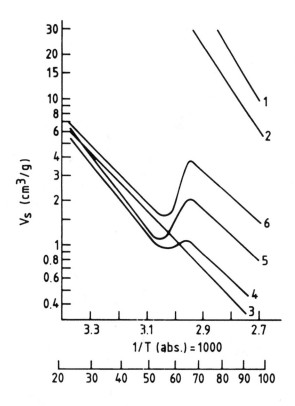

FIGURE 2
Log V_S vs. $1/T_c$ for ethanol on column packings consisting of Chromaton N AW coated with various amounts of PEG 20M. (1) Uncoated support; (2) $\tau = 0.0004$; (3) $\tau = 0.001$; (4) $\tau = 0.005$; (5) $\tau = 0.02$; (6) $\tau = 0.04$. (From Surowiec, K. and Rayss, J., *Chromatographia*, 37, 1993, 444–450. With permission.)

coefficients and the contribution of the adsorption of solutes were different and depended differently on the column temperature (Tables 7 and 8).

The exact estimation of retention indices may be influenced by the large neighboring peak effect when a large nearly eluted analyte causes a temporary change in the polarity of the stationary phase.[67] Adsorptions occurring both at the gas–liquid and the liquid–solid interfaces may considerably modify the accuracy of the determination of retention indices. The net retention volume[68] V_N can be described by

$$V_N = K_L \cdot V_L + K_{g,l} \cdot S_{g,l} + K_{l,s} \cdot S_{l,s} \tag{149}$$

where K_L is the partition coefficient of the analyte between the gas and the liquid phases, V_L is the volume of the liquid stationary phase, $K_{g,l}$ and $K_{l,s}$ are the adsorption coefficients of the analyte at the gas–liquid and the liquid–solid interfaces, respectively; $S_{g,l}$ and $S_{l,s}$ are the surfaces of the gas–liquid and liquid–solid interfaces ($S_{g,l} \approx S_{l,s}$). The retention index for pure gas–liquid chromatography (I_{0i}) can be calculated from the measured retention index (I_i) by taking into consideration the adsorption effect:[69]

$$I_i = I_{0i} + a_i/k_{st} \tag{150}$$

The measured retention index (I) can be divided:

$$I = I_{GLC} + I_{ads.g,l} + I_{ads.l,s} \tag{151}$$

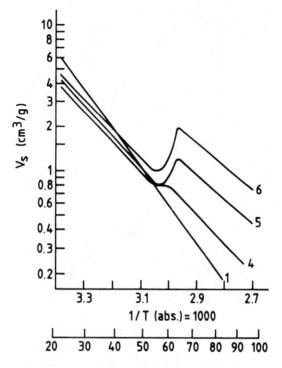

FIGURE 3
Log V_S vs. $1/T_c$ for *n*-octane on column packings consisting of Chromaton N AW coated with various amounts of PEG 20M. (1) Uncoated support; (4) $\tau = 0.005$; (5) $\tau = 0.02$; (6) $\tau = 0.04$. (From Surowiec, K. and Rayss, J., *Chromatographia*, 37, 1993, 444–450. With permission.)

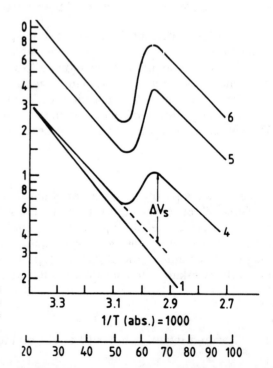

FIGURE 4
Log V_S vs. $1/T_c$ for chloroform on column packings consisting of Chromaton N AW coated with various amounts of PEG 20M. (1) Uncoated support; (4) $\tau = 0.005$; (5) $\tau = 0.02$; (6) $\tau = 0.04$. (From Surowiec, K. and Rayss, J., *Chromatographia*, 37, 1993, 444–450. With permission.)

TABLE 7

Values of the Partition Coefficient (K) on PEG 20M Coated on Chromaton N AW at Different Temperatures

Solute	12.0°C	29.9°C	39.3°C	49.4°C	71.7°C	84.0°C
n-Octane	20.23	19.2	12.6	9.6	44.5	30.5
Benzene	144.0	92.5	59.6	37.7	127.0	88.1
$CHCl_3$	312.8	182.4	109.3	70.9	206.1	130.1
CCl_4	69.3	52.7	35.7	24.1	80.6	58.0
Ethanol	?	43.0	46.7	40.1	91.6	63.7

From Surowiec, K. and Rayss, J., *Chromatographia*, 37, 1993, 444–450. With permission.

TABLE 8

Contribution of Partition ($V_g^0 \cdot \tau$) and Adsorption (A_g) to the Retention Volumes (V_S) Determined for a Column Packing Consisting of Chromaton N AW and Peg 20M for $\tau = 0.04$ at Different Temperatures

Solute	29.9°C			49.4°C		
	$V_g^0 \cdot \tau$	A_g	A%	$V_g^0 \cdot \tau$	A_g	A%
n-Octane	0.618	2.600	80.8	0.294	0.900	75.4
Benzene	2.976	0.594	16.6	1.158	0.315	21.4
$CHCl_3$	5.868	1.629	21.7	2.178	0.489	18.3
CCl_4	1.969	0.554	24.6	0.740	0.216	22.6
Ethanol	1.384	3.350	70.8	1.232	0.671	35.3
	71.7°C			84.0°C		
n-Octane	1.300	0.290	18.2	0.868	0.212	19.6
Benzene	3.712	0.055	1.5	2.512	0.035	1.4
$CHCl_3$	6.024	0.075	1.2	3.708	0.105	2.8
CCl_4	2.356	0.069	2.8	1.652	0.013	0.8
Ethanol	2.676	0.557	17.2	1.816	0.349	16.1

Note: A% was calculated as $A_g \cdot (100/V_S)$.

From Surowiec, K. and Rayss, J., *Chromatographia*, 37, 1993, 444–450. With permission.

where I_{GLC} is the retention index related to pure gas liquid chromatography; $I_{ads.g,l}$ and $I_{ads.l,s}$ are the contributions of adsorption effects at the gas–liquid and the gas–solid interfaces, respectively.

The influence of the adsorption effect on the retention index of six hydroxy compounds (Figure 5) was investigated using to two HP-5 fused-silica capillary columns (dimethylsilane with 5% diphenyl groups) at 80 to 150°C.[70] The film thicknesses were 0.11 and 0.33 µm. The retention indices and the standard deviation of retention indices calculated from ten independent parallel determinations are compiled in Tables 9 and 10. The data in Tables 9 and 10 clearly show that the retention indices are higher on the column with lower film thickness, that is, this column parameter has a marked effect on the retention indices. The differences decrease with increasing column temperature. The type of analyte also has a considerable effect on the differences. It has been assumed that both the steric accessibility of the primary hydroxy group as well as its acidity may influence the adsorption capacity.

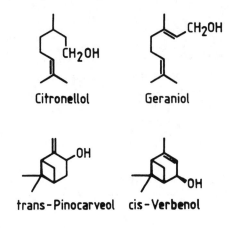

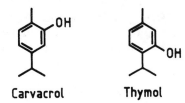

FIGURE 5
Structures of alcohols and phenols investigated. (From Hennig, P. and Engewald, W., *Chromatographia*, 38, 1994, 93–97. With permission.)

IV. DESCRIPTORS OF GLC RETENTION AND THEIR CORRELATION WITH MOLECULAR PARAMETERS

A. Retention Time

The retention time is the chromatographic parameter that can be easily determined. It has been used many times as a dependent variable to elucidate the influence of various molecular characteristics on the retention behavior. The retention behavior of alkylsilanes and alkylgermanes has been extensively studied.[71,72] Thus, the retention times of 26 tetra-n-alkylsilanes and 26 tetra-n-alkylgermanes were determined[73] and they were correlated by the molecular connectivity indices,[74,75] the new total topological indexes (TTV),[76] and the electrotopological state index (S)[77] using stepwise regression analysis. The chemical structure, logarithm of retention times, the topological indices $^1\chi$ and $^3\chi_c$, S, and TTV of tetra-n-alkylsilanes and tetra-n-alkyl-germanes are compiled in Table 11. The results of stepwise regression analysis proved that the retention behavior of both tetra-n-alkylsilanes and tetra-n-alkylgermanes can be related to the various topological indices. The equations for tetra-n-alkylsilanes were

$$\log T = -(0.455 \pm 0.040) + (0.514 \pm 0.007) \cdot {}^1\chi$$
$$n = 26, \quad r = 0.998, \quad s = 0.045, \quad F = 5448 \tag{152}$$

$$\log T = -(0.165 \pm 0.053) + (0.493 \pm 0.006) \cdot {}^1\chi - (0.193 \pm 0.031) \cdot {}^3\chi_c$$
$$n = 26, \quad r = 0.999, \quad s = 0.028, \quad F = 6959 \tag{153}$$

TABLE 9

Retention Indices (I) and Standard Deviations (SD) on HP-5 Capillary Column (Film Thickness 0.11 μm) at Various Temperatures

Analyte		Temperature (°C)			
		80	90	100	110
Citronellol	I	1228.71	1228.51	1228.46	1228.53
	SD	0.08	0.11	0.08	0.03
Geraniol	I	1254.32	1254.47	1254.78	1255.32
	SD	0.06	0.06	0.03	0.05
trans-Pinocarveol	I	1138.13	1143.22	1148.24	1153.54
	SD	0.08	0.09	0.09	0.06
cis-Verbenol	I	1140.98	1144.56	1148.35	1152.46
	SD	0.08	0.13	0.07	0.07
Carvacrol	I	1301.51	1301.31	1301.52	1302.21
	SD	0.04	0.03	0.04	0.03
Thymol	I	1293.12	1291.18	1291.73	1291.85
	SD	0.07	0.12	0.05	0.04

Analyte		Temperature (°C)			
		120	130	140	150
Citronellol	I	1228.88	1229.44	1230.12	1231.01
	SD	0.02	0.03	0.03	0.05
Geraniol	I	1256.06	1256.99	1258.04	1259.35
	SD	0.02	0.03	0.03	0.04
trans-Pinocarveol	I	1158.91	1164.53	1170.19	1176.26
	SD	0.04	0.05	0.06	0.06
cis-Verbenol	I	1156.63	1161.00	1165.47	1170.24
	SD	0.03	0.04	0.04	0.04
Carvacrol	I	1303.33	1304.79	1306.73	1308.84
	SD	0.05	0.04	0.05	0.12
Thymol	I	1292.31	1293.15	1294.32	1295.91
	SD	0.04	0.02	0.04	0.04

From Hennig, P. and Engewald, W., *Chromatographia*, 38, 1994, 93–97. With permission.

$$\log T = -(0.979 \pm 0.056) + (0.349 \pm 0.004)\cdot {}^0\chi^v - (0.391 \pm 0.040)\cdot {}^4\chi_c^v$$

$$n = 26, \quad r = 0.999, \quad s = 0.029, \quad F = 6610 \tag{154}$$

$$\log T = -(0.670 \pm 0.053) + (0.349 \pm 0.004)\cdot {}^0\chi^r - (0.954 \pm 0.098)\cdot {}^4\chi_c^r$$

$$n = 26, \quad r = 0.999, \quad s = 0.029, \quad F = 6610 \tag{155}$$

The equations for tetra-*n*-alkylsilanes were

$$\log T = -(0.029 \pm 0.052) + (0.490 \pm 0.006)\cdot {}^1\chi - (0.174 \pm 0.031)\cdot {}^3\chi_c$$

$$n = 26, \quad r = 0.999, \quad s = 0.027, \quad F = 7180 \tag{156}$$

$$\log T = -(1.210 \pm 0.055) + (0.348 \pm 0.004)\cdot {}^0\chi^v - (0.214 \pm 0.021)\cdot {}^4\chi_c^v$$

$$n = 26, \quad r = 0.999, \quad s = 0.026, \quad F = 7772 \tag{157}$$

TABLE 10

Retention Indices (I) and Standard Deviations (SD) on HP-5 Capillary Column (Film Thickness 0.33 µm) at Various Temperatures

Analyte		Temperature (°C)			
		80	90	100	110
Citronellol	I	1227.79	1227.67	1227.82	1228.07
	SD	0.02	0.02	0.06	0.04
Geraniol	I	1253.35	1253.63	1254.15	1254.82
	SD	0.02	0.02	0.02	0.04
trans-Pinocarveol	I	1137.89	1142.78	1147.86	1153.12
	SD	0.01	0.02	0.01	0.02
cis-Verbenol	I	1140.37	1144.02	1147.81	1151.89
	SD	0.03	0.02	0.02	0.02
Carvacrol	I	1300.36	1300.26	1300.72	1301.43
	SD	0.04	0.02	0.07	0.04
Thymol	I	1291.98	1291.18	1290.90	1291.04
	SD	0.03	0.03	0.06	0.05

Analyte		Temperature (°C)			
		120	130	140	150
Citronellol	I	1228.51	1229.12	1229.88	1230.31
	SD	0.03	0.01	0.02	0.02
Geraniol	I	1255.64	1256.60	1257.71	1258.98
	SD	0.03	0.04	0.02	0.03
trans-Pinocarveol	I	1158.62	1164.27	1169.94	1175.90
	SD	0.02	0.02	0.03	0.03
cis-Verbenol	I	1156.22	1160.68	1165.21	1169.98
	SD	0.03	0.05	0.03	0.03
Carvacrol	I	1302.60	1304.13	1306.02	1308.20
	SD	0.05	0.04	0.03	0.03
Thymol	I	1291.56	1292.53	1293.53	1295.40
	SD	0.04	0.05	0.03	0.03

From Hennig, P. and Engewald, W., *Chromatographia*, 38, 1994, 93–97.

$$\log T = -(0.529 \pm 0.049) + (0.347 \pm 0.004) \cdot {}^0\chi^r - (0.880 \pm 0.089) \cdot {}^4\chi_c^r$$

$$n = 26, \quad r = 0.999, \quad s = 0.027, \quad F = 7615 \tag{158}$$

These results show that the molecular size, the skeletal branching, and the structural and electronic environment of the silicone or the germanium atoms have a considerable impact on their retention. When the tetra-*n*-alkylsilanes and tetra-*n*-alkyl-germanes are simultaneously included in the calculations the equations are slightly modified:

$$\log T = -(0.468 \pm 0.022) + (0.485 \pm 0.004) \cdot {}^1\chi + (0.021 \pm 001) \cdot {}^3\chi^v$$

$$n = 52, \quad r = 0.999, \quad s = 0.033, \quad F = 9810 \tag{159}$$

$$\log T = -(0.290 \pm 0.040) + (0.476 \pm 0.004) \cdot {}^1\chi - (0.183 \pm 0.023) \cdot {}^3\chi_c$$
$$-(0.277 \pm 0.016) \cdot S$$

$$n = 52, \quad r = 0.999, \quad s = 0.029, \quad F = 8850 \tag{160}$$

$$\log T = -(1.265 \pm 0.030) + (0.372 \pm 0.003) \cdot {}^0\chi^r + (0.040 \pm 0.002) \cdot {}^4\chi_{pc}^v$$

$$n = 52, \quad r = 0.999, \quad s = 0.034, \quad F = 9478 \tag{161}$$

TABLE 11

Chemical Structure, Logarithm of Retention Times, and Topological Indices $^1\chi$, $^3\chi_c$, S, and TTV of Tetra-n-alkylsilanes and Tetra-n-alkylgermanes

No.	Compound[a]	$^1\chi_c$	$^3\chi_c$	S	TTV	log T[b] Meas.	log T[b] Calc.
1	Me$_3$BuSi	3.5607	1.5607	−0.6782	14.3584	1.29	1.29
2	MeEt$_3$Si	3.6820	0.9286	−0.6713	14.1734	1.43	1.49
3	Me$_2$Pr$_2$Si	4.1213	1.2071	−0.7033	17.2755	1.60	1.64
4	Me$_2$EtBuSi	4.1213	1.2071	−0.6982	17.2958	1.61	1.64
5	MeEt$_2$PrSi	4.1820	0.9268	−0.6973	17.2185	1.70	1.73
6	Et$_4$Si	4.2426	0.7071	−0.6914	17.1717	1.80	1.80
7	Me$_2$PrBuSi	4.6213	1.2071	−0.7243	20.3554	1.90	1.88
8	MeEtPr$_2$Si	4.6820	0.9268	−0.7234	20.2999	1.95	1.97
9	Et$_3$PrSi	4.7426	0.7071	−0.7174	20.2666	2.05	2.05
10	Me$_2$Bu$_2$Si	5.1213	1.2071	−0.7453	23.4573	2.15	2.12
11	MePr$_3$Si	5.1820	0.9268	−0.7494	23.4178	2.20	2.21
12	Et$_2$Pr$_2$Si	5.2426	0.7071	−0.7434	23.3979	2.29	2.29
13	Et$_3$BuSi	5.2426	0.7071	−0.7384	23.3913	2.32	2.29
14	MePr$_2$BuSi	5.6820	0.9268	−0.7704	26.5620	2.47	2.45
15	MeEtBu$_2$Si	5.6820	0.9268	−0.7653	26.5546	2.49	2.46
16	EtPr$_3$Si	5.7426	0.7071	−0.7695	26.5656	2.52	2.53
17	Pr$_4$Si	6.2426	0.7071	−0.7955	29.7698	2.74	2.77
18	EtPr$_2$BuSi	6.2426	0.7071	−0.7905	29.7462	2.77	2.77
19	Et$_2$Bu$_2$Si	6.2426	0.7071	−0.7854	29.7254	2.82	2.77
20	MeBu$_3$Si	6.6820	0.9268	−0.8123	32.9167	2.99	2.93
21	Pr$_3$BuSi	6.7426	0.7071	−0.8165	32.9782	2.99	3.02
22	EtPrBu$_2$Si	6.7426	0.7071	−0.8114	32.9489	3.03	3.01
23	Pr$_2$Bu$_2$Si	7.2426	0.7071	−0.8375	36.2088	3.24	3.25
24	EtBu$_3$Si	7.2426	0.7071	−0.8324	36.1737	3.30	3.25
25	PrBu$_3$Si	7.7426	0.7071	−0.8584	39.4616	3.49	3.49
26	Bu$_4$Si	8.2426	0.7071	−0.8794	42.7365	3.72	3.73
27	Me$_3$BuGe	3.5607	1.5607	−1.1086	13.4109	1.42	1.41
28	MeEt$_3$Ge	3.6820	0.9286	−1.1250	13.0229	1.57	1.61
29	Me$_2$Pr$_2$Ge	4.1213	1.2071	−1.1597	16.1384	1.77	1.77
30	Me$_2$EtBuGe	4.1213	1.2071	−1.1503	16.1933	1.77	1.76
31	MeEt$_2$PrGe	4.1820	0.9268	−1.1632	15.9902	1.84	1.86
32	Et$_4$Ge	4.2426	0.7071	−1.1667	15.8485	1.94	1.94
33	Me$_2$PrBuGe	4.6213	1.2071	−1.1885	19.1749	2.04	2.01
34	MeEtPr$_2$Ge	4.6820	0.9268	−1.2014	18.9886	2.09	2.10
35	Et$_3$PrGe	4.7426	0.7071	−1.2049	18.8569	2.18	2.18
36	Me$_2$Bu$_2$Ge	5.1213	1.2071	−1.2172	22.2309	2.31	2.25
37	MePr$_3$Ge	5.1820	0.9268	−1.2396	22.0178	2.35	2.35
38	Et$_2$Pr$_2$Ge	5.2426	0.7071	−1.2431	21.8963	2.42	2.43
39	Et$_3$BuGe	5.2426	0.7071	−1.2336	21.9310	2.45	2.42
40	MePr$_2$BuGe	5.6820	0.9268	−1.2683	25.1095	2.61	2.59
41	MeEtBu$_2$Ge	5.6820	0.9268	−1.2589	25.1429	2.64	2.59
42	EtPr$_3$Ge	5.7426	0.7071	−1.2813	24.9666	2.66	2.67
43	Pr$_4$Ge	6.2426	0.7071	−1.3494	28.0678	2.89	2.92
44	EtPr$_2$BuGe	6.2426	0.7071	−1.3100	28.0891	2.91	2.91
45	Et$_2$Bu$_2$Ge	6.2426	0.7071	−1.3006	28.1125	2.95	2.91
46	MeBu$_3$Ge	6.6820	0.9268	−1.3258	31.3514	3.14	3.07
47	Pr$_3$BuGe	6.7426	0.7071	−1.3482	31.2146	3.13	3.16
48	EtPrBu$_2$Ge	6.7426	0.7071	−1.3388	31.2313	3.16	3.16
49	Pr$_2$Bu$_2$Ge	7.2426	0.7071	−1.3769	34.3810	3.38	3.40
50	EtBu$_3$Ge	7.2426	0.7071	−1.3675	34.3930	3.40	3.40
51	PrBu$_3$Ge	7.7426	0.7071	−1.4057	37.5670	3.61	3.64
52	Bu$_4$Ge	8.2426	0.7071	−1.4344	40.7725	3.85	3.89

[a] Me = methyl, Et = ethyl, Pr = n-propyl, and Bu = n-butyl.
[b] Logarithm of retention time taken from Reference 73.

From Kupchik, E. J., *J. Chromatogr.*, 630, 1993, 223–230.

The results demonstrate again that the topological indices can be successfully used for the prediction of the retention behavior of nonhomologous series of tetra-n-alkylsilanes and tetra-n-alkylgermanes.

Retention time as dependent variable was also used in the gas chromatographic identification of flavonoid aglycones.[78] To facilitate the identification of unknown flavonoid aglycones the increment method has been used. The retention times of 51 flavonoid aglycones were determined on an OV-1 DF capillary column. The relative retention times were calculated with hispidulin (5,7,4'-tri-hydroxy-6-methoxyflavone) as the standard. The structure-retention increments (i) were calculated from the differences in the retention times of the corresponding analytes I and II. The increment values are compiled in Table 12. It has been concluded from the data in Table 12 that the increments can be calculated with acceptable reliability and they allow the identification of unknown flavonoid aglycones in complex mixtures.

TABLE 12

Influences of Substitution Patterns on GC Retention Times of Flavonoid Aglycones

General structure

I	II	i ± S.D.	n
Structure of C ring			
2,3-Dihydro	Δ2,3	0.590 ± 0.013	7
3-OH	3-Unsubstituted	1.099 ± 0.021	10
3-OH	3-OCH$_3$	1.028 ± 0.022	2
Methylation of OH Group			
4'-OH	4'-OCH$_3$	1.121 ± 0.049	9
3'-OCH$_3$, 4'-OH	3'4'-OCH$_3$	0.910 ± 0.044	4
7-OH	7-OCH$_3$	1.145 ± 0.018	9
6-OCH$_3$, 7-OH	6,7-OCH$_3$	0.763 ± 0.009	4
Introduction of Additional Substituents			
6-Unsubstituted	6-OCH$_3$	1.027 ± 0.039	6
6-Unsubstituted, 7-OCH$_3$	6,7-OCH$_3$	0.653 ± 0.046	3
6-Unsubstituted, 3'5'-OCH$_3$	6-OCH$_3$, 3'5'-OCH$_3$	0.54	1
3'-Unsubstituted, 4'-OH	3'-OCH$_3$, 4'-OH	0.829 ± 0.003	9
3'-Unsubstituted, 4'-OCH$_3$	3'-OH, 4'-OCH$_3$	0.665 ± 0.036	7
3'-Unsubstituted, 4'-OCH$_3$	3'4'-OCH$_3$	0.689 ± 0.011	5
Pairs of Isomers			
3'-OCH$_3$ 4'-OH	3'-OH, 4'-OCH$_3$	0.861 ± 0.012	7
6-OCH$_3$	6-Unsubstituted		
8-Unsubstituted	8-OCH$_3$	0.895 ± 0.004	2

From Schmidt, T. J., Merfort, I., and Willuhn, G., *J. Chromatogr. A*, 669, 1994, 236–240. With permission.

The retention time and the response factor (flame ionization detector) of 152 analytes were determined on a cross-linked methyl silicone capillary column and were correlated with 37 quantum-chemical and conventional molecular descriptors.[79] The parameters of the best two- to six-parameter correlations for retention times are compiled in Table 13. The parameters included in the calculation are significantly correlated with the retention behavior of the analytes, indicating that these characteristics have a considerable influence on the retention. The statistical data for the best six-parameter equation are given in Table 14. The data in Table 14 indicate that analytes with higher polarizability bind more strongly to the stationary phase. The significant contribution of the minimum valency of an H atom suggests the importance of hydrogen bonds in the interaction of analytes with the stationary phase. The parameters of the best two- to six-parameter correlations for response factors are compiled in Table 15. Similarly to the retention time, the response factor can also be related to the quantum-chemical and conventional molecular descriptors. The equations fit well to the experimental data, the significance level being in each instance over 95%. The statistical parameters for the best six-variable equation are given in Table 16. The process of flame ionization involves the thermal decomposition of the analyte, which depends considerably on the chemical structure and physicochemical characteristics of the analyte. The marked impact of the relative weight of "effective" carbon atoms in the analyte may be due to the capacity of nonoxidized carbon atoms to produce detector response.

TABLE 13

Best Two- to Six-Parameter Correlations of Gas Chromatographic Retention Times t_R of 152 Compounds with a Combined Set of 37 Molecular Descriptors

Eq	N[a]	D[b]	R^2 [c]	R^2_{cv} [d]	s[e]
1	2	8,24	0.9134	0.9092	0.7386
2	3	5,8,24	0.9318	0.9271	0.6576
3	4	8,13,21,24	0.9465	0.9419	0.5841
4	5	8,23,24,25,32	0.9539	0.9496	0.5445
5	6	5,8,23,24,25,32	0.9590	0.9550	0.5152

Note: 5 = Maximum atomic orbital electronic population; 8 = minimum valency of a H atom; 13 = maximum exchange energy for a C–H bond; 21 = transitional entropy of the molecule at 300 K; 23 = total entropy of the molecule at 300 K per number of atoms; 24 = α polarizability; 25 = molecular weight; 32 = relative number of C–H bonds.

[a] Number of parameters in the correlation.
[b] Descriptors involved in the correlation.
[c] Square of the correlation coefficient.
[d] Square of the cross-validated correlation coefficient.
[e] Standard deviation.

From Katritzky, A. R. et al., *Anal. Chem.*, 66, 1994, 1799–1807. With permission.

B. Capacity Factor

Not only retention time, but also capacity factor has been frequently used for the elucidation of the relationship between retention behavior and molecular characteristics of the analytes. It is well established that the partitioning of analytes between the gas and stationary phase is related to their retention in gas–liquid chromatography:

$$k' = K \cdot \Phi \tag{162}$$

TABLE 14

Details of the Best Six-Parameter Correlation of the Retention Times t_R with the Combined Set of 37 Descriptors

Descriptor	Coefficient[a]	Error[b]	t[c]	R_n^2 [d]
Intercept	26.50	5.98	6.942	
5	0.929	0.218	4.256	0.959
8	−21.55	4.02	−5.362	0.953
23	−0.871	0.102	−8.543	0.696
24	0.046	0.005	8.389	0.906
25	0.018	0.003	5.869	0.939
32	−6.912	0.746	−9.269	0.122

[a] Partial regression coefficient.
[b] Error of partial regression coefficient.
[c] t-Value for the regression term.
[d] R^2 for the correlation.

Note: 5 = Maximum atomic orbital electronic population; 8 = minimum valency of a H atom; 23 = total entropy of the molecule at 300 K per number of atoms; 24 = α polarizability; 25 = molecular weight; 32 = relative number of C–H bonds.

From Katritzky, A. R. et al., *Anal. Chem.*, 66, 1994, 1799–1807. With permission.

TABLE 15

Best Two- to Six-Parameter Correlations of Gas Chromatographic Response Factors RF of 152 Compounds with a Combined Set of 37 Molecular Descriptors

Eq	N[a]	D[b]	R^2 [c]	R_{cv}^2 [d]	s[e]
1	2	15,28	0.7630	0.7531	0.0795
2	3	11,15,28	0.8256	0.8160	0.0684
3	4	10,15,28,32	0.8490	0.8833	0.0639
4	5	8,10,15,36,37	0.8869	0.8763	0.0555
5	6	4,8,10,15,36,37	0.8924	0.8810	0.0543

Note: 4 = Total hybridization component of the molecular dipole; 8 = minimum valency of an H atom; 10 = minimum total bond order (>0.1) of a C atom; 11 = minimum total bond order (>0.1) of an H atom ; 15 = total molecular one-center electron–electron repulsion energy; 28 = relative weight of C atoms; 36 = relative number of "effective" C atoms; 37 = relative weight of "effective" C atoms.

[a] Number of parameters in the correlation.
[b] Descriptors involved in the correlation.
[c] Square of the correlation coefficient.
[d] Square of the cross-validated correlation coefficient.
[e] Standard deviation.

From Katritzky, A. R. et al., *Anal. Chem.*, 66, 1994, 1799–1807. With permission.

where k′ is the capacity factor of the analyte; K is the partition coefficient; and Φ is the phase ratio. The exact knowledge of the partition coefficients may facilitate the prediction of the retention time of the analytes and the efficiency of separation in a given GC system. The gas-*n*-hexadecane partition coefficients of 105 analytes were determined on fused-silica open tubulary capillary columns[80] and the results were compared with the values previously published.[81] The gas-*n*-hexadecane partition coefficients were calculated by

$$K_x = K_{n\text{-hexane}} \cdot (t_{R.x} - t_m)/(t_{R.n\text{-hexane}} - t_m) \tag{163}$$

where K_x is the partition coefficient of the analyte; $K_{n\text{-hexane}}$ is the partition coefficient of *n*-hexane; $t_{R.x}$ and $t_{R.n\text{-hexane}}$ are the retention times of the analyte and *n*-hexane, respectively;

TABLE 16

Details of the Best Six-Parameter Correlation of the Response Factors RF t_R with the Combined Set of 37 Descriptors

Descriptor	Coefficient[a]	Error[b]	t[c]	$R_n^{2\,d}$
Intercept	−2.327	0.458	−6.409	
4	−0.030	0.011	−2.733	0.892
8	3.316	0.376	8.827	0.886
10	−0.206	0.019	−10.602	0.824
15	−0.0028	0.0002	−11.343	0.752
36	−1.160	0.102	−11.343	0.777
37	−0.958	0.046	−20.762	0.465

Note: 4 = Total hybridization component of the molecular dipole; 8 = minimum valency of an H atom; 10 = minimum total bond order (>0.1) of a c atom; 15 = total molecular one-center electron–electron repulsion energy; 36 = relative number of "effective" C atoms; 37 = relative weight of "effective" C atoms.

[a] Partial regression coefficient.
[b] Error of partial regression coefficient.
[c] t-Value for the regression term.
[d] R^2 for the correlation.

From Katritzky, A. R. et al., *Anal. Chem.*, 66, 1994, 1799–1807. With permission.

and t_m is the dead time of the column. The gas-*n*-hexadecane partition coefficients are compiled in Table 17. The comparison of the results compiled in Table 17 with those found in the literature[81] clearly show that fused-silica capillary columns are also suitable for the determination of partition coefficients. The differences observed in the case of polar analytes suggest that in these instances the interactions between the silica wall and the polar substructures of analytes may influence the retention. The retention increases in each series of analytes with the increasing hydrophobicity of the analyte. This finding indicates that the interaction between the stationary phase and the analytes is of hydrophobic character and analytes with different hydrophobicity can be successfully separated on apolar stationary phases.

Two models developed for the description of the retention behavior of analytes in GLC[82,83] have been compared[84] using capacity factor as gas-chromatographic descriptor. According to the models the solvation of an analyte is subdivided into three consecutive steps: (1) the creation of a cavity in the solvent to accommodate the analyte; (2) reorganization of the solvent molecules around the cavity; and (3) formation of various solvent–analyte interactions. To compare the reliability of both methods a stationary phase (bis[3-allyl-4-hydroxyphenyl]sulphone) with marked hydrogen bond acidity (further H10) has been synthesized and the retention of a wide variety of analytes have been determined. The specific retention volume V_G has been subdivided into the contribution of the gas–liquid partition and interfacial adsorption (Table 18). It has been proved that the retention of the analytes is related to their physicochemical parameters:

$$\log K(121.2°C) = -(0.568 \pm 0.060) - (0.051 \pm 0.044) \cdot R_2 + (1.323 \pm 0.052) \cdot \pi_2^H$$
$$+ (1.266 \pm 0.052) \cdot \alpha_2^H + (1.457 \pm 0.070) \cdot \beta_2^H + (0.418 \pm 0.010) \cdot \log L^{16}$$

$$n = 58, \quad r = 0.9940, \quad S.D. = 0.069, \quad F = 865 \tag{164}$$

$$\log K(176.2°C) = -(0.749 \pm 0.123) + (0.165 \pm 0.088) \cdot R_2 + (1.160 \pm 0.096) \cdot \pi_2^H$$
$$+ (0.808 \pm 0.105) \cdot \alpha_2^H + (1.290 \pm 0.139) \cdot \beta_2^H + (0.332 \pm 0.020) \cdot \log L^{16}$$

$$n = 54, \quad r = 0.9738, \quad S.D. = 0.134, \quad F = 176 \tag{165}$$

TABLE 17
n-Hexadecane Gas–Liquid Partition Coefficients

Analyte	K	Analyte	K
n-Pentane	2.163	Acetophenone	4.458
n-Hexane	2.668	Acetonitrile	1.537
n-Heptane	3.173	Propionitrile	1.978
n-Octane	3.677	Benzonitrile	3.913
n-Nonane	4.176	Benzyl cyanide	4.570
n-Decane	4.685	Acetaldehyde	1.240
Cyclopentane	2.426	Propionaldehyde	1.770
Cyclohexane	2.906	Benzylaldehyde	3.935
2,4-Dimethylpentane	2.812	Tetrahydrofuran	2.521
2,5-Dimethylhexane	3.309	p-Dioxane	2.788
Cycloheptane	3.543	Diisopropyl ether	2.561
2-Methylpentane	2.507	Diethyl ether	2.066
Ethylcyclohexane	3.767	Dipropyl ether	2.971
2,3,4-Trimethylpentane	3.401	Dibutyl ether	3.954
Hexene	2.571	Anisole	3.916
Benzene	2.792	Methylene chloride	1.997
Toluene	3.343	Chloroform	2.478
Ethylbenzene	3.785	Carbon tetrachloride	2.822
Propylbenzene	4.239	1,2-Dichloroethane	2.572
Butylbenzene	4.714	Chlorobutane	2.716
p-Xylene	3.858	Chloropentane	3.232
m-Xylene	3.868	Fluorobenzene	2.785
o-Xylene	3.947	Chlorobenzene	3.630
Ethanol	1.556	Bromobenzene	4.022
n-Propanol	1.975	Iodobenzene	4.505
n-Butanol	2.539	o-Dichlorobenzene	4.453
n-Pentanol	3.057	p-Dichlorobenzene	4.405
n-Hexanol	3.550	Dimethylformamide	2.922
n-Heptanol	4.067	Dimethylacetamide	3.357
n-Octanol	4.569	Dimethylsulfoxide	3.110
2-Propanol	1.750	Ethylamine	1.646
Benzyl alcohol	4.162	Propylamine	2.083
2,2,2-Trifluoroethanol	1.315	Butylamine	2.575
tert-Butanol	1.994	Hexylamine	3.612
2-Methyl-1-propanol	2.381	Triethylamine	3.008
sec-Butanol	2.322	Diethylamine	2.386
Isopentanol	2.885	Pyridine	2.969
Cyclopentanol	3.107	Aniline	3.934
Cyclohexanol	3.594	N-Methylaniline	4.492
1,1,1,3,3,3-Hexafluoroisopropanol	1.370	N,N-Dimethylaniline	4.753
		Methyl formate	1.454
2-Phenylethanol	4.552	Methyl acetate	1.946
3-Phenylpropanol	4.663	Ethyl acetate	2.359
4-Phenylbutanol	5.049	Propyl acetate	2.861
Phenol	3.641	Acetic acid	2.331
m-Cresol	4.187	Propionic acid	2.978
p-Cresol	4.254	Butyric acid	3.427
o-Cresol	4.183	Ethyl propionate	2.860
Acetone	1.766	Nitromethane	1.839
2-Butanone	2.269	Nitroethane	2.313
2-Pentanone	2.726	Nitropropane	2.773
Cyclopentanone	3.093	Nitrobenzene	4.433
Cyclohexanone	3.580		

From Zhang, Y., Dallas, A. J., and Carr, P. W., *J. Chromatogr.*, 638, 1993, 43–56. With permission.

TABLE 18

Examples of Contributions from Gas–Liquid Partition and Interfacial Adsorption to V_G Values (cm³g⁻¹) for Solutes on H10 at 121°C (Phase Loading 8.9% [w/w] H10)

Solute	$V_{G(exp)}$	$V_{G(partition)}$	$V_{G(adsorption)}$	%ᵃ
n-Dodecane	64.59	29.91	34.68	53.7
n-Tridecane	112.01	55.06	56.95	50.8
n-Tetradecane	194.72	99.71	95.01	48.8
n-Pentadecane	567.17	286.67	280.50	49.5
cis-Hydrindane	27.79	25.70	2.09	7.5
Oct-2-yne	27.25	19.44	7.81	28.7
Heptan-2-one	312.59	269.79	42.80	13.7
Methyl nonanoate	828.23	652.13	176.10	21.3
1-Nitropentane	333.49	323.93	9.57	2.9
N,N-Dimethylformamide	2159.34	2115.21	44.09	2.0
Heptan-1-ol	764.64	609.60	137.04	18.4
Toluene	31.66	25.54	6.12	19.3
Anisole	161.55	143.11	18.44	11.4
Bromobenzene	94.49	90.37	4.12	4.4
Acetophenone	1528.49	1363.25	165.24	10.8
Phenol	1059.18	993.24	65.94	6.2
4-Chlorophenol	4585.81	4413.33	172.48	3.8

ᵃ This is the percent contribution of adsorption to $V_{G(exp)}$ with a phase loading of 8.9%. The $V_{G(partition)}$ values used to obtain the log K values are those extrapolated to zero contribution from adsorption.

From Abraham, M. H. et al., *J. Chromatogr.*, 646, 1993, 351–360. With permission.

where K is the partition coefficient of a solute between the mobile and the stationary phases; R_2 is an excess molar refraction reflecting general dispersion interactions; π_2^H is a dipolarity/polarizability parameter; α_2^H and β_2^H are the hydrogen bond acidity and basicity of the analyte, respectively; log L^{16} is the gas–liquid partition coefficient on hexadecane at 25°C.

Equations 164 and 165 clearly show that the retention behavior of analytes and the separation capacity of a given GLC support can be precisely predicted when the physicochemical parameters are included. It has been further established that the application of both methods leads to compatible and precise prediction of the retention behavior of these analytes.

The capacity factor k′ and the solvation enthalpy ΔH^0 in GLC were correlated by the various physicochemical parameters of stationary phases and analytes.[85] The linear solvation energy relationships equations were

$$\log k' = SP_0 + l \cdot \log L^{16} + s \cdot \pi_2^{*,C} + a \cdot \alpha_2^C + b \cdot \beta_2^C \tag{166}$$

$$\Delta H^0 = SP_0^H + l^H \cdot \log L^{16} + s^H \cdot \pi_2^{*,C} + d^H \cdot \delta_2 + a^H \cdot \alpha_2^C + b^H \cdot \beta_2^C \tag{167}$$

where SP_0 is a solute-independent, column-dependent constant; L^{16} is the partition coefficient of the analyte from the gas phase to n-hexadecane at 298 K; $\pi_2^{*,C}$ is a GC-based analyte dipolarity/polarizability parameter; δ_2 is an empirical polarizability correction factor (0 for aliphatics, 0.5 for polyhalogenated compounds, and 1 for aromatics); α_2^C and β_2^C are hydrogen bond donor acidity and hydrogen acceptor basicity of analytes, respectively. The retention indices of 53 analytes with highly different chemical structures were determined on eight capillary columns, and Eq. 167 was used for the elucidation of the relationships between retention behavior and physicochemical parameters. The results are compiled in Table 19. The data in Table 19 prove that the solute parameters can be used

TABLE 19

Regression Results for Log k' at All Temperatures

Column	T(°C)	SP$_0$	l	s	d	a	b	S.D.	r
DB-1	150	−2.120	0.438	0.217	0.070	−0.035	ns	0.057	0.996
		0.026	0.007	0.030	0.023	0.035			
	115	−2.013	0.513	0.281	0.026	0.058	ns	0.044	0.998
		0.020	0.006	0.023	0.018	0.027			
	80	−1.957	0.627	0.323	−0.015	0.215	ns	0.029	0.999
		0.013	0.004	0.015	0.012	0.018			
	45	−1.877	0.769	0.401	−0.104	0.372	ns	0.038	0.999
		0.017	0.005	0.020	0.016	0.023			
DB-5	150	−2.180	0.446	0.356	0.052	−0.046	ns	0.050	0.997
		0.023	0.006	0.026	0.020	0.031			
	115	−2.095	0.517	0.414	0.019	−0.017	ns	0.043	0.998
		0.019	0.005	0.022	0.019	0.026			
	80	−2.030	0.620	0.451	−0.023	0.146	ns	0.034	0.999
		0.015	0.004	0.018	0.014	0.021			
	45	−1.961	0.760	0.523	−0.093	0.309	ns	0.042	0.999
		0.019	0.005	0.022	0.017	0.026			
DB-1301	115	−2.293	0.526	0.636	−0.031	0.360	ns	0.056	0.997
		0.026	0.007	0.030	0.023	0.035			
	80	−2.149	0.621	0.723	−0.073	0.557	ns	0.048	0.998
		0.022	0.006	0.025	0.019	0.029			
	60	−2.093	0.695	0.842	−0.139	0.698	ns	0.051	0.998
		0.024	0.007	0.027	0.021	0.031			
	45	−2.083	0.765	0.902	−0.156	0.884	ns	0.056	0.998
		0.027	0.007	0.029	0.023	0.035			
DB-1701	150	−2.294	0.427	0.824	−0.035	0.267	−0.133	0.033	0.998
		0.018	0.004	0.022	0.014	0.021	0.027		
	115	−2.233	0.507	0.931	−0.077	0.463	−0.085	0.024	0.999
		0.013	0.003	0.016	0.010	0.016	0.020		
	80	−2.156	0.616	1.071	−0.133	0.669	−0.086	0.032	0.999
		0.017	0.004	0.021	0.013	0.021	0.026		
	60	−2.102	0.687	1.157	−0.163	0.838	−0.069	0.038	0.999
		0.021	0.005	0.025	0.016	0.025	0.031		
	45	−2.016	0.744	1.239	−0.214	0.963	−0.111	0.044	0.999
		0.023	0.006	0.028	0.018	0.028	0.035		
DB-17	150	−2.420	0.427	0.827	0.081	−0.068	ns	0.047	0.997
		0.021	0.006	0.025	0.019	0.039			
	115	−2.354	0.506	0.960	0.058	−0.003	ns	0.041	0.998
		0.019	0.005	0.022	0.017	0.034			
	80	−2.266	0.600	1.121	0.031	0.034	ns	0.043	0.999
		0.020	0.006	0.023	0.018	0.036			
	45	−2.152	0.723	1.343	−0.017	0.171	ns	0.049	0.999
		0.022	0.006	0.026	0.020	0.040			
DB-210	115	−2.149	0.399	1.454	0.220	−0.319	ns	0.049	0.997
		0.022	0.006	0.025	0.020	0.030			
	80	−2.052	0.489	1.667	0.285	−0.274	ns	0.057	0.997
		0.026	0.007	0.030	0.023	0.035			
	60	−1.996	0.551	1.815	0.332	−0.224	ns	0.063	0.997
		0.028	0.008	0.033	0.025	0.038			
	45	−1.938	0.606	1.930	−0.367	−0.181	ns	0.054	0.998
		0.025	0.007	0.029	0.022	0.034			
DB-225	150	−2.367	0.371	1.512	0.001	0.436	−0.096	0.053	0.997
		0.028	0.007	0.034	0.022	0.034	0.043		
	115	−2.287	0.445	1.618	−0.021	0.584	0.004	0.036	0.999
		0.019	0.005	0.023	0.015	0.023	0.029		
	80	−2.194	0.537	1.794	−0.044	0.837	0.073	0.036	0.999
		0.019	0.005	0.023	0.015	0.023	0.029		
	45	−2.060	0.654	2.006	−0.091	1.128	0.144	0.038	0.999

TABLE 19 (CONTINUED)

Regression Results for Log k' at All Temperatures

Column	T(°C)	SP_0	l	s	d	a	b	S.D.	r
		0.020	0.005	0.025	0.016	0.025	0.031		
DB-WAX	115	−2.245	0.416	1.819	0.095	1.365	ns	0.077	0.996
		0.037	0.010	0.041	0.031	0.048			
	80	−2.195	0.505	2.127	0.058	1.953	ns	0.052	0.998
		0.027	0.007	0.028	0.022	0.051			
	60	−2.119	0.559	2.325	0.022	2.198	ns	0.052	0.998
		0.028	0.007	0.030	0.024	0.052			
	40	−2.062	0.606	2.501	0.004	2.414	ns	0.059	0.998
		0.033	0.008	0.034	0.027	0.060			

Note: Second rows refer to the standard deviation of the variable; ns = not significant.

From Li, J. and Carr, P. W., *J. Chromatogr. A*, 659, 1994, 367–380. With permission.

for the exact prediction of the retention indices on any column and at any temperature. The magnitude of coefficient l is an indicator of the capacity of the stationary phases for dispersion interactions with the analytes, coefficient s increases with the increasing polarity of the stationary phase, and coefficient a reflects the capacity to accept hydrogen bonds. Because the parameters in Table 19 depend linearly on 1/T, the retention indices can be predicted at any temperature, thus facilitating considerably the optimization of any GLC separation. The infinite dilution enthalpy of solution (H^0) was calculated from the temperature dependence of the capacity factor. The enthalpy values are compiled in Table 20. The enthalpy values differ considerably both among the stationary phases and analytes, indicating that the retention is equally influenced by the physicochemical characteristics of the analyte and the stationary phase. The apparent Gibbs free energies ($\Delta G'$) were also calculated[86] for the same analytes under the same gas-chromatographic conditions.[87] The energy values are shown in Table 21. Linear solvation energy calculations proved that not only the enthalpy, but also the Gibbs free energy significantly depends on the various physicochemical parameters such as dispersive interaction, dipolarity, and hydrogen bond donor and hydrogen bond acceptor strength. This finding emphasizes again the importance of these processes in the retention mechanism.

In order to find the physicochemical characteristic of analyte that accounts for their retention behavior the log k' value of 87 analytes were determined on five stationary phases.[88] The phases were selected in an attempt to find strong hydrogen donor stationary phases: stearic acid (SA), N-tetradecyl-1,1,1-trifluoroacetamide (TFAMIDE), 3-pentadecylphenol (PDP), α-trifluoromethyl alcohols (TFOH), and α,α-bis(trifluoromethyl)-p-dodecyl benzyl alcohol (HFOH). The physicochemical parameters of the analytes included in the calculations were L^{16} (partition coefficient of the solute from the gas phase to n-hexadecane at 298 K), $\pi_2^{*,c}$ (dipolarity/polarizability parameter), δ_2 (polarizability correction parameter), a_2^c and β_2^c (hydrogen bond donor acidity and hydrogen bond acceptor basicity of the analyte, respectively).[89] In each instance highly significant linear correlations were found between the retention behavior and physicochemical characteristics of analytes:

$$\log k'(\text{HFOH}) = (-1.47 \pm 0.01) + (0.693 \pm 0.003) \cdot \log L^{16} + (0.67 \pm 0.02) \cdot \pi_2^{*,c}$$
$$+ (-0.26 \pm 0.01) \cdot \delta_2 + (-0.09 \pm 0.02) \cdot \alpha_2^c + (2.22 \pm 0.02) \cdot \beta_2^c$$
$$\text{SD} = 0.027, \quad r = 0.9995, \quad n = 85 \tag{168}$$

$$\log k'(\text{TFOH}) = (-1.51 \pm 0.02) + (0.742 \pm 0.005) \cdot \log L^{16} + (0.56 \pm 0.03) \cdot \pi_2^{*,c}$$
$$+ (-0.18 \pm 0.02) \cdot \delta_2 + (0.48 \pm 0.03) \cdot \alpha_2^c + (1.58 \pm 0.03) \cdot \beta_2^c$$
$$\text{SD} = 0.046, \, r = 0.998, \, n = 80 \tag{169}$$

TABLE 20
Enthalpy (–ΔH) of the Retention Process (kJ/mol)

Compound	Stationary Phase							
	A	B	C	D	E	F	G	H
Cyclohexane	27.9	26.2	27.0	nd	25.9	nd	26.8	nd
1-Hexene	25.0	23.3	25.2	24.7	23.2	23.9	25.9	22.7
Pentane	21.6	20.5	24.3	21.0	19.3	18.5	19.1	16.1
Hexane	29.3	24.5	28.1	25.6	24.1	22.6	23.0	18.9
Octane	34.1	33.0	36.4	32.9	31.2	29.8	30.2	26.0
Decane	43.3	40.8	43.6	42.1	39.3	38.5	39.8	34.8
Undecane	47.7	45.0	47.6	46.4	43.2	41.7	42.9	37.8
Tetradecane	61.3	58.5	61.0	59.4	56.5	52.1	53.3	48.0
Pentadecane	66.0	63.0	65.5	63.9	59.0	55.9	57.1	51.5
Ethyl acetate	27.2	25.5	29.7	29.2	28.5	29.9	30.9	29.6
Propyl acetate	31.1	28.8	33.4	33.0	33.4	33.5	35.0	32.7
Diethyl ether	23.7	21.2	23.4	24.3	22.3	22.6	25.4	19.6
Dipropyl ether	30.3	27.8	31.3	30.9	29.1	28.0	31.2	25.7
Dibutyl ether	38.4	35.3	38.8	39.2	36.8	35.0	37.3	34.1
Acetonitrile	nd	20.3	24.8	26.2	24.9	27.6	28.5	32.3
Propionitrile	24.7	23.1	29.3	28.8	29.0	30.7	31.7	32.8
Acetone	20.1	19.3	28.1	24.4	24.9	26.8	27.1	26.2
2-Butanone	25.6	23.8	31.1	28.8	27.5	30.3	31.8	29.0
2-Pentanone	28.9	27.1	32.2	32.2	30.9	33.8	34.4	31.7
Dimethylformamide	32.3	31.4	37.4	38.9	37.5	41.2	41.4	42.4
Dimethylacetamide	35.6	33.3	40.9	42.0	41.2	45.3	44.8	45.2
Dimethylsulfoxide	34.6	32.3	40.9	41.1	41.3	44.6	45.2	49.6
Propionaldehyde	22.5	19.4	nd	25.9	23.9	25.3	24.6	25.7
Tetrahydrofuran	26.3	24.1	32.5	27.8	28.0	27.9	29.4	30.3
Triethylamine	32.3	27.5	31.1	33.4	28.1	30.3	33.5	nd
Nitromethane	25.7	25.8	27.9	29.7	28.4	30.6	32.1	nd
Nitroethane	27.8	25.0	31.9	32.3	31.2	33.6	33.8	37.7
Nitropropane	30.5	30.2	35.7	35.3	34.4	36.5	36.5	39.5
Methanol	16.5	19.9	22.2	24.1	nd	20.5	27.7	31.0
Ethanol	nd	23.8	nd	27.1	23.1	22.5	28.0	33.5
1-Propanol	24.0	27.5	29.3	29.9	25.4	28.0	nd	37.2
2-Propanol	23.2	22.9	25.0	27.1	21.5	24.0	30.0	33.6
2-Methyl-2-propanol	23.6	22.1	26.6	27.2	24.5	25.7	29.8	32.8
Trifluoroethanol	27.8	23.7	35.1	34.6	23.9	25.2	33.1	44.1
Hexafluoroisopropanol	34.0	30.1	43.3	43.5	nd	29.9	44.2	nd
Acetic acid	nd	30.0	40.8	nd	nd	nd	nd	nd
Aniline	37.9	36.9	44.4	44.3	43.8	42.9	49.0	nd
N-Methylaniline	41.4	38.5	47.0	46.6	47.3	45.4	50.6	nd
Phenol	41.1	40.0	52.1	50.6	43.8	41.2	53.6	nd
Benzyl alcohol	40.4	38.5	48.7	47.3	45.8	43.4	51.9	nd
m-Cresol	44.1	nd	55.6	54.8	47.3	45.1	56.9	nd
Ethylamine	19.3	18.6	nd	23.3	19.5	19.8	26.8	nd
Propylamine	25.7	22.3	nd	26.8	23.5	23.8	28.9	nd
Butylamine	29.0	24.6	32.2	32.3	29.3	28.6	32.2	nd
Benzene	27.1	25.4	28.5	28.4	28.1	26.4	29.4	29.2
Toluene	31.1	29.3	32.9	32.4	32.2	31.3	32.9	32.5
Ethylbenzene	34.7	32.8	36.4	36.1	35.8	34.6	36.8	35.4
Propylbenzene	38.5	36.1	40.1	40.3	39.3	37.8	40.2	38.2
p-Xylene	35.6	33.3	37.0	36.3	36.0	35.4	37.1	35.7
Benzaldehyde	37.3	36.1	41.4	41.3	42.5	41.7	43.3	47.6
Benzonitrile	39.0	37.0	43.3	42.4	44.6	43.9	45.2	nd
N,N-Dimethylaniline	43.2	38.6	45.9	44.8	46.3	45.8	47.8	nd
Carbon tetrachloride	27.1	26.9	27.8	27.1	27.3	nd	26.5	27.8

Note: nd = not determined; A = DB-1; B = DB-5; C = DB-1301; D = DB-1701; E = DB-17; F = DB-210; G = DB-225; H = DB-WAX.

From Li, J. and Carr, P. W., *J. Chromatogr. A*, 659, 1994, 367–380. With permission.

TABLE 21

Apparent Gibbs Free Energy (ΔG', kJ/mol) at the Harmonic Temperature

	Stationary Phase							
	A	B	C	D	E	F	G	H
Compound	93°C	93°C	73°C	86°C	93°C	73°C	93°C	73°C
Cyclohexane	1.78	2.23	nd	nd	4.00	nd	5.28	nd
1-Hexene	3.39	3.87	3.53	4.79	6.41	5.92	7.47	6.28
Pentane	5.34	6.04	5.73	7.11	8.95	7.97	10.14	9.45
Hexane	3.27	3.87	3.39	4.93	6.77	6.06	8.11	7.34
Octane	−1.02	−0.27	−1.07	0.56	2.57	2.32	4.07	3.43
Decane	−5.14	−4.51	−5.52	−3.55	−1.47	−1.23	0.48	−0.35
Undecane	−7.20	−6.53	−7.71	−5.71	−3.60	−3.08	−1.48	−2.30
Tetradecane	−13.43	−12.71	−14.20	−12.05	−9.85	−8.34	−7.06	−7.99
Pentadecane	−15.49	−14.75	−16.36	−14.15	−11.93	−10.20	−8.88	−9.86
Ethyl acetate	3.46	3.83	2.43	3.29	4.35	1.72	3.61	1.42
Propyl acetate	1.31	1.49	0.12	1.12	2.33	−0.19	1.72	−0.27
Diethyl ether	5.39	5.96	5.24	6.55	7.70	6.71	8.15	6.69
Dipropyl ether	1.52	2.03	1.14	2.66	3.93	3.57	4.87	3.70
Dibutyl ether	−2.51	−2.03	−3.16	−1.47	0.02	0.05	1.04	0.14
Acetonitrile	nd	6.28	4.23	4.22	5.24	1.35	2.49	−0.91
Propionitrile	4.45	4.32	2.39	2.37	3.65	−0.28	0.97	−1.36
Acetone	5.89	6.15	5.01	5.09	6.56	nd	4.49	2.71
2-Butanone	3.82	3.90	2.68	2.92	4.32	0.71	2.68	1.05
2-Pentanone	1.84	1.99	0.40	1.03	2.44	−0.96	0.98	−0.42
Dimethylformamide	0.23	−0.10	−2.56	−2.54	−1.46	−5.87	−4.67	−7.05
Dimethylacetamide	−1.68	−1.74	−4.48	−4.36	−3.24	−7.46	−6.25	−8.52
Dimethylsulfoxide	−0.84	−1.02	−4.31	−4.61	−3.50	−8.25	−7.66	−11.29
Propionaldehyde	6.26	6.03	nd	5.39	6.47	3.43	5.00	3.22
Tetrahydrofuran	3.11	3.10	nd	2.87	3.78	2.13	3.10	1.67
Triethylamine	1.77	2.22	1.08	2.67	4.14	2.77	4.79	nd
Nitromethane	5.08	5.01	2.42	2.51	3.67	0.01	0.64	nd
Nitroethane	2.96	2.68	0.64	0.73	1.90	−1.66	−0.90	−3.78
Nitropropane	1.03	1.00	−1.23	−0.99	0.23	−3.18	−2.34	−4.71
Methanol	8.50	9.21	7.34	7.42	nd	7.74	6.32	1.25
Ethanol	nd	7.80	nd	6.03	8.03	6.13	5.03	0.49
1-Propanol	4.45	5.32	3.08	3.65	5.73	4.05	nd	−1.50
2-Propanol	5.94	6.55	nd	5.23	7.15	5.29	4.71	0.69
2-Methyl-2-propanol	5.26	5.61	4.07	4.83	7.01	4.62	4.51	1.30
Trifluoroethanol	8.53	7.95	4.02	4.07	9.15	5.63	2.54	−4.16
Hexafluoroisopropanol	2.28	6.82	0.47	0.60	nd	nd	nd	nd
Acetic acid	nd	5.14	1.85	nd	nd	nd	nd	nd
Aniline	−4.03	−4.06	−6.61	−6.32	−5.64	−5.70	−8.44	nd
N-Methylaniline	−5.80	−5.69	−8.27	−7.75	−7.14	−7.02	−9.14	nd
Phenol	−3.96	−3.98	−8.15	−8.09	−4.83	−4.38	−10.52	nd
Benzyl alcohol	−5.09	−5.00	−8.08	nd	−6.44	−6.16	−9.88	nd
m-Cresol	−5.95	nd	−10.09	−9.93	−6.77	−6.35	−12.09	nd
Ethylamine	6.90	7.76	nd	7.05	8.45	5.96	7.68	nd
Propylamine	5.23	5.36	nd	5.17	6.29	3.82	5.60	nd
Butylamine	2.87	2.94	1.46	2.81	4.33	1.62	3.21	nd
Benzene	1.92	2.21	1.28	2.21	2.67	2.52	2.50	0.16
Toluene	−0.24	0.08	−1.01	0.01	0.59	0.55	0.55	−1.69
Ethylbenzene	−2.15	−1.81	−3.05	−1.92	−1.37	−1.02	−1.18	−3.29
Propylbenzene	−4.01	−3.66	−5.00	−3.77	−3.10	−2.62	−2.75	−4.77
p-Xylene	−2.32	−2.01	−3.21	−2.08	−1.38	−1.13	−1.12	−3.42
Benzaldehyde	−3.66	−3.82	−5.81	−5.34	−4.94	−6.22	−6.53	−10.26
Benzonitrile	−4.13	−4.26	−6.60	−6.26	−5.62	−7.65	−7.77	nd
N,N-Dimethylaniline	−6.52	nd	−8.25	−7.57	−7.15	−7.04	−7.78	nd
Carbon tetrachloride	1.94	2.29	1.49	2.68	3.23	nd	3.72	nd

Note: nd = not determined; A = DB-1; B = DB-5; C = DB-1301; D = DB-1701; E = DB-17; F = DB-210; G = DB-225; H = DB-WAX.

From Li, J. and Carr, P. W., *J. Chromatogr. A*, 670, 1994, 105–116. With permission.

$$\log k'(\text{PDP}) = (-1.32 \pm 0.02) + (0.602 \pm 0.005) \cdot \log L^{16} + (0.66 \pm 0.03) \cdot \pi_2^{*,c}$$
$$+ (-0.13 \pm 0.02) \cdot \delta_2 + (0.29 \pm 0.03) \cdot \alpha_2^c + (1.46 \pm 0.03) \cdot \beta_2^c$$
$$\text{SD} = 0.044, \quad r = 0.998, \quad n = 80 \tag{170}$$

$$\log k'(\text{TFAMIDE}) = (-1.66 \pm 0.02) + (0.67 \pm 0.006) \cdot \log L^{16} + (0.90 \pm 0.03) \cdot \pi_2^{*,c}$$
$$+ (-0.19 \pm 0.02) \cdot \delta_2 + (0.78 \pm 0.03) \cdot \alpha_2^c + (0.66 \pm 0.04) \cdot \beta_2^c$$
$$\text{SD} = 0.053, \quad r = 0.997, \quad n = 80 \tag{171}$$

$$\log k'(\text{SA}) = (-1.46 \pm 0.02) + (0.702 \pm 0.005) \cdot \log L^{16} + (0.43 \pm 0.02) \cdot \pi_2^{*,c}$$
$$+ (1.23 \pm 0.05) \cdot \alpha_2^c + (0.39 \pm 0.03) \cdot \beta_2^c$$
$$\text{SD} = 0.040, \quad r = 0.999, \quad n = 75 \tag{172}$$

The significant relationships between the retention parameters and physicochemical characteristics of analytes prove again that the retention behavior of analytes on these new stationary phases can be described by the parameters listed above.

The retention of 8 *para*-hydroxy benzoic acid esters was determined on an OV-17 capillary column at different temperatures and the enthalpy (ΔH°) and entropy (ΔS°) of the transfer of the analyte from the carrier gas to the stationary phase were calculated from the temperature dependence of the retention.[90] The physicochemical parameters are compiled in Table 22. Significant linear correlations were found between the physicochemical parameters and the first connectivity index ^1X of analytes:

$$-\Delta H^\circ = 51.89 + 4.73 \cdot {}^1\chi \quad r^2 = 0.95 \tag{173}$$

$$-S^\circ = 74.69 + 6.08 \cdot {}^1\chi \quad r^2 = 0.94 \tag{174}$$

Equations 173 and 174 allow the prediction of the retention by using the temperature and the first connectivity index:

$$\ln k' = {}^1\chi \cdot (5.68 \cdot 100/T - 0.73) + 6.24 \cdot 100/T - 14.97 \tag{175}$$

It has been stated that the equation above predicts with acceptable accuracy the retention of this homologous series of analytes.

TABLE 22

$-\Delta H^\circ$, $-\Delta S^\circ$, and $-T \cdot \Delta S^\circ$ Values for All Solutes for T = 443 K and Correlation Coefficient r

Ester	$-\Delta H^\circ$ (kJ/mol)	$-\Delta S^\circ$ (J/mol/K)	$-T \cdot \Delta S^\circ$ (kJ/mol)	r
Methyl	66.98	94.02	41.65	0.998
Ethyl	68.89	96.99	42.97	0.997
Propyl	70.94	101.00	44.74	0.999
Butyl	74.00	103.77	45.97	0.996
i-Propyl	71.44	98.98	43.85	0.999
i-Butyl	73.98	101.97	45.17	0.998
s-Butyl	73.85	104.00	46.07	0.998
t-Butyl	73.02	100.00	44.30	0.999

From Guillaume, Y. and Giunchard, C., *Chromatographia*, 39, 1994, 438–442. With permission.

C. The Kováts Index

Absolute and relative retention data have been extensively used in gas chromatography (GC) both for practical and theoretical purposes. It is fairly easy to obtain relative retention data, therefore they are mainly used in the solution of various practical separation problems. The exact knowledge of flow rate and the loading of stationary phase are not necessary for the calculation of relative retention indices. However, the absolute retention data related to the thermodynamic parameters are needed for theoretical works.[91–93] The Kováts index is the most frequently used retention parameter in gas chromatography. Its application and importance in the up-to-date gas chromatographic theory and practice has been evaluated many times.[94–97] Many good relationships were found between the retention index and the various physicochemical parameters of analytes. Thus, it has been found that boiling point,[98] density, molecular weight, molar volume,[99] refractive index,[100] various thermodynamic parameters,[101] van der Waals volume,[102] dipole moment,[103] capacity factor, electric interactions,[104] molecular orbitals,[105] ionization potential,[106] Taft and Palm constants,[107] molecular polarizability,[108] solubility factors,[109] various parameters derived from molecular structure,[110,111] and connectivity indices[112] have a marked influence on the retention of various analytes. However, the relationships mentioned above refer to different sets of analytes and the physicochemical parameters included in the calculations were not always the same, therefore it is not surprising that the conclusions also are different. Moreover, it has been recently established that the Kováts index is not entirely an additive property of the analytes.[113] The use of the results of correlation analysis instead of the additivity rule has been strongly advocated:

$$I(Cpd_i) = a \cdot I(Std_i) + b \tag{176}$$

where I is the Kováts retention index, Cpd is the compound, and Std is the reference compound.

1. Isocratic Separation Mode

a. Relationship with Physicochemical Parameters

Retention indices have been extensively used also for the tentative identification of volatile environmental pollutants. It has been recently proposed that the retention indices of various possible pollutants can be calculated relative to polycyclic aromatic hydrocarbons[114] and they can be related to the boiling point of the analytes. The measured retention indices of various pollutants relative to naphthalene, phenantrene, chrysene, and benzo[ghi]perylene on J&W Scientific DB-5 fused capillary column and the calculated retention indices using the relationship between boiling point and retention indices are compiled in Tables 23 and 24. It has been concluded that the I values listed in Tables 23 and 24 may facilitate the identification of these environmental pollutants. Good linear correlations were found between the measured I values and the boiling point (bp) of the various types of analytes.

Alkyl-substituted phenols:

$$bp = 0.815 \cdot I + 56.93, \quad r^2 = 0.935 \tag{177}$$

n-Alkylcarboxylic acids:

$$bp = 0.888 \cdot I + 65.78, \quad r^2 = 0.976 \tag{178}$$

TABLE 23
Calculated (Calc. I) and Measured Retention Indices (Exptl. I) of Various Environmental Pollutants

	Calc. I	Exptl. I	%Diff.
Alkyl-substituted phenols			
2-Methylphenol	164.42	170.82	3.79
4-Methylphenol	177.79	175.32	−1.41
3-Methylphenol	178.16	175.41	0.56
2,6-Dimethylphenol	190.18	182.78	−4.05
2-Ethylphenol	184.04	188.28	2.25
2,4-Dimethylphenol	187.72	190.86	1.65
2,5-Dimethylphenol	189.56	190.94	0.72
4-Ethylphenol	198.76	194.51	−2.19
3-Ethylphenol	192.63	194.63	1.03
2,3-Dimethylphenol	197.53	196.83	−0.46
3,4-Dimethylphenol	206.12	199.63	−3.25
2,4,6-Trimethylphenol	201.21	202.51	0.64
2-n-Propylphenol	199.99	204.88	2.39
2,3,5-Trimethylphenol	212.25	214.69	1.14
2,3,5,6-Tetramethylphenol	233.10	231.28	−0.79
n-Alkylcarboxylic acids			
Pentanoic acid	135.37	137.02	1.20
Hexanoic acid	156.76	158.12	0.86
Heptanoic acid	177.03	177.85	0.46
Octanoic acid	195.38	196.58	0.61
Nonanoic acid	227.69	214.27	−6.26
Decanoic acid	229.94	231.58	0.71
Undecanoic acid	241.20	247.95	2.72
Hexadecanoic acid	320.02	322.15	0.66

From Donnelly, J. R. et al., *J. Chromatogr.*, 642, 1993, 409–411. With permission.

n-Alkyl esters:

$$bp = 1.104 \cdot I - 9.838, \quad r^2 = 0.982 \tag{179}$$

n-Alkyl-substituted benzenes:

$$bp = 1.041 \cdot I + 3.821, \quad r^2 = 1.00 \tag{180}$$

Cyclic alkenes:

$$bp = 1.045 \cdot I + 2.472, \quad r^2 = 0.957 \tag{181}$$

Primary alkenes:

$$bp = 0.951 \cdot I + 17.18, \quad r^2 = 0.998 \tag{182}$$

Aldehydes:

$$bp = 0.741 \cdot I + 49.49, \quad r^2 = 0.974 \tag{183}$$

It has been further stated that the good correlations between the boiling point and retention indices make possible the prediction of the retention indices for additional members of analytes in each type of compounds.

TABLE 24

Calculated (Calc. I) and Measured Retention Indices (Exptl. I) of Various Environmental Pollutants

	Calc. I	Exptl. I	%Diff.
n-Alkyl esters			
Propyl butirate	138.44	135.60	−2.09
Methyl hexanoate	145.69	141.95	−2.64
Ethyl hexanoate	161.09	159.04	−1.29
Methyl heptanoate	164.71	164.64	−0.04
Methyl octanoate	184.64	185.69	0.57
Ethyl octanoate	196.41	200.00	1.80
Methyl nonanoate	201.85	205.26	1.66
Ethyl decanoate	230.83	235.83	2.12
Methyl dodecanoate	246.23	256.91	4.16
Methyl tetradecanoate	301.49	286.83	−5.11
n-Alkyl-substituted benzenes			
Propylbenzene	149.20	149.31	0.07
Hexylbenzene	213.43	213.07	−0.17
Heptylbenzene	232.07	231.69	−0.16
Octylbenzene	248.01	249.13	0.45
Nonylbenzene	265.30	265.56	0.10
Decylbenzene	282.59	280.92	−0.60
Cyclic alkenes			
4-Methylcyclohexene	95.91	95.57	−0.36
1,3,5-Cycloheptatriene	108.64	108.93	0.27
4-Vinylcyclohexene	120.98	120.10	−0.73
1,5-Cyclooctadiene	141.94	144.40	1.70
d-Limonene	167.97	166.62	−0.81
Primary alkenes			
Decene	157.66	157.55	−0.07
Dodecene	199.16	199.34	0.09
Hexadecene	267.83	267.76	−0.03
Aldehydes			
Hexanal	105.97	112.67	5.95
Heptanal	139.44	137.01	−1.77
Octanal	164.01	160.75	−2.03
Nonanal	191.00	182.61	−4.59
Decanal	213.95	221.34	3.34

From Donnelly, J. R. et al., *J. Chromatogr.*, 642, 1993, 409–411. With permission.

As has been previously discussed, many physicochemical parameters of the analytes influence their retention in GLC.[115–118] The most important molecular parameters were the analyte excess molar refraction, the analyte dipolarity/polarizability, the effective or summation analyte hydrogen bond acidity, the effective or summation analyte hydrogen bond basicity, and the analyte gas–liquid partition coefficient on hexadecane at 298 K. Using multilinear regression analysis these parameters can be calculated from the existing retention data and the retention of new analytes can be predicted. The analyte dipolarity/polarizability (π_2^H) and the analyte gas–liquid partition coefficient on hexadecane at 298 K (log L^{16}) were calculated for more than 1000 analytes.[119] The group increments are compiled in Table 25. It was concluded from the data that the additivity of increments is valid only for noninteracting substituents, that is, the additivity rule cannot be generally used for the prediction of the log L^{16} values. This finding has been supported by the finding that the π_2^H values markedly differ according to the position of substituents on the benzene ring (Table 26).

The relationship between retention indices, the total solubility parameter (δ_T),[120] molar volume (V), and the number of carbon atoms in the molecule (N) has also been used to

TABLE 25

Substituent Values for log L^{16} and π_2^H in PhX Analytes

Substituent X	log L^{16}	X π_2^H
H	0.000	0.00
Me	0.539	0.00
Et	0.992	−0.01
Pr	1.444	−0.02
Ph	3.228	0.47
F	0.002	0.05
Cl	0.871	0.13
Br	1.255	0.21
I	1.716	0.30
OMe	1.104	0.23
CHO	1.222	0.48
COMe	1.715	0.49
CO_2Me	1.918	0.33
CN	1.253	0.59
NH_2	1.148	0.44
NO_2	1.771	0.59
OH	0.980	0.37
SH	1.325	0.28
NHMe	1.692	0.38
NMe_2	1.915	0.32

From Abraham, M. H., *J. Chromatogr.*, 644, 1993, 95–139. With permission.

predict retention in GLC.[121] The basic equations describing the correlation between the retention characteristics and physicochemical parameters were

$$I = a \cdot N + b \cdot \Delta(V \cdot \delta_T) + c \cdot \Delta V + d \qquad (184)$$

$$\Delta(V \cdot \delta_T) = V^i \cdot \delta_T^i - V^{al} \cdot \delta_T^{al} \qquad (185)$$

$$\Delta V = V^i - V^{al} \qquad (186)$$

where superscript i and al refer to the analyte and the *n*-alkane with the same carbon atom number, respectively. The total solubility parameter includes the capacity of the analyte for dispersive, dipole orientation, inductive dipole, and hydrogen bond interactions. It was calculated by

$$\delta_T^2 = -0.02085 \cdot T_b^2/V + 58.93 \cdot T_b/V - 6892.14/V - 26.76 \qquad (187)$$

where T_b is the boiling point, and V is the molar volume of the analyte. Equation 187 has been applied for the prediction of the retention of alkenes on nonpolar and moderately polar supports. The results are compiled in Table 27. Equation 187 fits well to the experimental data, the significance level being always over 99.9%. The ratio of variance explained is high in each instance proving that the number of carbon atoms in the analyte, its solubility, and its molar volume have the highest impact on the GLC retention when the probability of polar (hydrophobic) interactions between the stationary phase and the analyte are nonexistent or fairly low.

The SPARC computer program (U.S. Environmental Protection Agency and University of Georgia) has been tested for the capacity to predict the retention indices of 143 analytes on the apolar squalane stationary phase.[122] The principle of the program is the breakdown

TABLE 26

Observed and Calculated π_2^H Values for Difunctional Benzene with One or Two Halogen Substituents

Substituents	Ortho	Meta	Para
F/F	0.75		
	0.62	0.58	0.60
Cl/Cl	0.78	0.73	0.75
	0.78	0.73	0.75
Br/Br	0.96	0.88	0.86
	0.94	0.88	0.90
I/I	1.21	1.07	1.15
	1.12	1.05	1.08
F/OMe	0.79	0.72	0.74
	0.80	0.75	0.77
Cl/OMe	0.91	0.86	0.86
	0.88	0.83	0.85
Br/OMe			0.90
	0.96	0.90	0.92
I/OMe			0.99
	1.05	0.99	1.01
F/CO$_2$Me	0.89	0.88	0.89
	0.90	0.85	0.86
Cl/CO$_2$Me	0.99		0.92
	0.98	0.92	0.94
Cl/CN	1.24	1.14	1.18
F/NO$_2$		1.11	
Cl/NO$_2$	1.25	1.13	1.17
	1.24	1.17	1.19
Cl/NHMe	0.96		1.01
	1.03	0.97	0.99
Br/NMe$_2$			0.97
	1.05	0.99	1.01
F/NH$_2$	0.88	1.08	1.09
	1.01		
Cl/NH$_2$	0.92	1.10	1.13
	1.09		
Br/NH$_2$	0.98	1.19	1.19
	1.17		
I/NH$_2$	1.00	1.26	1.28
	1.26		
F/OH	0.69	0.98	0.95
	0.94		
Cl/OH	0.86	1.06	1.08
	1.04		
Br/OH	0.90	1.15	1.17
	1.10		
I/OH	1.00	1.20	1.33
	1.16	1.09	1.11

Note: First line observed, second line calculated.

From Abraham, M. H., *J. Chromatogr.*, 644, 1993, 95–139. With permission.

of the analyte at each essential single bond and the linear summation of the fragment contributions. A good agreement between the measured and calculated retention indices was observed (root mean square deviation was lower than seven Kováts units), which indicates the adequate prediction power of the program.

It has been recently established that the specific retention volumes and retention indices of many volatile compounds can be predicted from stationary phase constants, thus facilitating the separation of a wide variety of compounds.[123]

TABLE 27

Regression Coefficients and Standard Deviations in Various Phases for the Equation $I = a \cdot N + b \cdot \Delta(V \cdot \delta_T) + c \cdot \Delta V + d$

Phase	Temperature (°C)	a	b	c	d	n	R	S
SQ	80	97.21	1.19	−3.44	12.63	39	0.9990	3.9
1-Octadecene	25	95.79	1.21	−3.77	20.10	35	0.9991	2.7
OV-101	50	96.84	1.19	−3.19	23.23	46	0.9989	3.5
SQ	50	96.56	1.19	−3.33	16.45	46	0.9990	3.3
OV-101	70	96.97	1.17	−3.26	21.74	46	0.9988	3.7
SQ	70	97.13	1.17	−3.23	13.55	46	0.9988	3.6
OV-101	40	97.71	1.24	−3.61	16.16	61	0.9996	3.8
OV-101	60	97.87	1.21	−3.74	13.75	61	0.9996	3.7
OV-101	80	98.05	1.18	−3.91	10.91	61	0.9996	3.8
DB-1	40	98.11	1.26	−3.27	18.13	69	0.9991	3.7
HP-PONA	40	98.10	1.26	−3.27	18.27	69	0.9991	3.7
12 DB-5	40	98.17	1.29	−3.33	20.90	67	0.9990	3.8

Note: n = number of the alkene; R = regression coefficient; S = standard deviation; SQ = squalane; DB-1 = cross-linked and bonded methylsilicone phase; DB-5 = similar to DP-1 with 5% phenyl substitution; HP-PONA = cross-linked methylsiloxane phase.

From Hu, Z. and Zhang, H., *J. Chromatogr.*, 653, 1993, 275–282. With permission.

It is well known that the specific retention volume (V_g) of *n*-alkanes depends logarithmically on the carbon number of the *n*-alkanes (n):

$$\log V_g = A + B \cdot n \tag{188}$$

where A is the phase constant, and B is the logarithm of the retention ratio (r). The retention ratio can be calculated from the retentions of two neighboring *n*-alkanes with carbon numbers n and n + 1:

$$B = \log(t'_{n+1}/t'_n) = \log r \tag{189}$$

The phase constant can be determined by measuring the specific retention volume of one *n*-alkane:

$$A = \log V_{gn} - B \cdot n \tag{190}$$

The retention ratio of a compound RX relative to an *n*-alkane RH can be expressed by

$$\log r = \log V_g^{RX} - \log V_g^{RH} \tag{191}$$

The specific retention volume of the compound can be expressed by

$$\log V_g^{RX} = \log r^{RX/RH} + A + B \cdot n^{RH} \tag{192}$$

These equations were used for the calculation of the specific retention volume of dodecane and A values from the slope B and from the retention indices (I) of benzene (Table 28). Mathematical-statistical methods such as multivariate linear regression analysis indicated that the specific retention volume and retention indices of a wide variety of compounds can be predicted from the stationary phase constants. This calculation method may help in the exact selection of stationary phases for the separation of any given set of volatile analytes and in the study of the correlation between molecular structure and retention behavior.

TABLE 28

Calculation of log V_g (Dodecane) and of Intercept A from Slope B and Retention Index I (Benzene)

Phase	Intercept A	I (benzene)	Slope B	V_g (dodecane) meas.	V_g (dodecane) calc.	% dev. of A calc.
Diethylene glycol succinate	−1.209	1186	0.1941	13.2	14.0	−1.18
Triethylene glycol succinate	−1.074	1122	0.2024	22.6	21.7	2.33
Hyprose SP 80	−1.073	983	0.2163	33.3	49.1	−16.79
Ethylene glycol adipate	−1.049	1019	0.2198	38.8	49.4	−9.79
Diethylene glycol adipate	−1.045	1037	0.2191	38.4	46.0	−6.89
Triton X	−0.932	907	0.2450	102.0	147.2	−16.33
Carbowax 300	−0.921	998	0.2219	55.3	56.0	−0.61
Sucrose octacetate	−0.867	992	0.2131	49.1	43.4	4.56
Carbowax 1000	−0.854	975	0.2284	77.0	72.9	3.14
Carbowax 400	−0.852	991	0.2205	62.1	54.7	6.12
Diethylene glycol sebacate	−0.846	903	0.2443	122.0	145.7	−8.50
Carbowax 1540	−0.841	968	0.2291	81.0	76.0	3.50
Quadrol	−0.822	915	0.2299	86.5	90.4	−3.74
Zonyl E 7	−0.822	867	0.2156	58.3	66.6	−12.33
Carbowax 600	−0.819	975	0.2217	69.4	59.3	7.51
Neopentyl glycol succinate	−0.816	918	0.2316	91.8	94.5	−2.46
Carbowax 4000	−0.809	965	0.2290	87.0	76.4	7.05
Carbowax 20M	−0.809	961	0.2300	89.4	79.7	6.36
Ethylene glycol sebacate	−0.804	875	0.2476	147.0	174.5	−8.96
Pluronic F 68	−0.802	907	0.2477	148.0	160.0	−2.79
XF 1150	−0.793	966	0.2085	51.2	40.5	9.41
Carbowax 6000	−0.792	964	0.2292	90.8	77.1	9.10
Kroniflex THPP	−0.789	935	0.2436	136.0	130.3	4.06
Igepal CO 880	−0.776	906	0.2445	144.0	145.4	0.30
Poly phenyl ether 6 rings	−0.765	829	0.2643	255.0	332.1	−13.23
Neopentyl glycol adipate, td	−0.749	867	0.2494	175.0	188.6	−3.89
Neopentyl glycol adipate	−0.747	886	0.2414	141.0	139.8	0.20
Poly phenyl ether 5 rings	−0.744	821	0.2668	287.0	366.9	−12.36
DC FS-1265 Fluid (QF 1)	−0.711	780	0.2117	67.6	75.5	−16.82
Tricresyl phosphate	−0.705	824	0.2644	294.0	337.9	−6.85
Pluronic P 85	−0.699	845	0.2589	256.0	286.9	−1.59
Oronite NIW	−0.691	834	0.2571	248.0	262.4	−2.83
Pluronic F 88	−0.688	908	0.2384	149.0	119.8	13.58
Pluronic P 46	−0.683	868	0.2478	195.0	179.1	5.64
Pluronic F 77	−0.683	885	0.2426	169.0	145.5	9.36
Pluronic P 84	−0.680	838	0.2581	262.0	267.6	−0.34
Ethofate 60-25	−0.677	834	0.2590	270.0	278.2	−0.87
Polytergent 300	−0.666	816	0.2688	362.0	395.7	−3.29
Dibutyltetrachloro phthalate	−0.666	785	0.2875	620.0	767.9	−10.11
Ucon 50 HB-2000	−0.665	846	0.2518	227.0	215.5	3.59
Tergitol NPX	−0.664	844	0.2565	259.0	250.4	3.28
Bis(ethoxyethyl) phthalate	−0.651	856	0.2533	245.0	219.4	8.18
Ucon LB 1715	−0.644	774	0.2649	342.0	395.0	−9.65
Pluronic P 65	−0.634	847	0.2495	229.0	200.2	8.99
Pluronic L 61	−0.625	833	0.2546	269.0	243.6	7.18
Sucrose acetate isobutyrate	−0.621	829	0.2472	222.0	196.2	6.96
Pluronic L 44	−0.620	842	0.2519	253.0	218.6	10.18
Pluronic L 81	−0.612	792	0.2683	406.0	416.9	−0.46
Pluronic L 63	−0.609	833	0.2536	264.0	236.2	9.92
Pluronic L 72	−0.571	817	0.2566	322.0	270.6	13.19
Pluronic L 42	−0.565	811	0.2559	320.0	269.8	12.73
Diisodecyl phthalate	−0.564	739	0.2829	677.0	758.6	−6.02
Dioctyl phthalate	−0.560	750	0.2808	645.0	689.5	−2.41
DC 550 Fluid	−0.540	733	0.2643	429.0	435.1	−3.40
Flexol 8N8	−0.526	749	0.2757	605.0	590.9	3.58
Apiezon L	−0.522	683	0.2873	841.0	1016.1	−14.48
TMP Tripelargonate	−0.520	734	0.2844	782.0	805.7	0.45

TABLE 28 (CONTINUED)
Calculation of log V_g (Dodecane) and of Intercept A from Slope B and Retention Index I (Benzene)

Phase	Intercept A	I (benzene)	Slope B	V_g (dodecane) meas.	V_g (dodecane) calc.	% dev. of A calc.
Apiezon J	−0.508	685	0.2834	782.0	897.8	−11.46
Castorwax	−0.505	749	0.2722	578.0	530.5	7.97
Apiezon N	−0.503	685	0.2833	789.0	894.1	−10.56
Hallcomid M 18 OL	−0.472	747	0.2833	848.0	750.9	14.92
Apiezon M	−0.470	685	0.2830	843.0	885.8	−4.32
SE-30 terminated	−0.457	675	0.2633	382.0	496.7	−4.82
Isooctyl decyl adipate	−0.455	726	0.2870	975.0	892.8	12.20
Di(ethylhexyl) adipate	−0.450	728	0.2873	995.0	896.0	14.26
Dioctylsebacate	−0.443	724	0.2879	1028.0	923.0	14.62
Di(ethylhexyl) sebacate	−0.438	724	0.2872	1021.0	903.4	16.01
SE-31	−0.430	666	0.2509	381.0	347.8	−1.84
Versilube F 50	−0.428	670	0.2574	458.0	420.1	−0.02
SE-52	−0.427	681	0.2541	419.0	367.9	3.85
Hallcomid M-18	−0.420	730	0.2877	1080.0	902.0	23.08
SE-30	−0.384	678	0.2507	422.0	334.2	14.58
Mean	−0.684	843.3	0.2515	331.0	332.9	
Standard deviation						9.00

From Rohrschneider, L., *Chromatographia*, 37, 1993, 250–258. With permission.

b. Relationship with Geometrical and Structural Characteristics

A wide variety of physicochemical parameters and molecular descriptors have been used in the studies trying to find the best relationships between the Kováts index and the characteristics of the analytes. Many excellent correlations were established; however, the studies frequently used different sets of analytes and physicochemical parameters, which made generalization of the conclusions difficult.

Various connectivity indices as the best descriptors of any structural characteristic of the analytes have been frequently used in structure-retention studies. Thus, connectivity indices have also been used to find the relationship between the retention behavior of linear alkylbenzenes on SE-54 and DB-1 stationary phases and their chemical structure.[124] The best correlations selected by a multilinear regression analysis program were

$$I_{SE-54} = 27.5536 \cdot {}^1X - 32.1700 \cdot {}^5\chi_p - 48.6688$$

$$n = 29, \quad r = 0.9955, \quad F({}^1\chi) = 2110.96 \ (P > 0.0001), \quad r^2 = 0.9911,$$

$$F({}^5\chi_p) = 317.00 \ (P > 0.0001) \tag{193}$$

$$I_{DB-1} = 283.0253 \cdot {}^1X - 329.4509 \cdot {}^5\chi_p + 133.9781$$

$$n = 23, \quad r = 0.9923, \quad F({}^1\chi) = 1133.88 \ (P > 0.0001), \quad r^2 = 0.9847,$$

$$F({}^5\chi_p) = 212.88 \ (P > 0.0001) \tag{194}$$

where ${}^1X$ contains information about the number of atoms in the analyte molecule, and ${}^5X_p$ relates to the branching of the analyte.

$$I_{SE-54} = 31.8111 \cdot {}^1X - 10.5633 \cdot {}^3\chi_p - 28.3486 \cdot {}^5\chi_p - 4.9117$$

$$n = 29, \quad r = 0.9985, \quad F({}^1\chi) = 1996.83 \ (P > 0.0001), \quad r^2 = 0.9969,$$

$$F({}^3\chi_p) = 47.89 \ (P > 0.0001), \quad F({}^5\chi_p) = 546.91 \ (P > 0.0001) \tag{195}$$

$$I_{DB-1} = 331.3419 \cdot {}^1\chi - 121.1496 \cdot {}^3\chi_p - 285.4131 \cdot {}^5\chi_p + 180.7905$$

$$n = 23, \quad r = 0.9979, \quad F({}^1\chi) = 1622.25 \; (P > 0.0001), \quad r^2 = 0.9958,$$

$$F({}^3\chi_p) = 49.59 \; (P > 0.0001), \quad F({}^5\chi_p) = 433.82 \; (P > 0.0001) \tag{196}$$

It has been stated that the introduction of the third connectivity index also relating to the branching of the analyte molecule markedly improved the prediction power of the correlations, proving the considerable role of this parameter in the determination of the retention behavior. Molecular connectivity indices have been successfully used for the prediction of the retention behavior of pyrido[1,2-α]pyrimidin-4-one[125] and pyrido[1,2-α]pyrimidine derivatives.[126] In a recent study the relationship between retention indices and molecular connectivity indices has been demonstrated by using multivariate mathematical-statistical methods.[127] The list of compounds, their retention indices, both on OV-17 and Apiezon L stationary phases, and the retention index differences are compiled in Table 29. It has been established that one connectivity index cannot predict adequately the retention index, indicating the involvement of more than one structural parameters in the retention:

$$I_{OV-17} = 341.1620 \cdot {}^1\chi - 188.8422$$

$$r = 0.9751, \quad F = 541.92 \; (P > 0.0001), \quad r^2 = 0.9509 \tag{197}$$

$$I_{OV-17} = 347.5188 \cdot {}^2\chi + 27.6842$$

$$r = 0.9769, \quad F = 586.09 \; (P > 0.0001), \quad r^2 = 0.9544 \tag{198}$$

where ${}^1\chi$ and ${}^2\chi$ contain information about the number of atoms in the molecule and about branching, respectively. The prediction power of the calculations were lower for the apolar Apiezon L ($r = 0.9046$ and $r = 0.8932$) and I values. The application of multilinear regression analysis resulted in better correlations between the retention indices and connectivity indices:

$$I_{OV-17} = 425.0345 \cdot {}^2\chi - 422.5803 \cdot {}^4\chi_{pc}^v - 75.2959$$

$$r = 0.9882, \quad F({}^2\chi) = 633.63 \; (P > 0.0001), \quad r^2 = 0.9765,$$

$$F({}^4\chi_{pc}^v) = 36.46 \; (P > 0.0001) \tag{199}$$

$$I_{ApL} = 372.1000 \cdot {}^1\chi - 170.3856 \cdot {}^3\chi_p^v - 203.3994$$

$$r = 0.9548, \quad F({}^1\chi) = 144.26 \; (P > 0.0001), \quad r^2 = 0.9263,$$

$$F({}^3\chi_p^v) = 15.87 \; (P > 0.0001) \tag{200}$$

The results of multivariate regression analysis indicate that the retention of the analyte on OV-17 stationary phase mainly depends on the branching (presence and number of adjacent atoms), on the unsaturation, and on the number and orientation of substituents. The relationship is similar on Apiezon L stationary phase; the number of adjacent atoms, the unsaturation, and the presence of heteroatoms exert the highest influence on the retention.

It has been early recognized that the use of a large set of physicochemical parameters and molecular descriptors resulted in a better relationship between molecular characteristics and retention than the use of a narrow range of parameters. These data sets generally contain information not only about the structure of the analytes, but also about the other

TABLE 29

Experimental Retention Indices and ΔI Values at 170°C for Tetralones, Coumarins, and Structurally Related Compounds on Nonpolar (Apiezon L) and Polar (OV-17) Stationary Phases

| No. | Compound | Retention Index I | | ΔI |
		OV-17	Apiezon L	
1	Cyclohexane	757	—	—
2	Methylcyclohexane	793	—	—
3	Benzene	774	—	—
4	Toluene	887	—	—
5	Cyclohexanone	1080	—	—
6	Methoxybenzene	1085	—	—
7	Tetrahydro-4H-pyran-4-one	1088	—	—
8	δ-Valerolactone	1107	—	—
9	2-Methylcyclohexanone	1119	—	—
10	4-Methylcyclohexanone	1144	—	—
11	3-Methylcyclohexanone	1155	—	—
12	Tetrahydronaphtalene	1348	—	—
13	2-Coumaronone	1501	1289	212
14	4-Chromanone	1621	1390	231
15	β-Tetralone	1625	1503	122
16	α-Tetralone	1645	1458	187
17	2-Methyl-1-tetralone	1667	1492	175
18	1-Methyl-2-tetralone	1669	1479	190
19	Dihydrocoumarin	1682	1434	248
20	4-Methyl-1-tetralone	1688	1514	174
21	Coumarin	1758	1521	237
22	6-Methylcoumarin	1863	1618	245
23	7-Methoxy-1-tetralone	1897	1666	232
24	7-Methoxy-2-tetralone	1903	1680	223
25	5-Methoxy-1-tetralone	1905	1690	216
26	6-Methoxy-2-tetralone	1910	1711	199
27	6-Methoxy-1-tetralone	1970	1820	162
28	7-Methoxycoumarin	2056	1798	258
29	4-Methoxycoumarin	2085	1844	241
30	7-Methoxy-4-methylcoumarin	2214	2021	193

From Arruda, A. C., Heinzen, V. E. F., and Yunes, R. A., *J. Chromatogr.*, 630, 1993, 251–256. With permission.

parameters (hydrophobicity, polarity, etc.). Thus, the retention indices of 38 benzene derivatives were determined on hydrogenated Apiezon M/sililated Chromosorb W columns and they were correlated with 58 molecular descriptors (topological, geometric, electronic, and physicochemical descriptors).[128] The analytes were randomly divided into a training and a prediction set and a stepwise regression analysis program was used to select the descriptors having the highest impact on the retention indices of the analytes using the training set. The optimal equation was found to be

$$I = (137.114 \pm 3.359) \cdot \chi V_0 - (55.414 \pm 3.791) \cdot NOCH_3 + (2.321 \pm 0.344) \cdot VOL + (9.462 \pm 2.675) \cdot DIMO - (5.726 \pm 25.084)$$

$$n = 32, \quad r = 0.998, \quad F = 1835, \quad S.D. = 10.067$$

(201)

where I is the retention index; XV_0 is the zeroth-order valence term; $NOCH_3$ is the number of methyl groups in the analyte molecule; VOL is the van der Waals volume of the molecule; and DIMO is the dipole moment. The chemical structure of the analytes, the descriptors selected by the program, and the calculated and experimental retention indices

TABLE 30

Chemical Structure of Analytes

No.	Compound

Training Set

1	Benzene
2	Fluorobenzene
3	Chlorobenzene
4	Bromobenzene
5	Toluene
6	Anisole
7	p-Chloroanisole
8	p-Xylene
9	p-Fluorotoluene
10	p-Bromotoluene
11	p-Bromofluorobenzene
12	p-Chlorofluorobenzene
13	p-Chlorotoluene
14	p-Fluoroanisole
15	m-Chloroanisole
16	m-Methylanisole
17	m-Xylene
18	m-Chlorobromobenzene
19	m-Bromotoluene
20	m-Fluorotoluene
21	m-Chlorotoluene
22	m-Chlorofluorobenzene
23	m-Dibromobenzene
24	m-Dichlorobenzene
25	o-Methylanisole
26	o-Chloroanisole
27	o-Bromofluorobenzene
28	o-Fluorotoluene
29	o-Fluoroanisole
30	o-Xylene
31	o-Bromochlorobenzene
32	o-Chlorofluorobenzene

Prediction Set

1	p-Chlorofluorobenzene
2	p-Methylanisole
3	o-Chlorotoluene
4	o-Bromotoluene
5	m-Fluoroanisole
6	m-Bromofluorobenzene

From Jalali-Heravi, M. and Garkani-Nejad, Z., *J. Chromatogr.*, 648, 1993, 389–393. With permission.

are compiled in Tables 30 and 31. It has been concluded from the results that topological ($NOCH_3$ and XV_0), geometric (VOL), and electronic (DIMO) molecular parameters influence the retention of analytes in this GLC systems, which means that various molecular interactions are involved in the binding of analytes to the stationary phase.

The retention indices of 73 industrially important fragrance compounds were also determined on an apolar OV-101 and a polar Carbowax 20M capillary column and a wide set of molecular parameters (atom counts, bond counts, branching, moments of inertia, molecular volume, surface area, charge distribution throughout the molecule, hydrogen

TABLE 31

Experimental and Calculated Retention Indices and Descriptors Employed in the Selected Model

Compounds in Table 30	Descriptors				Retention Index	
	$NOCH_3$	χV_0	VOL	DIMO	Calc.	Exptl.
Training Set						
1	0	3.464	88.618	0.000	674.9	681.3
2	0	3.163	93.558	1.995	664.0	664.1
3	0	4.591	102.304	1.837	878.6	877.9
4	0	5.371	105.972	1.392	989.8	979.6
5	1	4.387	105.096	0.068	784.9	788.2
6	0	4.795	113.870	1.072	926.2	923.6
7	0	5.921	127.585	2.242	1123.5	1131.7
8	2	5.309	121.642	0.001	893.7	889.2
9	1	4.086	110.077	1.960	773.1	777.7
10	1	6.294	122.492	1.384	1099.3	1096.3
11	0	5.070	110.923	0.667	953.2	940.9
12	0	6.497	119.634	0.463	1167.2	1174.4
13	1	5.513	118.808	1.826	987.8	989.2
14	0	4.494	118.855	2.353	908.6	910.6
15	0	5.921	127.506	1.096	1112.4	1126.0
16	1	5.718	130.453	1.534	1040.2	1029.6
17	2	5.309	124.332	0.061	900.5	892.0
18	0	6.497	119.596	1.585	1177.7	1179.0
19	1	6.294	122.387	1.378	1098.9	1100.0
20	1	4.086	110.058	1.998	773.5	778.0
21	1	5.513	118.779	1.827	987.7	990.0
22	0	4.290	107.228	1.873	849.1	835.4
23	0	7.278	123.271	1.315	1290.7	1287.7
24	0	5.717	115.941	1.744	1063.8	1060.5
25	1	5.718	130.495	1.459	1039.6	1013.5
26	0	5.922	127.615	2.482	1125.9	1135.6
27	0	5.070	110.900	2.743	972.8	959.6
28	1	4.086	110.085	1.952	773.1	777.4
29	0	4.494	118.910	2.674	911.8	919.7
30	2	5.309	121.534	0.073	894.2	916.2
31	0	6.497	119.609	2.490	1186.3	1197.6
32	0	4.290	107.171	3.123	860.8	862.0
Prediction Set						
1	0	4.290	107.289	0.204	833.4	840.5
2	1	5.718	130.396	1.077	1035.7	1029.5
3	1	5.513	118.782	1.786	987.4	986.3
4	1	6.294	122.415	1.345	1098.7	1095.7
5	0	4.494	118.780	1.263	898.1	908.5
6	0	5.070	110.889	1.771	963.6	932.8

From Jalali-Heravi, M. and Garkani-Nejad, Z., *J. Chromatogr.*, 648, 1993, 389–393. With permission.

bonding capacity, dipole–dipole interactions, electron withdrawing power, etc.) were tested for the prediction of the retention behavior.[129] The relationships between retention indices and physicochemical parameters were calculated by using 90% of the retention data; the remaining 10% were used to test the prediction power of the equations. The parameters of the best fitting equations are compiled in Table 32. The prediction power of the equations was higher on the apolar OV-101 column than on the polar Carbowax 20M stationary phase indicating that the analyte–stationary phase interaction is more complicated on polar support. The inclusion of geometric descriptors also increased the predictive power. The electronic parameters exerted a considerable influence on the retention too. The determinative role of hydrogen bonding capacity can be explained by the fact that this parameter considerably influences the boiling point of the analyte. It has been supposed that the descriptor "number of carbonyl groups" is related to the dipole–dipole interactions between the analyte and the stationary phase. It has been further supposed that the conformation of an analyte influences its solvent-accessible surface area, which modifies the accessibility of its positive and negative charges.

A wide variety of molecular descriptors were used for the elucidation of the retention behavior of complex hydrocarbon mixtures both on apolar SE-30 and polar Carbowax 20M columns.[130] The retention data of 81 hydrocarbons were taken from Reference 131. The parameters of the regression models developed using topological, geometrical, and electronic descriptors are compiled in Table 33. The substructure path count and the third-order path are related to the side groups and branching of the analyte. The third geometric moment measures the width of the analyte. Together with the molecular weight, it is related to the bulkiness of the analyte (correlated with the polarizability of the analyte). These parameters indicate that the interaction between the apolar stationary phase and the hydrocarbons is governed mainly by dispersive forces and by the polarizability of the analyte. The partially negatively charged surface area of the analyte significantly influences the retention of hydrocarbons on the polar Carbowax 20M stationary phase. This finding suggests the increased role of polar interactions. The regression models are modified when the boiling point is included as an independent variable in the calculations. The parameters of the regression models developed using topological, geometrical, and electronic descriptors and boiling point are compiled in Table 34. The data in Table 34 indicate that the inclusion of boiling point considerably increases the prediction power of the correlations, that is, the boiling point seems to be the most important of the physicochemical parameters influencing retention. It has been assumed that these equations are suitable for the estimation of the retention behavior of similar analytes in instances when the retention values are not readily available.

2. *Temperature Programming*

The Kováts retention index system has been developed for isocratic gas chromatography.[132,133] As temperature programmed gas chromatography has found growing acceptance and application both in theoretical studies and in the solution of practical separation problems, it was necessary to extend the Kováts retention index system also for programmed temperature gas chromatography. Many different methods have been developed for the calculation and prediction of retention indices in temperature programmed GC. Thus, various computer simulation models based on linear elution strength approximation[134–137] or on cubic spline interpolation have been developed for the calculation of retention indices and peak widths in temperature programmed GC.[138–140]

A theoretical model[141] described the retention of an analyte depending on the column temperature by

$$T_R = T_0 + b \cdot t_R \quad (202)$$

where T_0 is the column temperature at the beginning of the temperature rise, and b is the rate of temperature rise.

TABLE 32

Retention Index Models

Descriptor	Coefficient	±SD
Model 1: OV-101 with Geometrical Descriptors		
Sum of molecular charges	−188.1	39.4
Total molecular surface area	4.147	0.243
Partially negative surface area-3	−13.92	1.15
Number of ring atoms	47.12	1.73
Number of donatable hydrogens	138.4	17.4
Intercept	−560.5	
$n = 65; r = 0.989; s = 45; F = 528$		
External predictions: $r = 0.984$; rms = 48		
Model 2: OV-101 with No Geometrical Descriptors		
Sum of molecular charges	−387.6	37.8
Cluster-3 molecular connectivity	−54.7	13.21
Simple $^1\chi$	316.4	12.4
Number of donatable hydrogens	265.8	21.0
Number of carbonyl groups	108.4	16.3
Intercept	185.7	
$n = 64; r = 0.987; s = 49; F = 431$		
External predictions: $r = 0.977$; rms = 56		
Model 3: Carbowax 20M with Geometrical Descriptors		
Sum of molecular charges	−774.9	59.8
Weighted negative surface area-3	−89.46	7.01
Number of bonds	180.9	8.4
Cluster-3 molecular connectivity	−248.0	32.3
Number of donatable hydrogens	549.6	32.6
Intercept	614.9	
$n = 61; r = 0.977; s = 93; F = 229$		
External predictions: $r = 0.992$; rms = 55		
Model 4: Carbowax 20M with No Geometrical Descriptors		
Sum of molecular charges	−742.7	106.8
Charge on the most positive atom	1335	320
Number of bonds	222.2	13.6
Cluster-3 molecular connectivity	−465.5	43.0
Number of donatable hydrogens	452.1	68.5
Number of carbonyl groups	210.5	46.4
Flexibility of OC=O compounds	−36.37	11.5
Intercept	455.1	
$n = 63; r = 0.958; s = 127\ F = 88$		
External predictions: $r = 0.993$; rms = 47		

Note: Coefficient = descriptor coefficient; SD = standard deviation of the descriptor coefficient; n = number of analytes used for the development of the model.

From Egolf, L. M. and Jurs, P. C., *Anal. Chem.*, 65, 1994, 3119–3126. With permission.

$$b \cdot t_M = \int_{T_0}^{T_R} [1 + A \cdot \exp(\Delta H_v^o / RT)]^{-1} \cdot dT \qquad (203)$$

where t_M is the retention time of a nonabsorbed analyte (methane); R is the gas constant; ΔH_v^o is the standard vaporization enthalpy of the analyte from the solution in the stationary

TABLE 33

Model Developed for Hydrocarbons Using Topological, Geometrical and Electronic Descriptors

Reg. Coeff.	S.D.	Descriptor
SE-30 Column		
4.7	±0.3	Molecular weight
4.6	±0.3	Substructure path count
68.2	±7.2	Third-order path
−142.8	±19.3	Third major geometric axis
161.2		Intercept
	r = 0.983; s = 18.6; n = 67; F = 454	
	Std. error of the mean: 1.8%	
Carbowax 20M Column		
3.2	±0.2	Partially negatively charged surface area
2.8	±0.4	Substructure path count
156.1	±9.5	Third-order path
−207.2	±25.5	Third major geometric axis
3.4	±0.4	Molecular weight
106.1		Intercept
	r = 0.987; s = 23.3; n = 65; F = 432	
	Std. error of the mean: 1.93%	

From Woloszyn, T. F. and Jurs, P. C., *Anal. Chem.*, 65, 1993, 582–587. With permission.

TABLE 34

Model Developed for Hydrocarbons Using Topological, Geometrical, and Electronic Descriptors and Boiling Point

Reg. coeff.	S.D.	Descriptor
SE-30 Column		
3.7	±0.1	Boiling point
4.6	±1.1	Number of single bonds
340.1		Intercept
	r = 0.998; s = 6.4; n = 40; F = 4994	
	Std. error of the mean: 0.6%	
Carbowax 20M Column		
6.6	±0.2	Boiling point
−27.0	±3.8	Molecular polarizability
−20.6	±3.1	Number of single bonds
660.7		Intercept
	r = 0.994; s = 13.0; n = 40; F = 1056	
	Std. error of the mean: 1.1%	

From Woloszyn, T. F. and Jurs, P. C., *Anal. Chem.*, 65, 1993, 582–587. With permission.

phase; and A is the entropy. The modification of the gas viscosity by temperature can be included in the equation:

$$b \cdot t_{MT_0} \cdot (T_0)^{-0.7} = \int_{T_0}^{T_R} T^{-0.7} \cdot [1 + A \cdot \exp(\Delta H_v^o / RT)]^{-1} \cdot dT \tag{204}$$

TABLE 35

Solution Thermodynamic Parameters of Standard n-Paraffins, Cz, with Chain Length Z, and Their Measured (m) and Calculated (c) Retention Values

n-Alkane	ΔH_v^o (kcal/mol)	$A \times 10^6$	t_R (min)	T_{Rm} (°C)	T_{Rc} (°C)	T_{Rc} [a] (°C)	t_{Rc3} [a] (min)
C6	2.416	4978	4.72	89	89.2	89.3	4.65
C7	2.856	4375	5.27	91	90.4	90.6	5.30
C8	6.587	42.960	6.20	92	91.8	92.1	6.05
C9	8.510	6.507	8.15	96	95.7	96.0	8.00
C10	9.034	6.059	11.13	102	101.4	101.8	10.90
C11	10.440	2.045	15.93	112	111.5	111.9	15.95
C12	11.311	1.017	21.10	122	120.7	121.9	20.95
C13	11.858	0.856	27.03	135	131.6	133.3	26.65
C14	12.413	0.719	33.08	148	143.3	145.6	32.80
C15	13.311	0.464	40.90	162	159.3	162.4	41.20
C16	14.483	0.193	47.62	174	172.2	175.9	41.95
C17	15.149	0.132	52.85	185	182.0	186.7	53.35
C18	15.626	0.112	58.08	196	191.8	196.5	58.25

[a] Data obtained by taking into account the effect of the temperature dependence of carrier gas viscosity.
From Messadi, D. and Ali-Mokhnache, S., *Chromatographia*, 37, 1993, 264–270. With permission.

where t_{MT_0} is the retention time of the nonabsorbed analyte at temperature T_0. The enthalpy and entropy can be determined by measuring the retention time of analyte or analytes at different temperatures:

$$(t_R - t_M)/t_M = A \cdot \exp(\Delta H_v^o/RT) \tag{205}$$

The thermodynamic parameters and the measured and calculated retention times for some n-alkanes and organochlorine compounds are compiled in Tables 35 and 36, respectively. It was found that the relative standard error was under 4% for ΔH_v^o, and under 8% for A values. The retention behavior of n-alkanes indicates that the entropy decrease of solution is related to the heat of solution, which means that the higher the heat of interaction with the solvent, the higher the degree of order of the analyte molecules in the solution. Excellent linear correlations were found between the measured and calculated retention time values. $F_{calculated}$ were 57,208 and 2557 for n-alkanes and organochlorine compounds, respectively, proving the high predictive power of the theoretical equations. The retention index of the analytes in programmed GC was calculated by both linear and polynomial approximations.

Linear approximation

$$I_X^P = 100 \cdot Z + 100 \cdot i \cdot (T_{R(X)} - T_{R(Z)})/(T_{R(Z+i)} - T_{R(Z)}) \tag{206}$$

where I_X^P is the retention index of the analyte; Z and Z + i are the carbon numbers of n-alkanes between the analyte that is eluted; and $T_{R(X)}$, $T_{R(Z)}$, and $T_{R(Z+i)}$ are the retention temperatures of the corresponding compounds.

Polynomial approximation

$$I_P = \sum_{i=0}^{k'-2} a_i' \cdot t_R^i + \sum_{i=k'-1}^{m} a_i \cdot t_R^i \tag{207}$$

TABLE 36

Solution Thermodynamic Parameters of Some Organochlorine Species, and Their Measured (m) and Calculated (c) Retention Values

Analyte	ΔH_v° (kcal/mol)	$A \times 10^6$	t_R (min)	T_{Rm} (°C)	T_{Rc} (°C)	T_{Rc} [a] (°C)	t_{Rc3} [a] (min)
MCB	7.425	21.516	7.05	94	93.8	94.0	7.02
2-CP	9.049	5.505	10.23	101	100.6	100.9	10.45
1,4-DCB	7.603	46.263	11.28	103	102.3	102.8	11.40
1,2-DCB	9.206	5.935	12.62	105	103.8	104.3	12.15
1,3,5-TCB	9.440	7.996	17.20	114	112.7	113.5	16.77
2,4-DCP	10.642	1.904	18.78	117	115.9	116.9	18.45
1,2,4-TCB	9.810	5.987	18.83	118	116.4	117.4	18.67
4-CP	10.936	1.520	19.70	120	118.9	120.0	20.0
2,4-DCP	10.050	4.853	20.33	120	118.4	119.5	19.76
2,4,6-TCP	11.251	2.354	29.47	138	136.3	138.3	29.15

Note: MCB = chlorobenzene; 2-CP = 2-chlorophenol; 1,4-DCP = 1,4-dichlorobenzene; 1,2-DCP = 1,2-dichlorobenzene; 1,3,5-TCP = 1,3,5-trichlorobenzene; 2,4-DCP = 2,4-dichlorophenol; 1,2,4-TCB = 1,2,4-trichlorobenzene; 4-CP = 4-chlorophenol; 2,4-DCP = 2,4-dichlorophenol; 2,4,6-TCP = 2,4,6-trichlorophenol.

[a] Data obtained by taking into account the effect of the temperature dependence of carrier gas viscosity.

From Messadi, D. and Ali-Mokhnache, S., *Chromatographia*, 37, 1993, 264–270. With permission.

TABLE 37

Comparison of Programmed Temperature Retention Indices Determined by Linear (I_L) and Polynomial (I_P) Interpolations

n-Alkane	I_L		I_P			
			m = 7		m = 9	
			k' = 4	k' = 6	k' = 4	k' = 6
C6	792	<u>600.0</u>	599.9	<u>600.0</u>	600.0	<u>600.0</u>
C7	818	685.7	699.7	700.4	700.0	700.0
C8	834	728.6	799.0	799.1	799.9	800.0
C9	<u>900.0</u>	<u>900.0</u>	<u>900.0</u>	<u>900.0</u>	<u>900.0</u>	<u>900.0</u>
C10	960	975.0	999.9	1000.0	999.9	999.9
C11	<u>1100.0</u>	<u>1100.0</u>	<u>1100.0</u>	<u>1100.0</u>	<u>1100.0</u>	<u>1100.0</u>
C12	<u>1200.0</u>	<u>1200.0</u>	<u>1200.0</u>	<u>1200.0</u>	<u>1200.0</u>	<u>1200.0</u>
C13	1290	1300.0	1300.6	1300.5	1300.0	1299.8
C14	1390	1392.3	1400.2	1400.2	1400.0	1400.2
C15	1510	1507.7	1499.6	1499.6	1500.1	1500.0
C16	<u>1600.0</u>	<u>1600.0</u>	<u>1600.0</u>	<u>1600.0</u>	<u>1600.0</u>	<u>1600.0</u>
C17	1690	1700.0	1700.5	1700.4	1700.0	1700.0
C18	1780	<u>1800.0</u>	1799.8	<u>1800.0</u>	1800.0	<u>1800.0</u>

Note: I_L values were obtained either graphically (Column 2) or by means of calculations (Column 3). Underlined values have been used as references and are those given by definition.

From Messadi, D. and Ali-Mokhnache, S., *Chromatographia*, 37, 1993, 264–270. With permission.

The results of the two computational methods are listed in Tables 37 and 38. The data in Table 37 prove that the polynomial equations have higher predictive power than the traditional linear one. It can be further concluded that the predictive power of the polynomial equations increases with increasing order of polynomials. The data in Table 38 prove again that the polynomial equations are more suitable for the prediction of the retention behavior of organochlorine environmental pollutants than the linear approximation. It has been stated that the retention temperatures and retention times of analytes

TABLE 38

Mean and Standard Deviations of the Retention Indices Computed by Linear (I_L) and Polynomial (I_P) Techniques

Solute	I_L	δ_L	I_P	δ_P	$(I_L)_c$	$(I_P)_c$
MCD	840.7	6.73	845.7	1.07	850.1	842.5
2-CP	932.6	3.39	952.0	1.54	947.2	947.5
1,4-DCB	955.1	2.87	974.4	2.29	960.3	969.6
1,2-DCB	967.1	2.58	994.7	1.06	975.6	992.0
1,3,5-TCB	1084.5	3.20	1094.9	1.97	1100.2	1096.1
2,4-DCP	1114.5	2.32	1134.6	1.78	1124.6	1135.0
1,2,4-TCB	1131.3	3.68	1144.3	1.56	1141.6	1145.4
4-CP	1145.1	2.90	1166.3	0.60	1161.1	1167.2
2,6-CP	1151.1	2.86	1177.5	1.49	1148.9	1178.7
2,4,6-TCP	1325.6	3.96	1320.3	1.04	1335.8	1325.3

Note: c-Indexed values were obtained by using calculated retention data. For abbreviations see Table 36.

From Messadi, D. and Ali-Mokhnache, S., *Chromatographia*, 37, 1993, 264–270. With permission.

can be calculated and the retention indices can be determined accordingly. The dependence of peak width (S) of the analyte i on the column temperature can determined by

$$_iS_A(T) = {_iA_S} \cdot \exp(_iB_S/T) + {_iC_S} \tag{208}$$

where $_iS_A$ is the calculated peak width; $_iA_S$ and $_iB_S$ are constants determined from the experimental values of peak width $_iS_R$ measured at different temperatures; and $_iC_S$ is determined to minimize the sum

$$S = \sum_{j=1}^{n} (_iS_R - {_iS_A})^2 \tag{209}$$

The measured and calculated values of peak widths for various analytes are listed in Table 39. The calculation method allows the estimation of the optimal GC conditions (short analysis time and acceptable separation) for the analysis of any mixture of analytes when the isothermal retention times and peak widths have been previously determined.

The effect of various gas chromatographic conditions on the temperature programmed gas chromatographic retention indices of nine chemical warfare agents was studied using

TABLE 39

Comparison of Measured ($_iS_R$) and Calculated ($_iS_A$) Peak Widths at Half Height for a Six-Component Model Mixture

Component	$_iA_S$	$_iB_S$	r	$_iC_S$	$_iS_R$	$_iS_A$	$(\Delta s/s)\%$
C9	35.560	1711.092	0.8707	0.020	3.68	3.9	5.64
2-CP	0.147	3963.534	0.9834	0.013	5.87	6.3	6.82
1,4-DCP	3.592	2512.145	0.9803	0.020	3.66	4.0	8.50
1,2-DCB	11.184	2389.591	0.9894	−0.168	6.02	6.5	7.38
1,3,5-TCB	1.034	3291.839	0.9928	0.010	5.34	5.8	7.93
C12	7.341	4576.188	0.9871	−0.150	8.11	8.9	9.18

Note: For abbreviations see Tables 36 and 38; r = correlation coefficient; $(\Delta s/s)\%$ = relative standard deviation.

From Messadi, D. and Ali-Mokhnache, S., *Chromatographia*, 37, 1993, 264–270. With permission.

n-alkane (I_C) and n-alkylbis(trifluoromethyl)phosphine sulfide standard series (I_M)[142]. The impact of the following chromatographic conditions was determined:

1. carrier gas flow rate
2. temperature programming rate
3. starting point of the temperature program
4. multistep temperature programming
5. injection volume
6. injection mode
7. sample solvent
8. background
9. column length
10. inside diameter of the column
11. film thickness
12. repeated use of the same column
13. individual columns
14. columns from different manufacturers
15. make of instrument
16. different operators
17. different laboratories
18. one or two retention index standard series in the same run
19. reduced number of retention index standards

The chemical warfare agents included in the study were sarin (isopropyl methylphosphonofluoridate); soman (1,2,2-trimethylpropyl methylphosphonofluoridate); tabun (ethyl N,N-dimethylphosphoramidocyanidate); VX (O-ethyl S-2-diisopropylamino)ethyl methylphosphonothioate; mustard (bis(2-chloroethyl)sulphide); CS (2-chlorobenzalmalononitrile); CN (α-chloroacetophenone); and CR (dibenz[b,f]-1,4-oxazepin. The retention indices calculated at different carrier gas flow rates are compiled in Table 40. The data in Table 40 clearly show that the carrier gas flow rate significantly influences the retention indices: they decrease with increasing flow rate. The effect was higher for compounds with lower volatility; however, the reproducibility of the measurements was good in each instance. The retention indices calculated at different temperature programming rates are compiled in Table 41. The results in Table 41 reveal that not only the carrier gas flow rate, but also the temperature programming rate may have a considerable impact on the retention indices. The retention indices markedly increased with increasing programming rate. It has been further established that the starting point of the temperature program, the use of multistep temperature program, the quality and quantity of organic solvent, the column length and internal diameter, and the film thickness exert a negligible influence on the retention indices.

Many methods have been developed to calculate the retention temperature Θ_f at which the analyte elutes from the column during linear temperature programming. It has been stated that the retention temperature can be calculated by[143]

$$g = \int_{\Theta_0}^{\Theta_f} d\Theta / (A + a \cdot \exp[b/(273 + \Theta)]) = \int_{\Theta_0}^{\Theta_f} y(\Theta) \cdot d\Theta \qquad (210)$$

TABLE 40
Effect of Carrier Gas Flow Rate on Retention Indices

Compound	Parameter	10°C/min							
		0.5 ml/min[a]		1.0 ml/min[a]		1.5 ml/min		2 ml/min	
		I	S.D. (3 runs)	I	S.D. (3 runs)	I	S.D. (3 runs)	I	S.D. (3 runs)
Sarin	I_C	823.5	0.5	823.7	0.1	—	—	—	—
	I_M	344.5	1.4	344.0	0.1	326.9	0.1	326.7	0.1
Tabun	I_C	1132.9	0.8	1132.5	0.4	1129.2	0.0	1129.9	0.3
	I_M	667.9	0.7	666.9	0.4	662.3	0.1	662.0	0.4
Mustard	I_C	1179.0	0.6	1176.2	0.2	1172.9	0.2	1171.8	0.3
	I_M	716.4	0.6	712.5	0.4	708.4	0.2	706.2	0.3
CN	I_C	1291.4	0.5	1287.4	0.4	1283.6	1.0	1281.4	0.1
	I_M	834.6	0.9	829.2	0.4	823.7	0.4	821.0	0.0
CS	I_C	1564.2	0.8	1559.6	0.5	1554.4	0.2	1552.2	0.1
	I_M	1116.6	0.8	1110.6	0.6	1103.8	0.2	1103.2	0.1
VX	I_C	1716.3	1.0	1712.2	0.6	1705.6	0.4	1703.3	0.4
	I_M	1272.6	1.1	1267.7	0.7	1260.5	0.4	1257.2	0.2
CR	I_C	1814.8	1.4	1804.8	0.7	1793.2	0.2	1787.6	0.1
	I_M	1373.9	1.3	1361.0	1.2	1350.0	0.2	1343.5	0.3

Compound	Parameter	10°C/min		5°C/min					
		2.5 ml/min		1.5 ml/min		2 ml/min		2.5 ml/min	
		I	S.D. (3 runs)	I	S.D. (3 runs)	I	S.D. (5 runs)	I	S.D. (4 runs)
Sarin	I_C	—	—	—	—	—	—	820.2	0.1
	I_M	326.2	0.1	326.2	0.3	—	—	340.8	1.1
Tabun	I_C	1129.2	0.1	1130.3	0.7	1129.2	0.2	1129.0	0.1
	I_M	660.9	0.1	661.2	0.6	658.9	0.3	657.8	0.1
Mustard	I_C	1170.6	0.1	1170.1	0.6	1168.5	0.3	1167.3	0.0
	I_M	704.6	0.1	703.1	0.4	700.2	0.2	700.0	—
CN	I_C	1279.8	0.1	1278.8	0.2	1278.0	0.8	1274.5	0.2
	I_M	818.8	0.2	816.5	0.6	814.5	0.3	811.5	0.1
CS	I_C	1550.7	0.2	1548.3	0.7	1547.7	0.5	1543.9	0.1
	I_M	1100	—	1096.0	0.6	1095.7	0.3	1090.3	0.2
VX	I_C	1700.0	—	1700.0	—	1699.4	0.4	1695.2	0.3
	I_M	1255.0	1.1	1251.9	0.1	1250.3	0.1	1246.8	0.2
CR	I_C	1783.7	0.5	1777.5	0.4	1774.1	0.1	1767.1	0.1
	I_M	1338.7	1.3	1330.8	0.4	1328.4	0.1	1319.6	0.1

Note: Columns — cross-linked SE-54 (2), 15 m × 0.32 mm I.D., 0.25 µm; cross-linked SE-54 (3), 15 m × 0.32 mm I.D., 0.25 µm; temperature programs — 40 to 280°C at 10 and 5°C/min.

[a] Column (3).

From Kokko, M., *J. Chromatogr.*, 630, 1993, 231–249. With permission.

where g is the temperature programming rate; Θ_0 is the initial temperature; A is the column dead volume under isothermal conditions.[144] The equation can be solved either by using an empirical initial temperature value (Method A; see Tables 42 and 43) or by iteration (Method B). Method C consists of the transformation of the integral into

$$g = \sum_{i=1}^{N} A(i) \tag{211}$$

$$A(i) = \Delta\Theta \cdot [y(\Theta_{i-1}) + y(\Theta_i)]/2 \tag{212}$$

and Method D calculates the integral by

TABLE 41

Effect of Temperature Programming Rate on the Reproducibility of the Retention Indices

Compound	Parameter	I	S.D. (4 runs)	II	S.D. (4 runs)	III	S.D. (3 runs)	IV	S.D. (3 runs)
Sarin	I_C	821.5	0.3	822.0	0.2	821.0	0.0	820.8	0.1
	I_M	342.9	0.2	342.7	0.1	341.5	0.0	338.7	0.1
Tabun	I_C	1130.3	0.9	1131.5	0.4	1129.3	0.1	1130.0	0.1
	I_M	664.8	0.8	665.2	0.3	661.8	0.1	658.5	0.1
Mustard	I_C	1176.2	0.7	1175.8	0.3	1171.4	0.1	1167.6	0.0
	I_M	713.2	0.5	711.8	0.1	706.2	0.1	700.0	—
CN	I_C	1287.2	0.6	1285.7	0.6	1280.5	0.1	1275.0	0.2
	I_M	829.9	0.6	827.8	0.5	820.7	0.1	811.9	0.2
CS	I_C	1559.2	0.6	1558.0	0.5	1551.4	0.1	1544.4	0.3
	I_M	1111.6	0.6	1109.0	0.4	1100.0	—	1091.1	0.4
VX	I_C	1710.2	0.2	1708.2	0.2	1702.4	0.2	1695.0	0.2
	I_M	1266.7	0.3	1264.1	0.2	1257.0	0.1	1246.7	0.2
CR	I_C	1805.7	0.4	1800.0	—	1786.8	0.1	1766.5	0.2
	I_M	1364.4	0.4	1358.0	0.2	1342.9	0.1	1319.2	0.2

Note: Column — cross-linked SE-54, 25 × 0.32 I.D., 0.25 μm; temperature programs — (I) 40 to 260°C at 10°C/min, (II) 40 to 260°C at 8°C/min, (III) 40 to 260°C at 5°C/min, (IV) 40 to 260°C at 2°C/min.
From Kokko, M., *J. Chromatogr.*, 630, 1993, 231–249. With permission.

TABLE 42

Example of the Retention Values (min) Obtained through Different Programs Compared with the Experimental Data Obtained on the Nonpolar DB-1 Column

No.	Experimental t_r	Method A t_r	E%	Method B t_r	E%	Method C t_r	E%	Method D t_r	E%
1	4.279	4.230	1.15	4.230	1.15	4.230	1.15	4.230	1.15
2	4.430	4.378	1.18	4.378	1.18	4.378	1.18	4.378	1.18
3	5.048	4.988	1.19	4.988	1.19	4.988	1.19	4.988	1.19
4	6.436	6.364	1.11	6.366	1.09	6.362	1.14	6.375	0.95
5	8.344	8.322	0.27	8.331	0.15	8.337	0.08	8.325	0.23
6	8.897	8.889	0.09	8.899	0.02	8.912	0.17	8.925	0.31
7	10.418	10.446	0.26	10.462	0.42	10.462	0.43	10.475	0.55
8	10.571	10.615	0.41	10.629	0.55	10.612	0.39	10.625	0.51
9	10.915	10.934	0.17	10.961	0.42	10.987	0.66	10.975	0.55
10	11.003	11.029	0.23	11.056	0.48	11.062	0.54	11.075	0.65
11	11.598	11.649	0.44	11.673	0.65	11.687	0.77	11.675	0.66
12	11.732	11.782	0.43	11.811	0.67	11.812	0.69	11.825	0.79
13	12.036	12.089	0.44	12.121	0.71	12.112	0.64	12.125	0.74
14	12.220	12.273	0.43	12.308	0.72	12.312	0.76	12.325	0.86

Note: Numbers refer to analytes in the text.
From Castello, G., Moretti, P., and Vezzani, S., *J. Chromatogr.*, 635, 1993, 103–111. With permission.

$$g = \sum_{i=1}^{M} A(i) \tag{213}$$

$$A(i) = \Delta\Theta \cdot [y(\Theta_{2i-2}) + y(\Theta_{2i-1}) + y(\Theta_{2i})]/3 \tag{214}$$

where Θ is the increment of temperature values. Analytes for testing the prediction capacity of the various calculation methods have been chosen according to their capacity to interact with the stationary phase: 1,3-dichlorobenzene (Comp. 1), 1,4-dichlorobenzene (Comp. 2), 1,2-dichlorobenzene (Comp. 3), 1,3,5-trichlorobenzene (Comp. 6), 1,2,4-trichlo-

TABLE 43

Example of the Retention Values (min) Obtained through Different Programs Compared with the Experimental Data Obtained on the Polar DB-WAX Column

No.	Experimental t_r	Method A t_r	E%	Method B t_r	E%	Method C t_r	E%	Method D t_r	E%
1	5.310	5.207	1.94	5.195	2.17	5.242	1.29	5.250	1.13
2	5.782	5.752	0.52	5.762	0.35	5.792	0.17	5.785	0.02
3	6.475	6.552	1.18	6.593	1.82	6.592	1.80	6.585	1.67
4	10.776	10.755	0.19	10.801	0.23	10.792	0.15	10.785	0.07
5	14.382	14.366	0.11	14.458	0.53	14.458	0.53	14.450	0.47
6	6.954	6.883	1.03	6.923	0.44	6.908	0.66	6.920	0.54
7	9.115	9.077	0.42	9.131	0.18	9.108	0.07	9.120	0.02
8	10.776	10.756	0.19	10.801	0.23	10.792	0.15	10.785	0.07
9	18.312	18.312	0.00	18.296	0.09	18.442	0.71	18.450	0.76
10	18.164	18.161	0.02	18.151	0.07	18.292	0.70	18.285	0.66
11	10.562	10.538	0.23	10.586	0.23	10.592	0.28	10.585	0.20
12	13.609	13.572	0.27	13.640	0.23	13.642	0.24	13.650	0.30
13	14.301	14.223	0.54	14.304	0.02	14.308	0.05	14.320	0.11
14	15.357	15.550	1.28	15.645	1.87	15.675	2.07	15.685	2.12

Note: Numbers refer to analytes in the text.
From Castello, G., Moretti, P., and Vezzani, S., *J. Chromatogr.*, 635, 1993, 103–111. With permission.

robenzene (Comp. 7), and 1,2,3-trichlorobenzene (Comp. 11), all with π-electron availability influenced by the inductive effect of chloro substituents; nitrobenzene (Comp. 4), 1-chloro-3-nitrobenzene (Comp. 12), 1-chloro-4-nitrobenzene (Comp. 13), and 1-chloro-2-nitrobenzene (Comp. 14), all containing donor but not active hydrogen atoms; 2-chloroaniline (Comp. 5), 3-chloroaniline (Comp. 9), and 4-chloroaniline (Comp. 10), all containing both donor atoms and active hydrogen; and naphtalene (Comp. 8), with π-electron availability. Both polar (DB-WAX, J&W Scientific, Folsom, CA) and nonpolar (DB-1 J&W Scientific, Folsom, CA) capillary columns were used in the experiments. The calculated and experimental retention times and the relative error (E%) of the predictions are compiled in Tables 42 and 43. The data in Tables 42 and 43 indicate that each method is suitable for the prediction of the retention times of the analytes under investigation; however, the calculated values were a little lower for analytes with shorter retention times and higher for analytes with longer retention times. The percentage deviation (E values) was lower on the nonpolar than on the polar column.

The temperature programmed retention indexes of nonsubstituted polynuclear aromatic hydrocarbons (PAHs) were related to the various physicochemical characteristics of the analytes (first-order valence molecular connectivity ^1X, ionization potential IP, length L, height H, and quadrupole moment Q).[145] The results of calculations are compiled in Tables 44 to 46. The regression equations fit well to the experimental data in each instance, indicating that the physicochemical parameters included in the calculation are suitable for the prediction of the retention of PAHs. The first-order valence molecular connectivity index exerts the highest impact on the retention; however, the other descriptors also increase the prediction power of the equations.

The various methods used for the calculation of retention indices in temperature programmed GLC have recently been compared and critically evaluated in an excellent review.[146]

D. The Takács Retention Index

It has been assumed that the retention index RI depends on both on the molecular structure coefficient S_c (containing information about the physicochemical characteristics

TABLE 44

Regression Equations for Isomeric Groups

$$I = I_0 + k_1 \cdot IP + k_2 \cdot {}^1X + k_3 \cdot L + k_4 \cdot H + k_5 \cdot Q$$

Mol wt	252[a]	278	302	328
n	7	12	17	15
I_0	−2612.4	−957.18	−1256.8	5193.9
k_1(IP)	−15.65	9.717	−12.0	−15.36
	3.60%	5.33%	4.12%	2.62%
k_2(${}^1X$)	470.6	156.85	244.24	−468.67
	93.18%	80.43%	88.58%	90.79%
k_3(L)	0.849	2.190	0.370	1.625
	0.31	2.20	0.23	0.59
k_4(H)		−4.579	−7.157	−4.207
		1.27%	1.20%	0.38%
k_5(Q)	−0.957	1.427	−1.042	−1.947
	2.91%	10.77%	5.86%	5.62%
RMS(iu)	0.60	1.36	0.86	5.20
RMSP(iu)	1.73	2.65	1.27	6.01
RMS(%)	0.13	0.28	0.16	0.89
RMSP(%)	0.38	0.54	0.24	1.03

Note: RMS = root-mean-square error; iu = index unit; RMSP = root-mean-square error from prediction.

[a] H was not included in the calculation.

From Bemgard, A., Colmsjö, A., and Wrangskog, K., *Anal. Chem.*, 66, 1994, 4288–4294. With permission.

of the stationary phase and the analyte as well as about their interaction) and on the Kováts coefficient K_c:[147,148]

$$RI = S_c + K_c \tag{215}$$

The classical definition of the Kováts retention index is

$$RI = 100 \cdot Z + 100 \cdot (\log V_{g,i} - \log V_{g,Z})/b \tag{216}$$

$$b = \log V_{g,Z+i} - \log V_{g,Z} \tag{217}$$

where $\log V_{g,i}$ is the specific retention volume of the analyte i; $\log V_{g,Z+i}$ and $\log V_{g,Z}$ are the specific retention volumes of the n-alkanes with carbon numbers Z + 1 and Z; and b is the slope of the plot of $\log V_g$ vs. Z. Thus,

$$K_c = 100 \cdot (Z - \log V_{g,Z}/b) \tag{218}$$

$$S_{c,i} = 100 \cdot \log V_{g,i}/b \tag{219}$$

The molecular structural coefficients of some analytes were determined on Superox 20M, as well as on OV-105 and QF-1 stationary phases at different column temperatures, and the molecular structural coefficients were calculated.[149] The results are compiled in Tables 47 and 48, respectively. In each instance the molecular structure coefficients decrease with increasing column temperature. The polar substructure of analytes has a considerable impact on the molecular structure coefficient; it decreases in the order alcohol > ketone >

TABLE 45

Regression equations for molecular intervals

$$I = I_0 + k_1 \cdot IP + k_2 \cdot {}^1X + k_3 \cdot L + k_4 \cdot H + k_5 \cdot Q$$

Mol wt	178-350	178-300	300-350
n	70	33	38
I_0	180.63	80.457	226.96
k_1(IP)	−17.631	−9.693	−20.319
	15.21%	10.01%	15.16%
k_2(${}^1X$)	77.344	82.354	78.429
	66.20%	72.42%	64.54%
k_3(L)	1.041	0.497	0.822
	1.63%	0.84%	1.19%
k_4(H)	−9.116	−7.389	−9.069
	3.87%	3.60%	3.39%
k_5(Q)	−1.009	−0.994	−1.246
	13.10	13.12	15.72
RMS(iu)	5.09	3.77	4.40
RMSP(iu)	5.70	5.05	5.16
RMS(%)	0.99	0.84	0.78
RMSP(%)	1.11	1.13	0.92

Note: RMS = root-mean-square error; iu = index unit; RMSP = root-mean-square error from prediction.

From Bemgard, A., Colmsjö, A., and Wrangskog, K., *Anal. Chem.*, 66, 1994, 4288–4294. With permission.

TABLE 46

Regression Equations for Cata- and Pericondensed PAHs

$$I = I_0 + k_1 \cdot IP + k_2 \cdot {}^1X + k_3 \cdot L + k_4 \cdot H + k_5 \cdot Q$$

Mol wt	Cata	Peri
n	34	36
I_0	114.42	166.49
k_1(IP)	−14.038	−14.080
	7.38%	13.25%
k_2(${}^1X$)	118.193	73.924
	61.08	70.61
k_3(L)	1.873	0.146
	1.87%	0.24%
k_4(H)	−9.032	−6.750
	2.43%	3.02%
k_5(Q)	−3.576	−0.867
	27.24%	12.88%
RMS(iu)	5.80	1.62
RMSP(iu)	7.45	1.98
RMS(%)	1.13	0.31
RMSP(%)	1.45	0.38

Note: RMS = root-mean-square error; iu = index unit; RMSP = root-mean-square error from prediction.

From Bemgard, A., Colmsjö, A., and Wrangskog, K., *Anal. Chem.*, 66, 1994, 4288–4294. With permission.

TABLE 47

Molecular Structural Coefficients of n-Dodecane and Analytes with Four Different Functional Groups on Superox 20M

Analyte	Column Temperature (°C)			
	80	100	120	140
n-Dodecane	945	928	903	871
Methanol	633	614	591	551
Ethanol	668	646	613	577
n-Propanol	771	751	721	683
n-Butanol	879	860	832	795
n-Pentanol	986	967	940	906
Benzene	705	695	676	656
Methylbenzene	800	793	775	753
Ethylbenzene	883	877	861	839
n-Propylbenzene	961	955	940	918
Butanone	655	642	621	530
2-Pentanone	732	720	700	633
n-Heptanone	932	924	901	865
2-Octanone	1030	1020	1000	963
Methyl formate	504	489	463	444
Methyl acetate	576	562	539	472
Methyl butyrate	737	722	701	674
Methyl hexanoate	935	925	902	864
Methyl heptanoate	1032	1020	997	967

From Santiuste, J. M., *Chromatographia*, 38, 1994, 701–708. With permission.

ester > aromatic. The data in Table 48 indicate that the change of molecular structural coefficient with temperature is lower on the less polar OV-105 than on the QF-1 stationary phases, suggesting the marked role of polar intermolecular forces in the interaction of analytes with the stationary phases. To gain a more profound understanding of the role of various molecular substructures of the analyte in the retention behavior, the S_c increments (ΔS_c) were also calculated:

$$\Delta S_c = S_c(FG) - S_c(\text{equal } Z \text{ } n\text{-alkane}) \tag{220}$$

where FG denotes functional group. Some calculated S_c values are compiled in Table 49. The molecular structure increments increase with increasing polarity of the stationary phase and the functional group, proving again the considerable impact of polar interactions in the retention of these type of analytes.

The equations discussed above have been used for different predictive purposes:[151]

1. Prediction of V_g of any n-alkane with carbon number Z from the V_g of decane at 120°C.

$$\log V_{g,Z} = 0.01 \cdot b \cdot S_{c,Z} \tag{221}$$

$$\log V_{g,10} = 0.01 \cdot b \cdot S_{c,10} \tag{222}$$

where $\log V_{g,10}$ is the specific retention volume of decane, and $S_{c,10}$ is molecular structural coefficient of decane.

$$S_{c,Z} = S_{c,10} + 100(Z - 10) \tag{223}$$

TABLE 48

Molecular Structural Coefficients of 11 Analytes on OV-105 and 12 Analytes on QF-1

OV-105

	Column Temperature (°C)				
Analyte	100	110	120	130	140
Benzene	547	537	533	524	513
n-Butanol	557	543	536	522	507
2-Pentanone	580	567	561	550	537
1-Nitropropane	646	634	631	620	608
Pyridine	646	638	630	624	615
2-Methyl-2-pentanol	622	607	601	589	575
1-Iodobutane	701	691	689	681	670
2-Octyne	739	725	718	707	691
1,4-Dioxane	591	580	576	564	553
cis-Hydrindane	863	855	854	848	839
n-Dodecane	1063	1048	1043	1031	1017

QF-1

	Column Temperature (°C)			
Analyte	90	120	150	180
Benzene	551	497	469	370
Toluene	661	603	560	484
n-Octanol	999	943	891	844
Pyridine	772	719	688	623
2-Pentanone	743	693	660	585
n-Butanol	591	530	490	428
Ethyl acetate	600	530	487	412
Dimethyl aniline	1076	1031	1003	954
Propionitrile	710	666	643	604
Butyronitrile	803	763	735	699
Valeronitrile	912	874	848	814
n-Dodecane	962	898	844	774

From Santiuste, J. M., *Chromatographia*, 38, 1994, 701–708. With permission.

TABLE 49

Molecular Structural Coefficient Increments of Five Functional Groups on Five Stationary Phases at 120°C

Stationary Phase	RP	S_c				
		FG 1	FG 2	FG 3	FG 4	FG 5
PS-255	6.9	58	246	192	170	238
TFP26%	27.0		307		344	468
QF-1	46.6	202	439	432	446	673
OV-215	47.9	198	439	417	507	684
Superox 20M	71.9	368	731	506	502	

Note: RP = polarity of the stationary phase according to Reference 150; FG 1 = n-alkylbenzenes; FG 2 = n-alkanols; FG 3 = methyl ester; FG 4 = methyl-n-alkyl-ket-2-ones; FG 5 = unbranched alkyl nitriles.

From Santiuste, J. M., *Chromatographia*, 38, 1994, 701–708. With permission.

$$\log V_{g,z} = F \cdot \log V_{g,10} \qquad (224)$$

where F is the methylene group factor:

$$F = 1 + 100 \cdot (Z - 10)/(1000 - K_{c,Z}) \tag{225}$$

2. Prediction of V_g at any column temperature.

   The dependence of both $K_{c,Z}$ and $S_{c,Z}$ on the temperature can be described by

$$K_{c,Z}(T) = a_1 \cdot T + a_2 \tag{226}$$

$$S_{c,i}(T) = \alpha \cdot T + \beta \tag{227}$$

$$\log V_{g,i}(T) = 0.01 \cdot b(T) \cdot S_{c,i}(T) \tag{228}$$

for nonalkane analytes, and for n-alkanes

$$\log V_{g,Z}(T) = 0.01 \cdot b(T) \cdot K_{c,Z}(T) \tag{229}$$

3. Calculation of S_c and analyte chemical functions.

   The difference in the molecular structural coefficients (ΔS_c) of a series of analytes depends on the characteristics of the stationary phase RP:

$$\Delta S_c(\text{function}) = m \cdot RP + n \tag{230}$$

$$S_{c,i} = S_{c,Z} + \Delta S_c(\text{function}) = (m \cdot RP + n) + (100 \cdot Z - K_{c,Z}) \tag{231}$$

$$\log V_{g,i} = 0.01 \cdot b \cdot [(100 \cdot Z - K_{c,Z}) + (m \cdot RP + n)] \tag{232}$$

Retention polarities (RP) and Kováts coefficients ($K_{c,Z}$) for 21 stationary phases, molecular structural coefficient of n-decane ($S_{c,10}$), F factors, and V_g absolute deviations for dodecane (δ) are compiled in Table 50. The methylene group factor F increased with increasing polarity of the stationary phase. The differences between the calculated and measured values of the specific retention volumes were low, indicating that the theoretical equations fit well to the experimental results. The measured and calculated specific volumes of some n-alkanes on OV-22 stationary phase are compiled in Table 51. The low differences between the experimental and calculated specific retention volumes of n-alkanes prove the suitability of the theoretical treatment discussed above for the prediction of the retention of analytes at any less polar stationary phase.

E. Other Descriptors

Retention indices have been frequently used for the tentative identification of various volatile compounds;[152–155] however, the interlaboratory reproducibility of the determination of retention indices has been vigorously discussed.[156–159] The so-called unified retention index[160,161] has been recently used for the identification of alkylbenzenes on apolar OV-101 and SE-30 columns.[162] The unified retention indices (UI_o), their calculated temperature increment (dUI/dT), standard deviation (s), and number of experimental indices used in the calculation (n) are compiled in Table 52. It has been concluded from the comparison of the calculated and measured retention indices of alkylbenzene derivatives that the unified retention indices can be successfully used for their identification both on OV-101 and SE-30 columns. It has been further proposed that the dUI/dT values can be used for the calculation of the optimum column temperature for the efficient separation of any mixtures of alkylbenzenes.

TABLE 50

Retention Polarities (RP), Kováts Coefficients ($K_{c,Z}$) for 21 Stationary Phases, Molecular Structural Coefficient of n-Decane ($S_{c,10}$), F Factors, and V_g Absolute Deviations for Dodecane (δ)

Stationary Phase	RP	$K_{c,Z}$	$S_{c,10}$	F	δ
OV-101	6.8	170	830	1.241	0.0
PS-255	6.9	139	861	1.232	0.9
OV-3	13.2	173	827	1.242	0.7
OV-105	14.5	149	851	1.235	1.9
OV-7	18.4	193	807	1.248	−0.4
DC-550	19.2	193	807	1.248	0.7
OV-61	24.0	229	771	1.259	1.7
OV-11	24.4	223	777	1.257	1.1
TFPS26%	26.0	212	788	1.254	−0.6
DIOPH	26.0	219	781	1.256	−2.4
OV-17	27.4	244	756	1.264	0.9
OV-22	33.3	278	722	1.277	−0.2
OV-25	35.9	302	698	1.286	1.8
QF-1	46.6	303	697	1.287	0.5
OV-215	47.9	267	733	1.273	0.0
Ucon 50HB-2000	49.6	329	671	1.298	−1.1
OV-225	56.4	284	716	1.279	−0.3
Igepal Co-990	67.5	527	472	1.424	0.0
CW 6000	72.0	416	584	1.342	−0.6
CW 20M	72.0	430	570	1.351	−0.4
Superox 20M	71.9	354	646	1.309	0.0

Note: RP = retention polarities taken from Reference 150; DIOPH = diisooctyl phthalate; TFPS26% = trifluoropropylsiloxane.

From Santiuste, J. M., *J. Chromatogr. A*, 690, 1995, 177–186. With permission.

TABLE 51

Prediction of the Specific Retention Volume (V_g) of the n-Alkanes Z = 6 to 12 on OV-22 from Decane Data

Z	F	V_g(exptl.)	V_g(calc.)	Deviation, δ
6	0.4464	6.2	6.3	−0.1
7	0.5848	11.0	11.2	−0.2
8	0.7232	19.6	19.7	−0.1
9	0.8616	34.9	34.9	0.0
10	1.0	61.9		
11	1.1384	109.4	109.5	−0.15
12	1.2768	193.6	193.8	0.02
Mean				−0.12

From Santiuste, J. M., *J. Chromatogr. A*, 690, 1995, 177–186. With permission.

It has been observed many times that the adsorption of n-alkanes on the gas–liquid surface of a polar stationary phase highly depends on the film thickness, resulting in decreased reproducibility of Kováts retention indices in capillary GC.[163-167] The use of more polar reference compounds was proposed to increase the accuracy of the determination of retention indices. It has been found that unsaturated hydrocarbons can be successfully used for this purpose.[168] The retention indices and their temperature increments of n-tetradecenes and n-tetradecynes were determined on capillary columns coated with PEG 20M (film thickness: column A, 0.12 μm; column B, 0.24 μm; and column C, 0.25 μm).

TABLE 52

The Unified Retention Indices (UI_o), Their Calculated Temperature Increment (dUI/dT), Standard Deviation (s), and the Number of Experimental Indices Used in the Calculation (n)

No.	Hydrocarbon	UI_o	s	n	dUI/dT
1	Benzene	643.72	0.23	14	0.1930
2	Methylbenzene	745.95	0.23	13	0.2004
3	Ethylbenzene	831.61	0.63	6	0.2719
4	1,3-Dimethylbenzene	842.04	0.42	6	0.2442
5	1,4-Dimethylbenzene	843.14	0.41	4	0.2479
6	Styrene	856.94	0.70	5	0.2845
7	1,2-Dimethylbenzene	857.03	0.77	6	0.3264
8	Isopropylbenzene	881.14	0.33	6	0.2889
9	Allylbenzene	909.53	0.64	3	0.29
10	n-Propylbenzene	922.62	0.82	5	0.2691
11	1-Methyl-4-ethylbenzene	930.04	0.78	7	0.2795
12	1-Methyl-3-ethylbenzene	931.52	0.55	5	0.2501
13	1,3,5-Trimethylbenzene	938.08	0.58	7	0.2491
14	1-Methyl-2-ethylbenzene	942.67	0.73	6	0.3099
15	α-Methylstyrene	948.06	0.68	5	0.2477
16	1,2,4-Trimethylbenzene	952.87	0.50	8	0.3453
17	tert-Butylbenzene	953.41	0.37	5	0.3374
18	2-Methylstyrene	954.87	0.35	4	0.3236
19	3-Methylstyrene	956.31	0.22	4	0.2950
20	sec-Butylbenzene	970.50	0.37	6	0.3472
21	1-Methyl-3-isopropylbenzene	976.45	0.99	5	0.3489
22	1,2,3-Trimethylbenzene	981.87	0.78	6	0.3452
23	2,3-Dihydroindene	982.17	0.36	4	0.4532
24	Indene	986.07	0.37	4	0.4925
25	1-Methyl-4-isopropylbenzene	989.88	0.54	6	0.2740
26	1-Methyl-2-isopropylbenzene	991.84	0.66	5	0.3802
27	1-Methyl-3-n-propylbenzene	1013.43	0.73	8	0.2916
28	1,3-Diethylbenzene	1014.57	0.41	5	0.2563
29	1-Methyl-4-n-propylbenzene	1017.06	0.69	9	0.2933
30	n-Butylbenzene	1018.48	0.39	6	0.2945
31	1,4-Diethylbenzene	1019.19	0.52	5	0.2862
32	1,2-Diethylbenzene	1022.03	0.44	5	0.3102
33	1-Methyl-2-n-propylbenzene	1023.05	0.37	6	0.3551
34	tert-Pentylbenzene	1043.04	0.31	4	0.4224
35	1-Methyl-4-tert-butylbenzene	1052.97	0.39	4	0.3175
36	1-Ethyl-2-isopropylbenzene	1060.83	0.76	4	0.3796
37	1-Ethyl-3-isopropylbenzene	1063.89	0.67	4	0.2918
38	1,2,4,5-Tetramethylbenzene	1071.54	0.56	7	0.3532
39	1,2,3,5-Tetramethylbenzene	1073.79	0.62	7	0.3626
40	1-Ethyl-4-isopropylbenzene	1078.72	0.71	4	0.2563
41	Isopentylbenzene	1080.73	0.20	3	0.3175
42	1-Ethyl-3-n-propylbenzene	1096.29	0.70	4	0.2869
43	1-Ethyl-2-n-propylbenzene	1102.78	0.53	4	0.3136
44	1-Methyl-3-n-butylbenzene	1112.72	0.27	4	0.2764
45	1-Methyl-4-n-butylbenzene	1115.15	0.30	4	0.3059
46	n-pentylbenzene	1115.80	0.31	5	0.2947
47	1-Methyl-2-n-butylbenzene	1117.47	0.62	4	0.3655
48	1-Ethyl-4-n-propylbenzene	1121.67	0.03	3	0.2025
49	1,3-Diisopropylbenzene	1121.84	0.14	4	0.2072
50	1,2-Diisopropylbenzene	1123.93	0.92	4	0.2746
51	1,4-Diisopropylbenzene	1134.43	0.87	6	0.2629
52	1,3,5-Triethylbenzene	1185.83	0.34	6	0.2093
53	Pentamethylbenzene	1209.07	0.17	4	0.5105
54	n-Hexylbenzene	1212.03	0.12	3	0.315
55	1,3,5-Triisopropylbenzene	1320.37	0.06	3	0.045
56	Hexamethylbenzene	1350.57	0.23	3	0.65

From Skrbic, B. D. and Cvejanov, J. Dj., *Chromatographia*, 37, 1993, 215–217. With permission.

Various saturated and unsaturated hydrocarbons were used as standard. To compare the retention indices the differences between columns A and C were calculated by

$$\Delta I = I^C - I^A, \quad \Delta I_P = I_P^C - I_P^A \qquad (233)$$

where ΔI and ΔI_P are the differences in the retention indices using apolar and polar reference standards, respectively. The retention indices and their temperature increments are compiled in Tables 53 and 54. The retention indices of compounds increased with increasing thickness of PEG 20M film. The differences are higher for 1- and 2-isomers as well as for cis-isomers, indicating the effect of isomerization on the adsorption and dissolution capacity of these type of analytes. Using 3-alkyne standards the reproducibility of retention indices was considerably enhanced. This result was explained by the supposition that alkynes form hydrogen bonds with PEG 20M. The data in Table 55 clearly show that the use of polar reference compounds markedly reduces the temperature increments, resulting in a higher reliability of identification.

TABLE 53

Retention Indices (I, I_P) and Temperature Increments ($10 \cdot (\delta I/\delta T)$) of n-Tetradecenes on PEG 20M Columns at 80°C

n-Alkane Standards

Hydrocarbon	I^A	$10 \cdot (\delta I/\delta T)$	I^B	$10 \cdot (\delta I/\delta T)$	I^C	$10 \cdot (\delta I/\delta T)$
1-Tridecene	1337.0	1.5	1344.1	1.0	1343.9	1.0
1-Tetradecene	1436.3	2.4	1443.6	1.0	1444.3	1.6
trans-2-Tetradecene	1449.0	2.1	1458.6	0.6	1458.2	1.7
cis-2-Tetradecene	1459.6	3.6	1467.8	1.8	1470.2	2.5
trans-3-Tetradecene	1431.8	1.7	1438.4	0.4	1439.0	1.0
cis-3-Tetradecene	1434.5	2.1	1440.5	1.3	1442.4	0.9
trans-4-Tetradecene	1423.1	1.4	1429.3	0.8	—	—
cis-4-Tetradecene	1427.0	2.6	1432.9	1.8	1434.3	1.8
trans-5-Tetradecene	1421.5	1.7	1427.1	1.0	1427.8	0.2
cis-5-Tetradecene	1420.5	2.6	1427.1	1.8	1428.0	2.1
trans-6-Tetradecene	1418.8	1.7	1424.8	1.0	1425.3	0.8
cis-6-Tetradecene	1417.4	2.7	1422.6	1.8	1424.1	2.2
trans-7-Tetradecene	1417.4	1.6	1423.5	1.2	1423.3	1.0
cis-7-Tetradecene	1414.8	2.3	1421.1	1.6	1422.4	2.4

1-Alkene Standards

Hydrocarbon	I_P^A	$10 \cdot (\delta I/\delta T)$	I_P^B	$10 \cdot (\delta I/\delta T)$	I_P^C	$10 \cdot (\delta I/\delta T)$
1-Tridecene	1300	—	1300	—	1300	—
1-Tetradecene	1400	—	1400	—	1400	—
trans-2-Tetradecene	1412.8	−0.4	1415.1	−0.4	1413.8	−0.1
cis-2-Tetradecene	1423.5	0	1424.3	0.8	1425.8	0.7
trans-3-Tetradecene	1395.5	−0.6	1394.8	−0.6	1394.7	−0.4
cis-3-Tetradecene	1398.2	−0.3	1396.9	0.3	1398.1	−0.9
trans-4-Tetradecene	1386.7	−0.9	1385.6	−0.2	—	—
cis-4-Tetradecene	1390.6	0.3	1389.2	0.8	1390.0	0.1
trans-5-Tetradecene	1385.1	−0.6	1383.4	0	1383.9	−1.3
cis-5-Tetradecene	1384.1	0.4	1383.4	0.8	1383.6	0
trans-6-Tetradecene	1382.4	−0.5	1381.1	0	1381.7	−0.1
cis-6-Tetradecene	1381.0	0.5	1378.9	0.8	1379.9	0.3
trans-7-Tetradecene	1381.0	−0.6	1379.8	−0.2	1379.1	−1.0
cis-7-Tetradecene	1378.3	0.1	1377.4	0.6	1378.2	0.7

From Orav, A., Kuningas, K., and Rang, S., *Chromatographia*, 37, 1993, 411–414. With permission.

TABLE 54

Retention Indices (I, I_p) and Temperature Increments ($10 \cdot (\delta I/\delta T)$) of n-Tetradecynes on PEG 20M Columns at 80°C

n-Alkane Standards

Hydrocarbon	I^A	$10 \cdot (\delta I/\delta T)$	I^B	$10 \cdot (\delta I/\delta T)$	I^C	$10 \cdot (\delta I/\delta T)$
3-Tridecyne	1479.7	3.2	1498.5	3.2	1497.9	0.5
1-Tetradecyne	1603.5	2.9	1626.2	1.1	1627.4	0.8
2-Tetradecyne	1628.3	4.6	1651.8	3.0	1652.5	2.0
3-Tetradecyne	1576.4	4.2	1596.0	3.2	1596.5	0.8
4-Tetradecyne	1557.8	4.2	1576.0	2.9	1576.3	0.8
5-Tetradecyne	1551.0	4.1	1568.6	3.0	1569.0	1.4
6-Tetradecyne	1546.9	3.9	1564.3	2.9	1565.1	1.6
7-Tetradecyne	1545.1	3.8	1562.3	3.1	1563.0	1.6

1-Alkyne Standards

Hydrocarbon	I_p^A	$10 \cdot (\delta I/\delta T)$	I_p^B	$10 \cdot (\delta I/\delta T)$	I_p^C	$10 \cdot (\delta I/\delta T)$
3-Tridecyne	1300	—	1300	—	1300	—
1-Tetradecyne	1428.0	−1.7	1431.0	−2.0	1431.4	−0.1
2-Tetradecyne	1453.7	0.4	1457.2	−0.2	1456.8	1.0
3-Tetradecyne	1400	—	1400	—	1400	—
4-Tetradecyne	1380.6	0.2	1379.5	−0.3	1379.5	0.1
5-Tetradecyne	1373.6	0.1	1371.9	−0.2	1372.1	0.7
6-Tetradecyne	1369.5	0	1367.5	−0.3	1368.2	0.9
7-Tetradecyne	1367.7	−0.1	1365.4	−0.1	1366.0	0.9

From Orav, A., Kuningas, K., and Rang, S., *Chromatographia*, 37, 1993, 411–414. With permission.

TABLE 55

ΔI and ΔI_p Values of Unsaturated Hydrocarbons at 80°C

Hydrocarbon	ΔI	ΔI_p
1-Tetradecene	8.0	0
trans-2-Tetradecene	9.2	1.0
cis-2-Tetradecene	11.6	2.3
trans-3-Tetradecene	7.2	−0.8
cis-3-Tetradecene	7.9	−0.1
cis-4-Tetradecene	7.3	−0.6
trans-5-Tetradecene	6.3	−1.2
cis-5-Tetradecene	7.5	−0.1
trans-6-Tetradecene	6.5	−0.7
cis-6-Tetradecene	6.7	−1.1
trans-7-Tetradecene	5.9	−1.9
cis-7-Tetradecene	7.6	−0.1
1-Tetradecyne	23.9	3.3
2-Tetradecyne	24.2	3.1
3-Tetradecyne	20.1	0
4-Tetradecyne	18.5	−1.9
5-Tetradecyne	18.0	−1.5
6-Tetradecyne	18.2	−1.3
7-Tetradecyne	17.9	−1.7

From Orav, A., Kuningas, K., and Rang, S., *Chromatographia*, 37, 1993, 411–414. With permission.

A new approximation was used for the selection of parameters to describe the relationship between capacity ratio (k) and column temperature (T).[169] The principal equation was

$$t_r = t_0 \cdot (1 + k) \qquad (234)$$

With open tubular columns the dead time t_0 is given by

$$t_0 = L^2 \cdot \mu \cdot 16 / [r^2 \cdot j \cdot (P_0/P_i^2 - P_0^2)] \qquad (235)$$

where L and r are the length and radius of the column; μ is the viscosity of the carrier gas; P_i and P_0 are the inlet and outlet pressures; and j is the mobile-phase compressibility factor. When pressures are constant Eq. 235 becomes

$$t_0 = A_t \cdot (T)^{Bt} \qquad (236)$$

A_t and B_t can be calculated from the temperature dependence of the dead time. The capacity ratio depends on the partition coefficient (K) and the phase ratio (β):

$$k = K/\beta \qquad (237)$$

K depends only on the column temperature:

$$\ln K = -\Delta G^\circ / RT \qquad (238)$$

where ΔG° is the change of the Gibbs free energy for the evaporation of the analyte from the stationary phase. The characteristics of columns, chromatographic conditions, and analytes used in the experiment are compiled in Tables 56 and 57. It has been established that two or three parameters were sufficient to describe the relationship between the capacity ratio and column temperature. The best fitting equations and their errors are compiled in Table 58. Because the p_i parameters in Table 58 depend considerably on the experimental conditions, a new parameter T_{ik} has been proposed to describe the retention behavior of an analyte on a column. T_{ik} is defined as the isothermal column temperature at which analyte i has a capacity ratio k. It has been stated that this T_{ik} parameter can be calculated from the experimental data and can be used for the prediction of retention of any analyte.

Because the use of each reference series of compounds has some drawbacks, the application of more than one standard series of solutes for the determination of the retention indices has also been proposed.[170]

TABLE 56

Capillary Columns Used in the Experimental Determinations

Column	Stationary Phase	Length (m)	I.D. (mm)	d_f (μm)
A	OV-1	20	0.2	0.5
B	OV-1	22	0.22	0.14
C	SP-1000	25	0.22	0.25
D	OV-1	22	0.22	0.14

From De Frutos, M. et al., *Anal. Chem.*, 65, 1993, 2643–2649. With permission.

TABLE 57

Compounds Used in the Determination of k Values at Several Temperatures in Columns A to D and Temperature Ranges for Each Compound Class

Column	Compounds	Carbon no.	Temperature range (°C)
A (OV-1)	n-Alkanes	(9–19)	100–240
B (OV-1)	Methyl esters	(5–12)	60–200
	n-Acids	(6–12,14,16,18)	75–250
	2-Ketones	(8–13,15,17,19)	80–225
	Ethyl esters	(6–12,14,16)	75–250
	n-Alkanes	(10–20,22,24)	75–270
C (SP-1000)	Methyl esters	(5–11)	80–180
	n-Acids	(5–8)	120–180
	2-Ketones	(8–13)	80–80
	Ethyl esters	(4–12)	80–180
	n-Alkanes	(14–20)	80–180
D (OV-1)	n-Tetradecane		80–250

Note: Numbers in parentheses indicate carbon chain length.
From De Frutos, M. et al., *Anal. Chem.*, 65, 1993, 2643–2649. With permission.

TABLE 58

Equations used for correlating k and T values

Equation	Mean square residuals $(\Sigma(k_{exp} - k^2_{calc}/n \cdot t)^{1/2}$			
	A	B	C	D
$k = p_1 \cdot T \cdot e^{p_2/T}$	0.026	0.039	0.078	0.104
$k = p_1 \cdot e^{p_2/T}$	0.028	0.041	0.054	0.112
$k = p_1 \cdot T^{p_2}$	0.046	0.082	0.136	0.229
$k = p_1 \cdot (p_2 + T)^{p_3}$	0.022	0.019	0.013	0.019
$k = p_1 \cdot e^{p_2/T} + p_3$	0.031	0.023	0.013	0.090
$k = p_1 \cdot T^{p_2} \cdot e^{p_3/T}$	0.033	0.022	0.029	0.034

Note: For column notation see Table 57.
From De Frutos, M. et al., *Anal. Chem.*, 65, 1993, 2643–2649. With permission.

V. APPLICATION OF SOLUTE PARAMETERS RETENTION RELATIONSHIPS

The relationships between the retention characteristics of the analytes and their physicochemical parameters can be used not only for the prediction of the retention behavior and the optimization of separation process, but also for the determination of various physicochemical parameters of analytes from the measurement of the retention times or other chromatographic descriptors. The determination of these parameters (solubility of sparingly soluble components, infinite dilution activity coefficient, etc.) is generally difficult with other methods or the reproducibility of the classical physicochemical methods is lower than that of the gas chromatographic one.

The environmental fate of xenobiotics depends considerably on their partition coefficient between air and water (Henry's law constant). A new GC method has been developed for the determination of this molecular parameter.[171] The method is based on the measurement by GC of the equilibrium headspace peak areas of analytes from aliquots of the same solution in three separate vials with different headspace-to-liquid volume ratios. The slope of the linear correlation between the reciprocal peak area and the headspace-to-liquid ratio is related to the Henry's law constant:

$$P_i = H_p \cdot X_{i,j} \tag{239}$$

where P_i is the partial pressure of analyte i in the vapor phase (atm); H_p is the Henry's law constant of analyte i at a given temperature (atm); and $X_{i,j}$ is the mole fraction of analyte i in the liquid phase.

In ideal case

$$P_i = X_{i,g} \cdot P_t \tag{240}$$

where $X_{i,g}$ is the mole fraction of the analyte in the vapor phase, and P_t is the total pressure of the vapor phase. At low analyte concentration in the liquid phase,

$$P_i = N_{ihse} \cdot RT/V_{hs} \tag{241}$$

where N_{ihse} is the number of analyte molecules in the headspace vapor phase at equilibrium, and V_{hs} is the volume of the headspace vapor phase.

$$N_i = N_{iwe} + N_{ihse} \tag{242}$$

where N_i is the total number of moles of analyte in the container, and N_{iwe} is the number of moles of analyte in the water phase.

Dividing Eq. 242 by the volume of the water phase V_w gives

$$C_{iwo} = C_{iwe} + N_{ihse}/V_w \tag{243}$$

where C_{iwo} is the initial concentration of analyte in the water phase, and C_{iwe} is the equilibrium concentration of the analyte in the vapor phase.

$$C_{iwo} = C_{iwe} + C_{ihse}(V_{hs}/V_w) \tag{244}$$

where C_{ihse} is the equilibrium concentration of the analyte in the headspace phase, and V_{hs} is the volume of the headspace phase.

$$H_i = H_p \cdot k \cdot RT = H/RT \tag{245}$$

where H_i is Henry's constant in dimensionless terms, and H_p is Henry's constant in terms of pressure (atm).

$$H_i = C_{ihse}/C_{iwe} \tag{246}$$

$$C_{iwo} = C_{ihse} \cdot (1/H_i + V_{hs}/V_w) \tag{247}$$

Using a standard (std) the unknown analyte (unk) can be expressed by

$$(C_{iwo})_{unk}/(C_{iwo})_{std} = [(C_{ihse})_{unk}/(C_{ihse})_{std}] \cdot [(1/H_i + V_{hs}/V_w)_{unk}/(1/H_i + V_{hs}/V_w)_{std}] \tag{248}$$

When the Henry's law constant, the temperature, the sample matrix, and the ratios of the volumes of the headspace to the volume of the liquid phase are identical, Eq. 248 can be simplified to

$$(C_{iwo})_{unk}/(C_{iwo})_{std} = (C_{ihse})_{unk}/(C_{ihse})_{std} \tag{249}$$

TABLE 59

Dimensionless Henry's Law Constants at 25°C Determined by the Static Headspace Method

Analyte	Average	S.D.
Benzene	0.216	1.00×10^{-3}
Toluene	0.263	3.06×10^{-3}
Ethylbenzene	0.318	2.08×10^{-3}
m-Xylene and p-xylene	0.298	1.54×10^{-2}
o-Xylene	0.204	1.51×10^{-2}
1,1,1-Trichloroethane	0.718	6.45×10^{-2}
Trichloroethylene	0.420	3.64×10^{-2}
Tetrachloroethylene	0.697	1.75×10^{-2}
$Tert$-Butyl ether	0.0216	2.08×10^{-4}

From Robbins, G. A., Wang, S., and Stuart, J. D., *Anal. Chem.*, 65, 1993, 3113–3118. With permission.

$$1/C_{ihse} = (1/C_{iwo}) \cdot (1/H_i + V_{hs}/V_w) \tag{250}$$

$$1/PA = [(1/R \cdot C_{iwo}) \cdot (1/H_i)] + [(1/R \cdot C_{iwo}) \cdot (V_{hs}/V_w)] \tag{251}$$

where PA is the peak area of the GC response, and R is the response factor. The plot of $1/PA$ vs. V_{hs}/V_w is a straight line with a slope $1/R \cdot C_{iwo}$ and an intercept $1/(H_i \cdot R \cdot C_{iwo})$, and H_i = slope/intercept. The results are compiled in Table 59. The data showed excellent agreement with the values determined by other methods, proving the reliability and accuracy of this GC method.

The infinite dilution activity coefficients (τ^∞) of 31 analytes were determined by inert gas stripping and gas chromatography.[172] It has been proved many times that the infinite dilution activity coefficient of analytes be successfully used for the prediction and selectivity in chromatography.[173,174] The determination is based on the measurement of the decrease of the gas-phase analyte concentration as a function of time when an inert gas is passed through the solution. The concentration decrease can be described by

$$dn_2/dt = -\alpha \cdot n_2/(1 - \alpha \cdot n_2/P - p_1^0/P) \cdot F/RT \tag{252}$$

where n_2 is the number of analyte molecules in the cell; $\alpha = \tau_2^\infty \cdot p_2^0/N_1$; p_2^0 is the vapor pressure of the analyte (torr); N_1 is the number of water molecules; p_1^0 is the vapor pressure of the solvent (water) (torr); P is the pressure of the system (torr); T is the temperature (K); and F is the flow rate of the inert gas (ml/min at pressure P and temperature T). Neglecting the volatility of the solvent and the amount of analyte in the gas phase of the cell, Eq. 252 becomes

$$\ln A/A_0 = -\alpha \cdot F \cdot t/RT \tag{253}$$

where A is the peak area of the analyte at time t, and A_0 is the peak area of the analyte at the beginning of the experiment. When the number of the analyte molecules in the gas phase is not negligible:

$$\ln A/A_0 = -\alpha \cdot F/(RT \cdot \beta) \cdot (1 - n_2 \cdot \alpha/P \cdot \beta) \tag{254}$$

where $\beta = 1 + V \cdot p_2^0 \cdot \tau_2^\infty/RT \cdot N_1$, and V is the volume of the vapor phase in the cell. When the solvent loss is negligible:

TABLE 60

Limiting Activity Coefficients (τ_2^∞) and Gas-to-Water Partition Coefficients (K) of Some Nonelectrolytes

Analyte	τ_2^∞	K
Dichloromethane	253	9.33
Isopropyl ether	628	10.99
1,2-Dichloroethane	641	20.23
Bromoethane	679	3.23
Chloroform	903	5.85
2-Chloropropane	1.48×10^3	1.35
Chloropropane	1.75×10^3	1.71
2-Bromopropane	2.09×10^3	2.08
Iodoethane	2.19×10^3	3.48
Propyl ether	2.31×10^3	7.11
Benzene	2.48×10^3	4.36
Bromopropane	2.86×10^3	2.61
1,1,1-Trichloroethane	5.90×10^3	1.41
1-Chlorobutane	7.61×10^3	1.32
2-Bromobutane	8.32×10^3	1.91
Iodopropane	8.55×10^3	2.79
Trichloroethylene	8.75×10^3	2.49
Toluene	9.19×10^3	3.93
Bromobutane	1.22×10^4	2.04
o-Xylene	3.05×10^4	5.11
Chloropentane	3.21×10^4	1.03
Ethylenebenzene	3.27×10^4	3.28
m-Xylene	3.32×10^4	3.73
p-Xylene	3.33×10^4	3.56
Tetrachloroethylene	3.60×10^4	1.55
Butyl ether	4.72×10^4	3.24
Cumene	1.02×10^5	2.20
Mesitylene	1.17×10^5	3.52
Propylbenzene	1.36×10^5	2.25
Chlorohexane	1.41×10^5	0.77
Butylbenzene	5.66×10^5	1.65

From Li, J. et al., *Anal. Chem.*, 65, 1993, 3212–3218. With permission.

$$\ln A/A_0 = -\alpha \cdot F \cdot t/RT \cdot \beta \tag{255}$$

The results are compiled in Table 60. The relative standard deviation of the determination of the infinite dilution activity coefficients of nonelectrolytes was 0.2% when the flow rate was lower than 11 ml/min. The data are in good agreement with the results of other similar studies. The method is very accurate; however, it cannot be used for the determination of analytes with very low gas-to-water partition coefficients. It has been assumed that the data may facilitate the rational design of separations of hazardous wastes and pollution control.

A new headspace gas chromatographic method has been developed for the determinations of the water–hexadecane partition coefficients ($P_{w,16}$) of a wide variety of analytes.[175] The principle of the method is the measurement of the concentration of an analyte in a vapor phase in equilibrium with a dilute solution of the analyte in water upon addition of a known volume of hexadecane (Method 1) or the measurement of the concentration of an analyte in the vapor phase in equilibrium with a dilute solution of the analyte upon addition of a known volume of water (Method 2). The choice of the method depends on the lipophilicity (hydrophobicity) of the analyte or analytes under investigation. Considering a system in which the volume of water is V_w^0, the volume of the gas phase is V_g, and the total amount of analyte is n^0 mol, the amount of the analyte can be expressed by

$$[\text{analyte}]_w = n^0/V_w^0 \quad (256)$$

when the amount of the analyte is small in the gas phase. The amount of the analyte in the water is

$$[\text{analyte}]_g = n^0/(V_w^0 \cdot K_w) \quad (257)$$

where K_w is the partition coefficient of the analyte between gas and water. Adding a known volume (V_{16}) of n-hexadecane to the system will cause the analyte to partition between the water (amount = n_w) and the n-hexadecane phase (n_{16}):

$$n^0 = n_w + n_{16} = V_w^0 \cdot [\text{analyte}]_w + V_{16} \cdot [\text{analyte}]_{16} \quad (258)$$

$$n^0 = V_w^0 \cdot [\text{analyte}]_w + V_{16} \cdot P_{w,16} \cdot [\text{analyte}]_w \quad (259)$$

$$[\text{analyte}]_w = n^0/(V_w^0 + V_{16} \cdot P_{w,16}) \quad (260)$$

When the peak areas of the analyte are determined over pure water (A^0) and at various water–n-hexadecane mixtures (A^f) the partition coefficient can be calculated:

$$A^0/A^f = 1 + (P_{w,16}/V_w^0) \cdot V_{16} \quad (261)$$

The slope (S) of the equation above is given by

$$S = P_{w,16}/V_w^0 \quad (262)$$

When the partition of the analyte into the gas phase (n_g) is taken into consideration the equations are slightly modified:

$$n^0 = n_w + n_g + n_{16} \quad (263)$$

$$A^0/A^f = 1 + [P_{w,16}/(V_w^0 + V_g/K_w)] \cdot V_{16} \quad (264)$$

$$S = P_{w,16}/(V_w^0 + P_{w,16} \cdot V_g/K_{16}) \quad (265)$$

where K_{16} is the partition coefficient of gas to hexadecane:

$$P_{w,16,2}/P_{w,16,1} = 1 + V_g/(V_w^0 \cdot K_w) \quad (266)$$

where $P_{w,16,2}$ is the water-to-hexadecane partition coefficient when the analyte partitioned in the gas phase is ignored, and $P_{w,16,1}$ is the water-to-hexadecane partition coefficient when the analyte partitioned in the gas phase is taken into account.

$$K_{w,1} = K_{16}/(S \cdot V_w^0) \quad (267)$$

$$K_{w,2} = (K_{16}/(S - V_g)/\cdot V_w^0 \quad (268)$$

$$K_{w,1} - K_{w,2} = V_g/V_w^0 \quad (269)$$

where $K_{w,1}$ is the gas-to-water partition coefficient when the analyte partitioned in the gas phase is ignored, and $K_{w,2}$ is the gas-to-water partition coefficient when the analyte partitioned in the gas phase is taken into account. When the analyte is strongly hydrophilic (better solubility in water than in hexadecane) Method 2 has to be used. The calculation of the partition coefficients is similar:

$$A^0/A^f = 1 + V_w/(P_{w,16} \cdot V_{16}^0) \tag{270}$$

where V_w is the volume of water added to the initial hexadecane solution. When the amount of analyte n the gas phase has to be taken into account:

$$A^0/A^f = 1 + V_w/[P_{w,16} \cdot (V_{16}^0 + V_g/K_{16})] \tag{271}$$

The partition coefficients can be calculated from the equations listed above. The infinite dilution activity coefficient of an analyte in water τ^∞ was calculated by

$$\tau^\infty = RT \cdot \sigma_1/p_2^0 \cdot K_w \cdot MW_1) \tag{272}$$

where σ_1 is the density of water at 25°C; p_2^0 is the vapor pressure of the analyte; and MW_1 is the molecular weight of the water. The physicochemical parameters measured by this method are compiled in Table 61. The comparison of the physicochemical parameters in Table 61 with those measured with other methods[176] revealed that this method is suitable for measuring the partition coefficients of analytes between water and any immiscible organic solvents.

A GLC technique has been developed for the determination of the activity coefficient τ_1 of a nonelectrolytic analyte at molar fractions between 10^{-3} and 10^{-6} in a solvent.[177] The net retention volume V_N can be expressed by

$$V_N = J_3^2 \cdot F_C \cdot (t_R - t_M) \tag{273}$$

where F_C is the flow rate of the carrier gas at the column outlet; t_R and t_M are the retention times of an absorbed analyte vapor and a nonabsorbed carrier gas, respectively; and J_3^2, related to the pressure drop along the column, is defined by

$$J_3^2 = 3/4 \cdot [(p_i/p_0)^4 - 1]/[(p_i/p_0)^3 - 1] \tag{274}$$

where p_i and p_0 are the column inlet and outlet pressures, respectively. It has been shown that the correlation between the retention volume and inlet column pressure can be expressed by

$$\ln V_N = \ln(k^0 \cdot V_2) + \beta \cdot p_0 \cdot J_3^4 + \xi \cdot (p_0 \cdot J_3^4)^2 \tag{275}$$

where V_2 is the volume of the solvent on the GC support.

$$J_3^4 = 3/4 \cdot [(p_i/p_0)^4 - 1]/[(p_i/p_0)^3 - 1] \tag{276}$$

$$\ln k^0 = \ln(RT/\tau_1^\infty \cdot v_2^0 \cdot p_1^0) - (B_{11} - v_1^0) \cdot p_1^0/RT \tag{277}$$

TABLE 61

Log $P_{w,16}$, log K_w, and τ^∞ values of Various Analytes

Analyte	log $P_{w,16}$	log K_w	τ^∞
Methanol	−2.769	3.744	1.46
Ethanol	−2.200	3.662	3.80
Propanol	−1.555	3.539	14.17
Butanol	−0.941	3.451	53.33
Pentanol	−0.290	3.317	225.4
Hexanol	0.410	3.200	791.8
Acetonitrile	−1.500	3.060	10.9
Propionitrile	−0.810	2.788	37.56
Butyronitrile	−0.140	2.680	112.56
Isobutyronitrile	−0.110	2.368	134.69
Valeronitrile	0.510	2.547	413.0
Isopropyl ether	1.546	1.015	6.67×10^2
Propylether	2.099	0.872	2.21×10^3
Butyl ether	3.266	0.735	2.81×10^4
Anisole	2.025	1.901	3.65×10^3
Phenetole	2.509	1.631	1.57×10^4
Octanal	2.665	1.715	8.24×10^3
2-Nonanone	2.743	2.012	1.63×10^4
3-Nitrotoluene	2.130	2.840	7.09×10^3
1,1,2,2-Tetrachloroethane	1.975	1.851	3.46×10^3
Carbon tetrachloride	3.060	−0.237	1.54×10^4
1,2-Dichlorobenzene	3.334	1.071	6.82×10^4
Benzyl chloride	2.830	1.460	3.20×10^4
Fluorobenzene	2.404	0.436	4.80×10^3
Chlorobenzene	2.844	0.796	1.40×10^4
Bromobenzene	2.995	1.040	2.25×10^4
Iodobenzene	3.305	1.275	5.41×10^4
Benzene	2.173	0.630	2.53×10^3
Toluene	2.749	0.595	9.16×10^3
Ethylbenzene	3.258	0.507	3.34×10^4
Propylbenzene	3.860	0.361	1.33×10^5
Butylbenzene	4.442	0.244	5.33×10^5
o-Xylene	3.256	0.681	3.25×10^4
m-Xylene	3.362	0.502	3.90×10^4
p-Xylene	3.363	0.495	3.79×10^4
Cumene	3.751	0.354	9.91×10^4
Mesitylene	3.891	0.508	1.28×10^5

From Li, J. and Carr, P. W., *Anal. Chem.*, 65, 1993, 1443–1450. With permission.

where k^0 is the zero pressure partition coefficient; v_2^0 is the molar volume of pure solvent; p_1^0 is the vapor pressure of pure solute; B_{11} the solute second virial coefficient; v_1^0 is the molar volume of the pure solute.

$$\beta = (2B_{13} - v_{12}^\infty)/RT \qquad (278)$$

where B_{13} is the second virial coefficient between solute and carrier gas, and v_{12}^∞ is the partial molar volume of the solute infinitely dilute in the solvent.

$$\xi = (3C_{133} - 4B_{13} \cdot B_{33})/2(RT)^2 \qquad (279)$$

where C_{133} is the third virial coefficient between the solute and the carrier gas, and B_{33} is the second virial coefficient of the carrier gas. At low carrier gas pressures, at low solubility of the carrier gas in the solvent, and at low concentration of the analyte in the carrier gas the retention volume is given by

$$\ln V_N = \ln (k^0 \cdot V_2) + \beta \cdot p_0 \cdot J_3^4 \qquad (280)$$

By plotting $\ln (V_N/V_2)$ vs. $p_0 \cdot J_3^4$ both τ_1^∞ and B_{13} can be determined. The second virial coefficients and the derived Henry's law constants from the activity coefficients ($k_i = \tau_i^\infty \cdot p_i^0$) are listed in Table 62. It was concluded from the data that this GLC method is especially suitable for the determination of the molecular parameters listed above. The same method has been used for the determination of the second cross-virial coefficients and Henry's constant for brominated hydrocarbons.[178] The results are compiled in Table 63. The data indicated that each analyte contains only one type of group and their mixtures with water is a typical binary mixture.

TABLE 62

Second Virial Coefficients (B_{13}) for Benzene or Chlorinated Hydrocarbon (1) and Nitrogen (3) Mixtures, Group Assignments, and Henry's Law Constants (k_i) for Benzene or Chlorinated Hydrocarbon (1) and Water (2) Mixtures

Analyte	T (K)	B_{13} (cm^{-3}/mol)	Group Assignment	$k_i \cdot 10^{-4}$ (Pa)
Benzene	293.15	−118.95	6ACH	2512.65
	303.15	14.15		4013.57
	313.15	76.96		6065.12
	323.15	113.81		8914.52
Chlorobenzene	293.15	202.56	5ACH·1ACCl	1549.90
	303.15	175.38		2156.07
	313.15	142.07		2619.42
	323.15	−15.67		2809.97
Trichloroethylene	293.15	−66.23	3Cl(C=C)	4312.96
	303.15	−24.83	1CH=C	7640.23
	313.15	−10.15		11532.00
	323.15	102.60		17072.58
trans-1,2-Dichloroethene	293.15	−27.64	2Cl(C=C)	4418.50
	303.15	152.55	1CH=CH	7770.78
	313.15	169.30		11572.75
	323.15	177.51		16514.26

Khalfaoui, B. and Newsham, D. M. T., *J. Chromatogr. A*, 673, 1994, 85–92. With permission.

TABLE 63

The β Values, Second Virial Coefficients (B_{13}), and Henry's Constants (k_i) for Brominated Hydrocarbon (1) and Nitrogen (3) Mixtures

Compound	T (K)	β (atm^{-1})	B_{13} (cm^{3}mol^{-1})	k_i (MPa)
Tribromoethane	298.15	−0.0114	−96.76	1.06
	308.15	−0.0096	−78.16	1.90
	323.15	−0.0068	−46.61	3.16
1,2-Dibromoethane	298.15	−0.0104	−85.50	0.74
	308.15	−0.0092	−74.42	1.25
	323.15	−0.0061	−37.02	2.11
1,1,2,2-Tetrabromoethane	298.15	−0.0073	−31.79	0.98
	308.15	−0.0065	−24.43	1.48
	323.15	−0.0059	−20.21	2.87

Khalfaoui, I. and Newsham, D. M. T., *J. Chromatogr. A*, 688, 1994, 117–123. With permission.

The solubility parameters of 19 α,ω-diamino oligoethers were determined by means of inverse gas chromatography.[179] The chemical structures of the oligoethers are compiled in Figure 6. The oligoethers were used as liquid stationary phases and the retention of n-alkanes, aromatic hydrocarbons, ketones, nitropropane, alcohols, and pyridine was determined. The solubility parameters was calculated by[180,181]

$$\delta_1^2/R \cdot T - X^\infty/V_1^0 = 2 \cdot \delta_2 \cdot \delta_1/R \cdot T - (\delta_2^2/R \cdot T + X_S^\infty/V_1^0) \qquad (281)$$

where δ_1 and δ_2 are the solubility parameters of the analyte and the stationary phase, respectively; X^∞ and X_S^∞ are the Hildebrand–Schatchard interaction parameter and its entropic component, respectively; V_1^0 is the molar volume of the analyte; r is the gas constant; and T is the absolute temperature. The dispersive component of the solubility parameter (δ_d) was calculated from the slope of Eq. 281 applied for n-alkanes:

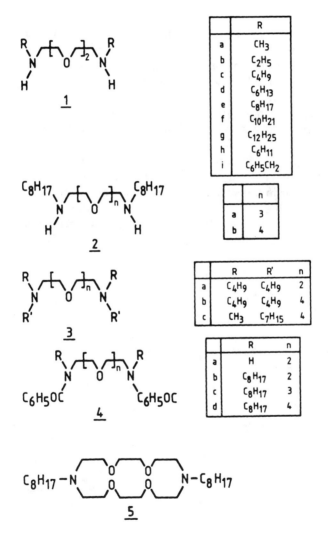

FIGURE 6
Chemical structure of α,ω-diamino oligoethers. (From Voelkel, A. and Janas, J., *J. Chromatogr. A*, 669, 1994, 89. With permission.)

$$\delta_d = \text{slope}_{n\text{-alkanes}} \cdot R \cdot T/2 \tag{282}$$

The polar increment of the solubility parameter (δ_p) was calculated from the difference of slope for polar analytes and n-alkanes:

$$\delta_p = (\text{slope}_{\text{polar}} - \text{slope}_{n\text{-alkanes}}) \cdot R \cdot T/2 \tag{283}$$

The hydrogen bonding increment (δ_h) was similarly calculated by

$$\delta_h = (\text{slope}_{\text{alcohols}} - \text{slope}_{n\text{-alkanes}}) \cdot R \cdot T/2 \tag{284}$$

The corrected solubility parameter (δ_T) was calculated

$$\delta_T^2 = \delta_d^2 + \delta_p^2 + \delta_h^2 \tag{285}$$

The binary parameters of polar (I_{12p}) and hydrogen bonding (I_{12h}) interactions were determined as described in Reference 182. These parameters are the measures for the deviation of a given system from the behavior of the regular solution. The solubility and binary parameters are listed in Table 64. Each parameter depends considerably on the chemical structure of the oligoethers. Thus, the solubility parameters δ_2, δ_p, δ_h, and δ_T decrease monotonously with the increasing length of the apolar hydrocarbon chain, while the dispersive component δ_d increases. The binary parameters approximate zero when the length of the hydrocarbon chain increases, indicating that the more hydrophobic oligoethers behave more regularly than the less hydrophobic ones.

Because the retention of an analyte depends on both its vapor pressure in the liquid state and its activity coefficient in the stationary phase, GLC can be used for the determination of the vapor pressure of the analyte when the activity coefficient is known.[183] The vapor pressure of an analyte (P_1) can be calculated from the comparison of its vapor pressure with that of a reference compound (P_2):

TABLE 64

Solubility and Binary Parameters for α, ω-Diamino Oligoethers at 90°C $[10^3(J/m^3)^{1/2}]$

Compound	δ_2	δ_d	δ_p	δ_h	δ_T	I_{12h}	I_{12p}
1a	18.1	11.1	5.1	10.2	15.9	−0.609	−0.453
1b	17.6	11.2	4.5	9.5	15.4	−0.587	−0.375
1c	17.2	11.4	4.0	9.1	15.1	−0.526	−0.342
1d	16.9	11.6	3.7	8.5	14.9	−0.468	−0.305
1e	16.7	11.8	3.6	8.2	14.8	−0.388	−0.215
1f	16.6	12.1	3.5	7.6	14.7	−0.382	−0.211
1g	16.2	12.5	3.0	7.0	14.6	−0.374	−0.183
1h	17.0	11.5	3.8	8.8	15.0	−0.472	−0.315
1i	16.8	11.8	3.7	8.1	14.8	−0.430	−0.292
2a	16.9	11.7	3.9	8.4	14.9	−0.444	−0.287
2b	17.1	11.7	4.1	8.8	15.2	−0.492	−0.317
3a	17.7	11.0	4.6	9.7	15.4	−0.537	−0.382
3b	18.2	10.5	5.3	10.1	15.8	−0.592	−0.433
3c	17.6	11.1	4.8	9.6	15.5	−0.555	−0.376
4a	18.3	10.9	6.0	10.0	16.0	−0.624	−0.448
4b	17.0	11.2	4.6	8.7	14.9	−0.472	−0.303
4c	17.3	11.1	5.1	8.9	15.1	−0.511	−0.320
4d	17.5	11.0	5.3	9.2	15.3	−0.547	−0.350
5	17.1	11.7	4.2	8.6	15.1	−0.500	−0.308

Note: Compounds correspond to those in Figure 6.
From Voelkel, A. and Janas, J., *J. Chromatogr. A*, 669, 1994, 89–95. With permission.

GAS CHROMATOGRAPHY

$$\ln P_1 = (\Delta H_1/\Delta H_2)\cdot \ln P_2 + C \tag{286}$$

where ΔH is the heat of vaporization.[184] With inclusion of the specific retention volumes Eq. 286 becomes:

$$\ln (V_{R1}/V_{R2}) = (1 - \Delta H_1/\Delta H_2)\cdot \ln P_2 - C \tag{287}$$

The Kováts retention index (I) can be correlated with the vapor pressure by

$$-\log P = B_0 + b_1 \cdot I \tag{288}$$

According to the equilibrium fugacity model the ratio of molar fractions y_i and x_i in the gas and the stationary phases is related to the vapor pressure and the activity coefficient (τ_i) of the analyte:

$$y_i/x_i = \tau_i \cdot P_i/P_t \tag{289}$$

where P_t is the pressure of the carrier gas. The relationship between the retention index and the activity coefficient can be expressed by

$$I_i = I_z + 100\cdot (\log \tau_z \cdot P_z - \log \tau_i \cdot P_i)/(\log \tau_z \cdot P_z - \log \tau_{z+1}\cdot P_{z+1}) \tag{290}$$

where subscripts z and z + 1 refer to the corresponding normal alkanes. Vapor pressure and the ratio of activity coefficients are given by

$$\log P_i = (d \log P_z/dI_z)\cdot I_i + \log P_{H_2} + \log (\tau_z/\tau_i) \tag{291}$$

$$\log (\tau_z/\tau_i) = \log [\tau_z(Sq)/\tau_i(Sq)] - (d \log P_z/dI_z)\cdot I \tag{292}$$

where P_{H_2} is the vapor pressure of liquid hydrogen; (Sq) refers to data determined on the apolar squalane stationary phase; and $I = I_i - I_i(Sq)$. The ratios of activity coefficients for some analytes are compiled in Table 65. The activity coefficient ratios are the highest on

TABLE 65

Calculated Activity Coefficient Ratios at 120°C of Compounds with Different Functional Groups on Squalane, SE-30, and Carbowax 20M (CW-20) Stationary Phases

Compound	log (τ_z/τ_i)		
	Squalane	SE-30	Carbowax-20M
Benzene	0.191	0.236	1.153
1-Butanol	−0.442	−0.284	1.158
2-Pentanone	−0.147	−0.016	0.952
Nitropropane	−0.436	−0.245	1.272
Pyridine	−0.093	0.030	1.430
2-Methyl-2-pentanol	−0.191	−0.098	0.964
1-Iodoputane	0.083	0.092	0.924
2-Octyne	0.060	0.126	0.720
1,4-Dioxane	−0.049	0.082	1.246

From Spieksma, W., Luijk, R., and Govers, H. A. J., *J. Chromatogr. A*, 672, 1994, 141–148. With permission.

TABLE 66

Vapor Pressure (P_i) Data for Chlorobenzenes (ca. 25°C) and Chlorophenols (160°C) Derived from Kováts Indices on SE-30

Chlorobenzene Congener	P_i
Mono-	1828
1,2-Di-	252
1,3-Di-	236
1,4-Di-	216
1,2,3-Tri-	44.1
1,2,4-Tri-	53.4
1,3,5-Tri-	62.4
1,2,3,4-Tetra-	12.1
1,2,3,5-Tetra-	15.1
1,2,4,5-Tetra-	15.1
Penta-	2.49
Hexa-	0.367
Phenol	79
2-Mono-	56
3-Mono	22
4-Mono-	22
2,4-Di-	20
2,6-Di-	17
2,4,5-Tri-	7.1
2,4,6-Tri-	7.5
2,3,4,6-Tetra	2.5
Penta-	0.91

From Spieksma, W., Luijk, R., and Govers, H. A. J., *J. Chromatogr. A*, 672, 1994, 141–148. With permission.

the strongly polar Carbowax 20M stationary phase, indicating the different retention characteristics of this stationary phase compared to the apolar squalane. The measured values of vapor pressure (P_i) for different analytes are compiled in Table 66. The vapor pressures calculated from Kováts retention indices differ considerably from the data determined by other methods (deviation between 3 to 51%; average of 19%). This result indicates that the relationship between the retention behavior of analytes (Kováts index) and the vapor pressure is not strong enough to predict the retention of the analytes in Table 66. This finding further suggests that physicochemical parameters other than vapor pressure may have a considerable impact on the retention even on a relatively apolar stationary phase such as SE-30.

Chapter 3

LIQUID CHROMATOGRAPHY

The term *liquid chromatography* summarizes chromatographic methods in which the mobile phase is liquid and the stationary phase is an inorganic or organic solid covered or not covered with ligands facilitating separation. The liquid chromatographic methods are generally divided into two parts: planar and column liquid chromatography. Because the same supports can be successfully used for both planar and column liquid chromatography, the characteristic and the main application parameters of various supports are discussed later in the chapter on high-performance liquid chromatography.

Due to the presence of a liquid phase consisting of one or more components the exact modeling of the partition and adsorption phenomena governing retention and separation is extremely difficult. It has to be borne in mind that each theoretical description of the chromatographic processes is necessarily an approximate one and uses simplifications that cannot be entirely verified either theoretically or practically. A liquid chromatographic separation process generally involves the adsorption of the analyte on the surface of the support and the desorption of the analyte from the surface with the components of the liquid mobile phase. This distribution process depends on the difference between adsorption strengths of the analyte and the molecules of the mobile phase on the active centers of the support and the dissolution capacity of the mobile phase for the analytes. However, this ideal condition is never reached in the practice of liquid chromatography. Even in the case of homologous mobile phase (which is improbable in the solution of concrete separation problems) the molecules of the eluent are adsorbed on the surface of the support. This adsorption results in the formation of a mono-, bi-, or multilayer on the surface, each bonded by different strengths. The dissolution capacity of the various adsorbed layers for the analytes can be different from that of the bulk phase and can vary according to the binding strength between the support and the adsorbed layers. Moreover, each analyte can form aggregates (dimers, trimers, tetramers, etc.) of undefined stoichiometry in the mobile phase that considerably influence the partition process. The stability of the aggregates may also have a marked impact on the efficiency of the separation.

The problems discussed above are multiplied in the case of eluents composed of two or more components. Each component can adsorb on the support surface forming layers of unknown composition and unknown adsorption and dissolution characteristics. Each analyte can interact with each component of the mobile phase resulting in the formation of complexes of analyte–analyte, analyte–one component, analyte–two components, etc. According to our knowledge the exact impact of the interactions discussed above on the liquid chromatographic separation processes has never been elucidated in detail.

We are well aware that this view of the molecular basis of liquid chromatography is a very pessimistic one. However, many excellent studies have been published on the various aspects of these separation problems and the agreement between the theoretical calculations and measured retention data sometimes was surprisingly good. The good agreement between theory and practice indicates that the main processes accounting for the efficiency of separation are well understood and the insufficient knowledge of the molecular interaction discussed above only modifies, not determines, the successful design of chromatographic separations. We strongly believe that a more profound understanding of the molecular basis of liquid chromatographic separation will be achieved in the future that will make possible the optimal and exact solution of any separation problems.

I. THIN-LAYER CHROMATOGRAPHY

A. Principles of Adsorption and Reversed-Phase Thin-Layer Chromatography

During the last few decades thin-layer chromatography (TLC) has been extensively used as a rapid and easily implemented analytical technique.[185] TLC is a very simple separation technique. The stationary phase is a thin-layer of sorbent and the sample is applied to the surface of the sorbent as a spot near the bottom end of the plate. An enclosed chamber is used for the separation, with the bottom end of the plate immersed in the mobile phase. The mobile phase moves through the stationary phase by capillary forces. The separation of compounds is produced by the different migration of the solutes. The particle size of the support is 10 to 60 μm, and the thickness of the layer is between 100 and 250 μm.[186] The sorbents most frequently used in TLC are silica, alumina, cellulose, and diatomaceous earth (Kieselguhr), silica being the most common sorbent. Because silica is slightly acidic the separation of basic solutes can be more easily carried out on an alumina layer that is neutral or slightly basic. The main advantages of TLC are[187]

1. More than one sample can be separated at one time
2. Great methodological simplicity
3. A wide variety of detection methods

However, the separation power and reproducibility of TLC is lower than that of gas chromatography (GC) or high-performance liquid chromatography (HPLC). TLC methods are generally less suitable for quantitative determination of residues, the standard deviation being higher than for the other chromatographic techniques. The separation can be carried out on unmodified sorbents (normal phase or direct phase separation, i.e., adsorptive TLC). In these instances the support is more polar and the mobile phase is relatively nonpolar; that is, the separation mechanism is based on adsorption processes. Reversed-phase thin-layer chromatography (RPTLC) uses apolar, hydrocarbonaceous stationary phases and polar mobile phases such as water and organic solvents miscible with water. The lipophilicity (hydrophobicity) of the solute governs its retention in RPTLC. The more hydrophobic the solute the lower is its mobility in reversed-phase chromatography. The theory and application of RPTLC has been recently reviewed.[188]

The retention of a solute in TLC and in RPTLC is characterized by the R_f value: the distance from the origin to the center of the separated zone divided by the distance from the origin to the solvent front. The capacity factor k can be defined as the ratio of retention times of a solute in the stationary (t_s) and mobile phase (t_m), respectively:

$$k = t_s/t_m \qquad (293)$$

The relationship between capacity factor and R_f is

$$R_f = t_m/(t_m + t_s) = 1/(k + 1) \tag{294}$$

$$k = (1 - R_f)/R_f = 1/R_f - 1 \tag{295}$$

In the case of equilibrium between the support and the mobile and vapor phase, the velocity of the solvent front is expressed by

$$(Z_f)^2 = K \cdot t \tag{296}$$

where Z_f is the migration distance of the solvent front in centimeters after t seconds; K is the mobile phase velocity constant (cm²/s). The velocity constant depends on the experimental conditions:

$$K = 2 \cdot K_0 \cdot d_p \cdot (\tau/\eta) \cdot \cos \Theta \tag{297}$$

where K_0 is the permeability constant (dimensionless); d_p is the mean particle diameter; τ is the surface tension of the mobile phase; η is the viscosity of the mobile phase; and Θ is the contact angle between support and mobile phase.

The velocity of the mobile phase (u) can be expressed by

$$u = dZ_f/dt = (K_0 \cdot d_p/\eta \cdot Z_f) \cdot \cos \Theta - Z_f \cdot (d_p \cdot p \cdot g/12) \tag{298}$$

where p is the density of the mobile phase and g is the gravitational field constant.

When the distribution of the sample within the spot is Gaussian, the number of theoretical plates (n) is given by:

$$n = 16 \cdot (R_f \cdot Z_f/w_b)^2 \tag{299}$$

where w_b is the peak width at the base.

The height equivalent of the theoretical plate (HETP) is given by

$$HETP = R_f \cdot Z_f/n = w_b^2/16 \cdot R_f \cdot Z_f \tag{300}$$

The average plate height (H) is given by

$$H = b \cdot (Z_f + Z_0) + a \cdot (Z_f^{2/3} - Z_0^{2/3})/(Z_f - Z_0) \tag{301}$$

where Z_0 is the distance of the sample spot above the solvent level in the tank.

The separation number (SN) is suitable for comparing the performance of various chromatographic systems. It can be calculated by the following equation:

$$SN = Z_f/(b_0 - b_1) - 1 \tag{302}$$

where b_0 and b_1 are the half-height peak widths of an imaginary compound at the sample origin and eluted at the solvent front, respectively.

The "real" plate number N_{real} can be calculated by:

$$N_{real} = 5.54 \cdot [Z_s/(b_s - b_0)]^2 \qquad (303)$$

where Z_s = migration distance of substance s and b_s = peak width at half-height for substance s.

The resolution (R_s) can be defined as:

$$R_s = (Z_{s1} - Z_{s2})/(w_{b1} - w_{b2}) \qquad (304)$$

where Z_{s1} is the migration distance of compound 1 with base width w_{b1} and Z_{s2} is the migration distance of compound 2 with base width w_{b2}.

A more profound theoretical treatment of the procedures governing TLC separation can be found in Reference 189.

B. Molecular Basis of Separation in Adsorption Thin-Layer Chromatography

The influence of eluent composition on retention characteristics has considerable theoretical and practical importance since mixed eluents are generally used for the successful separation of the analytes.

Many theories have been developed for the description of the relationship between retention parameters and the adsorption and partition processes in adsorption TLC. Each theoretical treatment highlights one or more aspects of the various interactions discussed above, resulting in different equations. The displacement theory[190-193] describes the retention as

$$\log(1/R_f - 1) = \alpha \cdot (S^0 - \epsilon_{AH} \cdot A_S) + \log(V_a \cdot W_a/V_m - 1) \qquad (305)$$

where α is the activity coefficient of the support; S^0 is the adsorption energy of the analyte; A_s is the cross-sectional area of the analyte molecule; ϵ_{AH} is the elution strength of the mobile phase consisting of alcohol–n-hexane; V_m is the volume of the mobile phase in the TLC system; V_a is the volume of the adsorbed mobile phase per unit mass of support; and W_m is the mass of the support. The elution strength can be expressed by

$$\epsilon_{AH} = \epsilon_H + [\log(N_A \cdot 10^{\chi \cdot n_A \cdot (\epsilon_A - \epsilon_H)} + 1 - N_A)]/\alpha \cdot n_A \qquad (306)$$

where N_A is the mole fraction of alcohol in the eluent; ϵ_A and ϵ_H are the elution strengths of the alcohol and n-hexane, respectively; and n_A is the area of the support surface occupied by an alcohol molecule.

The displacement theory outlined above concentrates on the sterical characteristics of the molecules in the mobile phase. However, parameters other than sterical also have a considerable impact on the retention. The acidic or alkaline nature of the eluent can markedly influence the retention because the acidic (silica) or alkaline (alumina) surface of supports facilitates the interaction with the eluent molecules of opposite charges.

These polar interactions can increase or decrease the role of sterical parameters in the determination of eluent strength.

As the adsorption energy is a direct measure of the strength of interaction between the mobile phase and the support, the heat capacities, temperature differences, and adsorption energies of some eluents on silica support were determined calorimetrically.[194] Eluent compositions were at the selectivity points of the "PRISMA" model.[195] The heat capacities are compiled in Table 67. The order of adsorption energies was tetrahydrofuran > ethyl

TABLE 67
Heat Capacities of Systems Comprising Silica Gel with Single Solvents and Ternary Solvents, c_p, and the Temperature Differences, ΔT, and Adsorption Energies, ΔH, at the Different Selectivity Points Studied

Selectivity Point					
Ethyl acetate	Chloroform	Tetrahydrofuran	c_p(J/gK)	ΔT(K)	ΔH(J/g)
0	0	10	1.98	0.39	0.772
0	10	0	1.21	0.37	0.446
10	0	0	1.78	0.35	0.622
1	1	8	1.86	0.24	0.446
1	3	6	1.68	0.21	0.352
1	6	3	1.44	0.19	0.274
1	8	1	1.31	0.13	0.170
3	1	6	1.83	0.25	0.455
3	3	4	1.64	0.25	0.410
3	4	3	1.56	0.27	0.421
3	6	1	1.41	0.20	0.282
4	3	3	1.63	0.25	0.406
6	1	3	1.76	0.34	0.599
6	3	1	1.58	0.19	0.301
8	1	1	1.72	0.37	0.637

From Merkku, P. et al., *J. Planar Chromatogr.*, 7, 1994, 305–308. With permission.

acetate > chloroform, suggesting that the solvent strengths of these solvents have to follow the same order according to the displacement theory.

A different model has been developed that takes into consideration the interaction between the components of the mobile phase.[196–198] This model divides a binary eluent system into three components. In the case of alcohol–*n*-hexane mixtures, the first component contains the molecules of alcohol (associated by hydrogen bonds), the second one contains the hydrocarbon molecules, and the third one is composed by the aggregates of alcohol–*n*-hexane molecules bonded by dispersive forces. The R_f values can be expressed by

$$R_f = A \cdot (x_1 + B \cdot x_2 + C)^{1/2} \tag{307}$$

where x_1 and x_2 are the volume ratios of alcohol and *n*-hexane, respectively. The constants of the equation above are

$$A = q_{alc} \cdot \Delta\mu_{alc/stph} \cdot (K_{alc}/c_{alc})^{1/2} \tag{308}$$

$$B = q_{hydr} \cdot \Delta\mu_{hydr/stph} \tag{309}$$

$$C = q_{mix} \cdot \Delta\mu_{mix/stph} \cdot (K_{mix}/c_{mix})^{1/2} \tag{310}$$

where q is an equation constant; subscripts "alc," "hydr," and "mix" indicate the alcohol, hydrocarbon, and alcohol–hydrocarbon parts of the eluents, respectively; $\Delta\mu$ is the chemical potential of the analyte partitioning between the mobile and the stationary phases (stph); K_{alc} is the dissociation constant of alcohol dimers; K_{mix} is the dissociation constant of the alcohol–*n*-hexane aggregates; c is the molar concentration of the eluent components.

It was concluded that A, B, and C are related to

$$A \sim \Delta\mu_{alc/stph} \tag{311}$$

$$B \sim \Delta\mu_{hydr/stph} \qquad (312)$$

$$C \sim \Delta\mu_{mix/stph} \qquad (313)$$

This theory has been extensively applied for the study of the retention mechanism in various TLC systems. The retention parameters of various analytes such as polycyclic aromatic hydrocarbons, ethers, ketones, higher fatty alcohols and acids, and mono- and dihydrixyaromatic compounds determined on silica TLC plates using 2-propanol–n-hexane eluent mixtures are compiled in Table 68. Due to the lack of polar substructures, polycyclic aromatic hydrocarbons are only weakly retained by the polar support. It has been assumed that their retention is mainly governed by the interaction of the delocalized π electrons with the acidic hydrogen atoms of the eluent molecules. This hypothesis was supported by the fact that the values of A were higher than those of B and C, indicating that these analytes show higher affinity to 2-propanol than to silica. In the case of the n-hexane–silica system, polycyclic aromatic hydrocarbons prefer the n-hexane phase (positive B values). The retention behavior of ethers and hydroxyaromatic compounds were similar to that of polycyclic aromatic hydrocarbons, indicating the similarity of their retention mechanisms. The data indicate that the presence of 2-propanol exerts the highest influence on the retention of higher fatty acids and alcohols. This fact was tentatively explained by the supposition that 2-propanol forms aggregates with analytes containing hydroxyl or carboxyl groups. Ketones have high A and B values. This fact may be due to the formation of hydrogen bonds between the carbonyl groups and the hydroxyl groups of the surface of the silica support. To elucidate the role of the length of the hydrocarbon

TABLE 68

Retention Parameters of Various Analytes on Silica Developed with 2-Propanol–n-Hexane

Analyte	A	B	C
Naphthalene	0.859	0.557	0.055
Phenanthrene	1.275	0.954	−0.454
Chrysene	0.872	0.488	0.093
Pyrene	0.818	0.502	0.100
Diphenyl ether	0.095	−0.219	1.033
Dibenzyl ether	1.334	0.762	−0.347
Acetophenone	3.062	1.852	−2.229
Benzophenone	3.329	2.196	−2.568
Methyl undecyl ketone	3.456	2.311	−2.731
Lauryl alcohol	1.985	0.082	−0.364
Myristyl alcohol	1.972	0.096	−0.388
Cetyl alcohol	1.899	0.085	−0.347
Stearyl alcohol	1.886	0.101	0.994
Arachidyl alcohol	1.887	0.103	−0.335
Lauric acid	1.747	0.062	−0.226
Myristic acid	1.793	0.080	−0.272
Palmitic acid	1.820	0.093	−0.291
Stearic acid	1.799	0.078	−0.253
Arachidic acid	1.814	0.102	−0.254
1-Naphthol	2.879	1.211	−1.617
2-Naphthol	2.263	0.477	−0.806
Pyrocatechol	1.406	−0.502	0.257
Resorcinol	0.906	−1.127	0.927
Hydroquinone	−0.039	2.159	2.098

From Kowalska, T. et al., *J. Planar Chromatogr.*, 5, 1992, 452–457. With permission.

TABLE 69
Retention Parameters of Various Analytes on Silica Developed with 2-Propanol–n-Heptane (I) and with 2-Propanol–n-Octane (II)

Analyte	Hydrocarbon	A	B	C
Naphthalene	I	0.110	−0.095	0.847
	II	−0.162	−0.281	1.125
Chrysene	I	−0.108	−0.347	1.139
	II	−0.318	−0.424	1.303
Phenanthrene	I	−0.152	−0.344	1.159
	II	−0.334	−0.425	1.317
Pyrene	I	0.006	−0.182	0.959
	II	−0.347	−0.435	1.322
Diphenyl ether	I	0.373	0.034	0.634
	II	−0.474	−0.548	1.473
Dibenzyl ether	I	1.113	0.454	−0.113
	II	−0.236	−0.645	1.324
Acetophenone	I	1.514	0.548	−0.574
	II	−0.654	−1.253	1.835
Benzophenone	I	1.648	0.906	−0.817
	II	−1.602	−1.898	2.851
Methyl undecyl ketone	I	1.365	0.715	−0.526
	II	−1.588	−1.872	2.832
Lauryl alcohol	I	0.355	−0.789	0.944
	II	−0.065	−0.893	1.248
Myristil alcohol	I	0.341	−0.790	0.946
	II	0.111	−0.747	1.049
Cetyl alcohol	I	0.298	−0.833	1.001
	II	0.190	−0.679	0.966
Stearyl alcohol	I	0.413	−0.728	0.869
	II	0.593	−0.348	0.521
Arachidyl alcohol	I	0.373	−0.762	0.914
	II	0.713	−0.267	0.400
Lauric acid	I	0.662	−0.810	0.785
	II	−1.084	−1.694	2.402
Myristic acid	I	0.691	−0.796	0.753
	II	−1.301	−1.858	2.628
Palmitic acid	I	0.800	−0.695	0.626
	II	−0.604	−1.282	1.848
Stearic acid	I	0.848	−0.621	0.551
	II	−0.508	−1.190	1.734
Arachidic acid	I	0.907	−0.556	0.476
	II	−0.432	−1.146	1.654
1-Naphthol	I	−0.063	−1.377	1.562
	II	−0.422	−1.201	1.629
2-Naphthol	I	−0.207	−1.516	1.714
	II	−0.503	−1.274	1.710
Pyrocatechol	I	−0.282	−1.639	1.768
	II	−0.043	−1.012	1.206
Resorcinol	I	−0.806	−2.266	2.429
	II	−0.510	−1.493	1.750
Hydroquinone	I	−1.083	−2.595	2.760
	II	−0.168	−1.315	1.394

From Kowalska, T. and Klama, B., *J. Planar Chromatogr.*, 7, 1994, 63–69. With permission.

chain the experiments listed above were extended by using n-heptane and n-octane as eluent components. It was assumed that the solubility of the analyte decreases with increasing molecular weight of the hydrocarbon component, resulting in the shift of the partition towards the stationary phase (reduction of B value). It was further supposed that larger hydrocarbon molecules inhibit the interaction of 2-propanol molecules with the stationary phase, resulting in reduced A values. The results are compiled in Table 69. The results in Table 69 support the theoretical considerations; in the majority of cases the numerical values of A and B are lower when the hydrocarbon component in the eluent is larger, proving that the affinity of solutes to the silica stationary phase increases. The individual contribution of methylene, hydroxyl, and carboxyl groups to the parameters of A, B, and C were calculated for higher fatty alcohols and acids. The results are compiled in Table 70. The data in Table 70 clearly show that the contribution of the hydrophobic apolar methylene groups to the retention processes in adsorption TLC is negligible; however, their contribution increases with increasing molecular weight of the hydrocarbon component. This fact supports the previous conclusions that the affinity of the polar substructures of analytes to silica becomes higher in the presence of larger hydrocarbon molecules. It was assumed that not only the type of the hydrocarbon component, but also the alcohol moiety may have considerable influence on the retention. Because alcohols show different capacities of self-association (n-butanol ≈ n-propanol < 2-propanol < t-butanol)[199-201] the energy of the alcohol–silica interaction may depend on the degree of self-association of alcohols. To verify this hypothesis the experiments discussed above were repeated using n-propanol, n-butanol, and t-butanol instead of 2-propanol. The results are compiled in Table 71. The data in Table 71 indicate that the character of the alcohol has a considerable influence on the retention parameters. Calculations proved that the contribution of methylene groups to the overall retention is negligible also in cases other than with 2-propanol and the character of polar substructures of the analytes determines their retention in these adsorption TLC systems.

The same model was used to study the retention behavior of azides and diazines in adsorption TLC[202] using n-propanol–n-hexane eluents. The calculated retention parameters are compiled in Table 72. It was concluded that the affinity of analytes to the alcohol–hydrocarbon mixtures is higher than to silica.

A thermodynamic model of liquid–solid chromatography separation assumes[203,204] that the R_M value of an analyte in adsorption TLC using binary eluent system can be expressed as

$$R_{M12} = X_1 \cdot R_{M1} + X_2 \cdot R_{M2} + (X_1^s - X_1) \cdot (\Delta R_{M12} + A_{12}) \tag{314}$$

$$(X_1 \cdot X_2)/(R_{M12} - X_1 \cdot R_{M1} - X_2 \cdot R_{M2}) = a \cdot X_1 + b \tag{315}$$

where R_{M12}, R_{M1}, and R_{M2} are the R_M values of the analyte determined in the binary eluent mixture and in components 1 and 2, respectively; $\Delta R_{M12} = R_{M1} - R_{M2}$; X_1 and X_2 are the molar fractions of components 1 and 2 in the mobile phase; X_1^s is the molar fraction of component 1 in the eluent adsorbed on the surface of support (surface phase); and A_{12} is related to the molecular interactions occurring in the mobile phase. The molar fraction of component 1 in the adsorbed phase can be calculated by

$$X_1^s = X_1 \cdot K_{12}/(X_2 + X_1 \cdot K_{12}) \tag{316}$$

where K_{12} is the adsorption equilibrium constant of the TLC system. The other physicochemical parameters can be calculated according to the following equations:

$$K_{12} = a/b + 1 \text{ and } A_{12} = 1/a - \Delta R_{M12} \tag{317}$$

The equations discussed above were used for the calculation of K_{12} and A_{12} values of some analytes on silica support. The results are compiled in Table 73. The same measurements were carried out in adsorption HPLC and the K_{12} and A_{12} values were calculated. A

TABLE 70
Numerical Values of Parameters A^{CH_2}, B^{CH_2}, C^{CH_2}, A^{OH}, B^{OH}, C^{OH}, A^{COOH}, B^{COOH}, and C^{OH} for Higher Fatty Alcohols and Acids in Adsorption TLC with Isopropanol–Hydrocarbon Mobile Phases

Homologous Series	Parameter	Hydrocarbon		
		n-Hexane	n-Heptane	n-Octane
Higher fatty alcohols	A^{CH_2}	−0.014	0.005	0.102
	A^{OH}	2.151	0.270	−1.322
	B^{CH_2}	0.002	0.006	0.082
	B^{OH}	0.056	−0.873	−1.908
	C^{CH_2}	0.005	−0.007	−0.111
	C^{OH}	−0.434	1.044	2.616
Higher fatty acids	A^{CH_2}	0.007	0.032	0.105
	A^{COOH}	1.690	0.296	−2.358
	B^{CH_2}	0.004	0.034	0.088
	B^{COOH}	0.024	−1.208	−2.757
	C^{CH_2}	−0.002	−0.041	−0.120
	C^{COOH}	0.231	1.253	3.846

From Kowalska, T. and Klama, B., *J. Planar Chromatogr.*, 7, 1994, 63–69. With permission.

good agreement was found between the values calculated by TLC and HPLC. It was suggested that TLC can be successfully used as a pilot technique for HPLC. It has been recently established that the conclusions discussed above are also valid in the case of sandwich chambers.[205]

The performances of the logarithmic[206,207] and quadratic relationships[208,209] for the description of the dependence of the R_M value of analytes on the volume fraction of the stronger component in binary eluent systems have been compared.[210] The dependence of the retention of 15 analytes with markedly different chemical structures was determined using eight eluent systems. Both logarithmic and quadratic correlations were calculated between the experimental R_M values and the molar fraction of the stronger component in the eluent. It was established that both equations fit well to the experimental data, the mean standard deviation being extremely low. However, the quadratic equation shows a better fit than the logarithmic one. This result was tentatively explained by the supposition that the logarithmic relationship assumes the presence of a monolayer of eluent on the support surface that is displaced by the analyte molecules. The displacement theory can be valid in the case of less polar eluents; however, it is less probable with highly polar eluents. The retention factor k can be defined by

$$k = (1 - R_f)/R_f = K_{th} \cdot \Phi = \Phi \cdot x_s/x_m = \Phi \cdot \tau_m/\tau_s \qquad (318)$$

where K_{th} is the thermodynamic equilibrium constant of an analyte; x_s and x_m are the molar fractions of an analyte in the stationary (s) and in the mobile phase (m), respectively; τ_m and τ_s are the corresponding activity coefficients of an analyte; and Φ is the phase ratio.

The activity coefficient in the mobile phase depends on the volume fraction of the stronger component in the eluent:[211,212]

$$\log \tau_m = a \cdot \Phi^2 + b \cdot \Phi + c \qquad (319)$$

In many instances the above equation can be simplified:

$$\tau_m = \text{const}/\Phi; \quad \log \tau_m = \text{const} - \log \Phi \qquad (320)$$

TABLE 71

Retention Parameters of Various Analytes on Silica Developed with
n-Propanol–n-Hexane (I), n-Butanol–n-Hexane (II), and t-Butanol–n-Hexane (III)

Analyte	Alcohol	A	B	C
Naphthalene	I	0.470	0.253	0.768
	II	0.289	1.670	2.230
	III	1.599	1.485	−0.967
Chrysene	I	1.172	0.116	0.082
	II	2.376	0.463	−0.691
	III	1.591	1.426	−0.919
Phenanthrene	I	1.369	0.604	−0.343
	II	0.263	−1.679	1.952
	III	2.820	2.961	−2.694
Pyrene	I	0.781	0.056	0.369
	II	0.251	−1.888	2.144
	III	1.847	1.816	−1.353
Diphenyl ether	I	1.772	1.060	−0.795
	II	0.729	0.765	1.001
	III	2.606	2.763	2.448
Dibenzyl ether	I	2.916	1.388	1.717
	II	6.316	5.466	−6.486
	III	7.420	6.802	−7.867
Acetophenone	I	0.823	1.087	1.073
	II	6.180	5.294	−6.273
	III	5.058	4.055	−4.758
Benzophenone	I	2.083	0.273	−0.537
	II	5.604	4.479	−5.368
	III	5.904	5.041	−5.855
Methyl undecyl ketone	I	2.889	1.214	−1.620
	II	5.618	4.531	−5.407
	III	7.013	6.260	−7.322
Lauryl alcohol	I	−0.017	−1.652	1.909
	II	4.790	3.074	−3.830
	III	5.210	3.604	−4.480
Myristyl alcohol	I	0.213	−1.416	1.613
	II	4.550	2.749	−3.468
	III	5.067	3.404	−4.261
Cetyl alcohol	I	0.218	1.417	1.612
	II	4.142	2.220	−2.874
	III	5.425	3.922	−4.831
Stearyl alcohol	I	0.444	−1.164	1.315
	II	4.219	2.354	−3.017
	III	5.333	3.787	−4.681
Arachidyl alcohol	I	0.070	−1.550	1.792
	II	4.668	2.970	−3.704
	III	5.415	3.908	−4.811
Lauric acid	I	−0.435	−2.197	2.548
	II	5.650	3.969	−4.877
	III	6.348	4.895	−5.951
Myristic acid	I	−0.422	−2.192	2.532
	II	4.976	3.088	−3.888
	III	6.340	4.860	−5.921
Palmitic acid	I	−0.449	−2.234	2.575
	II	4.813	2.888	−3.660
	III	5.793	4.078	−5.053
Stearic acid	I	−0.489	−2.283	2.634
	II	4.710	2.661	−3.435
	III	6.071	4.487	−5.506
Arachidic acid	I	−0.589	−2.348	2.732
	II	4.932	2.997	−3.801

TABLE 71 (CONTINUED)

Retention Parameters of Various Analytes on Silica Developed with
n-Propanol–n-Hexane (I), n-Butanol–n-Hexane (II), and t-Butanol–n-Hexane (III)

Analyte	Alcohol	A	B	C
	III	6.332	4.855	−5.914
1-Naphthol	I	1.922	0.260	−0.543
	II	4.595	2.606	−3.386
	III	4.790	3.134	−3.934
2-Naphthol	I	1.680	−0.045	−0.219
	II	4.476	2.386	−3.170
	III	3.932	1.933	−2.627
Pyrocatechol	I	1.516	−0.089	−0.188
	II	4.042	1.926	−2.664
	III	3.221	1.275	−1.895
Resorcinol	I	1.192	−0.531	0.259
	II	3.250	0.831	−1.477
	III	2.140	−0.047	−0.422
Hydroquinone	I	0.188	−1.599	1.473
	II	1.298	−1.930	1.565
	III	1.458	−0.757	0.399

From Kowalska, T. and Klama, B., *J. Planar Chromatogr.*, 7, 1994, 147–152. With permission.

TABLE 72

Retention Parameters of Various Analytes on Silica Developed with
n-Propanol–n-Hexane

Analyte	A	B	C
Pyrazine	0.456	−0.010	−0.005
2,2′-Bipyrazil	0.363	0.014	−0.001
Quinoxaline	0.744	−0.004	−0.003
2,2′-Biquinoxaline	1.022	−0.008	−0.002
Quinoline	0.490	0.026	−0.0003
2,2′-Biquinolyl	0.688	0.007	−0.001
2-Methylquinoxaline	0.515	0.014	−0.001
3,3′-Dimethyl-2,2′-biquinoxalyl	0.860	−0.015	−0.003
6,6′-Biquinolyl sulfide	0.258	0.017	−0.0002
8,8′-Biquinolyl disulfide	0.796	0.009	−0.001
τ-Picoline	0.218	0.010	−0.0006
τ,τ′-Bipicolil	0.162	0.009	−0.0005
Thioquinanthrene	0.823	−0.017	−0.002

From Baranowska, I. and Swierczek, S., *J. Planar Chromatogr.*, 7, 1994, 251–253. With permission.

TABLE 73

K_{12} and A_{12} Values Obtained from TLC Measurements for the following Mobile Phases:
I, Hexane–Carbontetrachloride; II, Hexane–Benzene; III, Hexane–Ethylene Chloride;
IV, Hexane–Ethyl Acetate

	I		II		III		IV	
Analyte	K_{12}	A_{12}	K_{12}	A_{12}	K_{12}	A_{12}	K_{12}	A_{12}
2,3-Dichlorophenol	1.8	−0.13	5.2	0.56	8.0	0.27	9.0	0.34
2,4-Dichlorophenol	1.8	0.13	5.4	0.49	8.0	0.28	9.0	0.29
2,5-Dichlorophenol	1.9	−0.26	5.8	0.50	8.0	0.23	8.7	−0.17
2,6-Dichlorophenol	1.9	−0.17	5.6	0.43	8.0	0.27	9.0	−0.44

From Rozylo, J. K. and Janicka, M., *J. Liquid Chromatogr.*, 14, 1991, 3197–3212. With permission.

The results discussed above suggest that the retention, both in adsorption and reversed-phase separation mode, is mainly governed by analyte–mobile phase interactions.

The decisive role of hydrophilic forces, especially hydrogen bond formation, in the determination of retention strength and retention selectivity has been many times demonstrated for a wide variety of analytes. Thus, a study dealing with the retention behavior of tris(β-diketonato) complexes of cobalt(III), chromium(III), and ruthenium(III) on silica established that the substitution of one oxygen atom by a sulfur atom decreased the retention.[213] This phenomena was explained by the supposition that the more electronegative oxygen atom forms a stronger hydrogen bond than sulfur with the silanol groups of the support, resulting in enhanced retention. The substitution of acetylacetonato ligand by a monothiobenzoylmethanato ligand also decreased the retention. It was stated that this effect may be due to the negative inductive effect of the phenyl group.

The retention of 10 N,N-disubstituted dithiocarbamates of nickel(II) and cobalt(II) was also determined on silica using various organic eluent systems.[214] It was concluded from the retention data that dithiocarbamate derivatives are bonded to the surface OH groups of silica by the N atom and the retention of the analytes decreased with increasing molecular area.

During the last few decades the number of papers dealing with the elucidation of the quantitative relationship between retention characteristics of analytes and their molecular parameters enormously increased. This interesting field facilitates the prediction of retention in chromatography, enables the measurements of a wide variety of molecular parameters, and promotes the more profound understanding of retention mechanisms. Some recent studies have tried to find a relationship between the retention characteristics of analytes and a more or less large set of physicochemical parameters. Thus, the roles of molar refraction, molar volume, electron polarizability, and dispersion component of the solubility parameters in the retention behavior of aliphatic aminoalcohols on silica were determined.[215] The analyte–solvent interaction parameter A_z related to the partition coefficient of the analyte between the two components of binary eluents was calculated as follows:

$$G(x_1) = a \cdot x_1 + b \quad (321)$$

$$G(x_1) = (x_1 \cdot x_2)/(R_{M12} - x_1 \cdot R_{M1} - x_2 \cdot R_{M2}) \quad (322)$$

$$a = \Delta R_M + A_z \quad (323)$$

$$\Delta R_M = R_{M1} - R_{M2} \quad (324)$$

where x_1 and x_2 are the volume fractions of the stronger and weaker components in the eluent, respectively; R_{M1}, R_{M2}, and R_{M12} are the R_M values of an analyte determined in pure components 1 and 2 and in the binary eluent ($x_1 + x_2$), respectively. The A_z values of amino alcohols determined in various eluents on silica support are compiled in Table 74. The data in Table 74 clearly show that both the character of the eluent components and the chemical structure of analytes considerably influence the association constant. Linear correlations were found between the association constant and the difference between the dispersion components of the solubility parameters of the solvents and between the association constant and the difference between the proton-acceptor component of solubility parameters of the solvents. It was assumed that in the given adsorption TLC system the proton-acceptor component has higher importance than does the dispersion. It was further suggested that the different degree of association may modify the shape of the analyte, resulting in changed accessibility of the adsortion centers on the silica surface. Similar methods have also been used for the determination of the A_z values of quinoline derivatives[216] and aromatic hydrocarbons.[217]

Another paper used four physicochemical parameters (hydrophobicity, π; electric effect, σ; field effect, F; and steric effect E_s) for the description of the retention behavior of 15 O-ethyl-O-aryl-N-isopropyl phosphoroamidothioate derivatives on silica.[218] The eluents were chloroform–n-hexane mixtures, the volume fraction of chloroform (X_s) being 0.2, 0.4, 0.5, and 0.7. Multilinear correlations were calculated between the R_f values and the physicochemical parameters at each eluent composition and the regression coefficients were related to the X_s value by a polynomial function. The parameters of the multilinear regressions are compiled in Table 75. The equations fit well to the experimental data; the significance level is high in each instance. This result indicates that the retention behavior of these analytes markedly depends on the molecular hydrophobicity, electric effect, field effect, and steric effect. Unfortunately, the path coefficients were not calculated, therefore the relative importance of the four physicochemical parameters in the retention cannot be evaluated. The polynomial functions relating a, b, c, d, and e to X_s were

$$a = -0.2652 + 4.6163 \cdot X_s - 11.3058 \cdot (X_s)^2 + 7.5283 \cdot (X_s)^3 \tag{325}$$

$$b = -0.5068 + 4.6699 \cdot X_s - 11.7418 \cdot (X_s)^2 + 8.6647 \cdot (X_s)^3 \tag{326}$$

$$c = 0.4462 - 4.2145 \cdot X_s + 10.9519 \cdot (X_s)^2 - 8.5090 \cdot (X_s)^3 \tag{327}$$

$$d = 0.5664 - 5.2270 \cdot X_s + 13.0850 \cdot (X_s)^2 - 9.7530 \cdot \cdot (X_s)^3 \tag{328}$$

$$e = 1.3697 - 9.6780 \cdot X_s + 22.6463 \cdot (X_s)^2 - 16.3527 \cdot (X_s)^3 \tag{329}$$

TABLE 74

Association Constants A_z for the Amino Alcohols Studied. Mobile Phase Composition: I, Methanol + $CHCl_3$; II, Methanol + CH_2Cl_2; III, Methanol + Acetone; IV, Acetone + $CHCl_3$

Analyte	I	II	III	IV
2-Aminoethanol	0.5730	1.6074		0.4270
N-(2-Aminoethyl)ethanolamine	0.3517	0.1129	0.1113	
1,1'-Iminodipropan-2-ol	0.2892	0.0283	–0.0285	
Diethanolamine	0.6395	0.5571		0.6095
N-Methyl-diethanolamine	1.1940	0.6474	0.8748	0.2399
N-Butyldiethanolamine	1.0890	0.6525	0.5778	0.2259
N-tert-Butyldiethanolamine	0.8100	0.6403	0.0852	–0.4227
N-Phenyldiethanolamine	0.5465	0.6039	0.3814	0.2761
N-o-Toluidinediethanolamine	0.5898	0.3698	0.2132	–0.0561
N-p-Toluidinediethanolamine	0.6371	0.6904	0.4348	0.1946
Tris-(2-hydroxyethyl)amine	0.0142	0.1308		0.2060
Tris-(hydroxymethyl)aminomethane	0.0909	0.0866	0.1643	

From Bazylak, G., *J. Planar Chromatogr.*, 5, 1992, 239–245. With permission.

TABLE 75

Results from the Use of Regression Analysis to Determine the Relationship $R_f = a \cdot \pi + b \cdot \sigma + c \cdot F + d \cdot E_s + e$ between R_f and π, σ, F and E_s

X_s	a	b	c	d	e	r
0.2	0.2660	0.0268	–0.0267	–0.0337	0.2092	0.9778
0.4	0.2542	0.0370	–0.0319	–0.0551	0.0754	0.9794
0.5	0.1575	–0.0243	0.0133	0.0050	0.1482	0.9400
0.7	0.0085	–0.0194	–00561	–0.0262	0.0828	0.9320

From Qin-Sun, W. et al., *J. Planar Chromatogr.*, 6, 1993, 144–146. With permission.

Using the equations listed above, good agreement was found between the measured and calculated R_f values, which proves the high predictive power on the calculations.

The effect of the various physicochemical parameters of eluents on the efficiency of separation was assessed in a series of papers.[219-222] The difference between the R_f values of two analytes is

$$\Delta R_f = (k_2 - k_1)/[(1 + k_1) \cdot (1 + k_2)] \qquad (330)$$

where k_1 and k_2 are the capacity factors of analyte 1 and analyte 2. The distance between the spot centers, S_D, for a migration distance L of the mobile phase is

$$S_D = \Delta R_f \cdot L \qquad (331)$$

Separation quality (PRF) was characterized by

$$PRF = \sum_{i=1}^{n} \ln S_D^{ij} / S_D^{SPEC} \qquad (332)$$

where n is the number of neighboring pairs; S_D^{ij} is the distance between spot centers of analyte pair i and j; and S_D^{SPEC} is a specified separation distance. Analyte pairs with $S_D^{ij} > S_D^{SPEC}$ have no contribution to PRF. The influence of various solvent parameters such as solvent strength, molar volume, density, surface area, polarizability/depolarizability, proton donor and acceptor properties, dipole moment, and density on PRF was evaluated. It has been found that the various parameters of solvent strength and molar volume are well correlated with the separation quality parameter PRF whereas density and dipole moment show only poor correlations.

The dependence of eluent strength on the character of homologous series of solvents has been vigorously discussed. It was established that eluent strength decreases with the increasing length of alkyl chain in the stronger component. This phenomena may be due to the fact that the relative concentration of the polar group decreases and the sterical hindrance of adsorption is enhanced.[223] A recent study proved that the elution strength of methyl alkyl ketones was considerably higher than that of isomeric dialkyl ketones. This result was explained by the sterical hindrances occurring during the adsorption process.[224]

Topological indices have also been used in adsorption TLC, either alone or in combination with other molecular parameters.[225,226] The topological indices used in the calculations were taken from References 227 to 229. Hydrophobicity parameters and the electron-withdrawing property of substituents were also included in the calculations. The best fitting equations in the case of eight benzoic acid derivatives were

$$R_M = (4.8219 \pm 0.8214) - (1.8195 \pm 0.1916) \cdot \sigma - (2.0612 \pm 0.2879) \cdot {}^0\chi^v$$
$$+ (0.4451 \pm 0.0570) \cdot D$$
$$r = 0.9787; \quad F = 30.26; \quad s = 0.1142 \qquad (333)$$

$$R_M = (8.2729 \pm 1.1553) - (0.5593 \pm 0.0799) \cdot \pi - (1.6094 \pm 0.2185) \cdot {}^1\chi^v$$
$$- (6.4740 \pm 1.0738) \cdot {}^2B_q$$
$$r = 0.9875; \quad F = 52.37; \quad s = 0.0876 \qquad (334)$$

$$R_M = (1.9533 \pm 0.4048) - (0.5236 \pm 0.0674) \cdot \log P - (3.0349 \pm 0.5043) \cdot {}^1\chi^v$$
$$+ (2.5590 \pm 0.4953) \cdot \chi_{012}$$
$$r = 0.9877; \quad F = 53.14; \quad s = 0.0870 \tag{335}$$

$$R_M = (6.0657 \pm 1.1415) - (0.7186 \pm 0.1270) \cdot \sigma + (2.9906 \pm 0.5710) \cdot \pi$$
$$- (2.9742 \pm 0.4734) \cdot \log P - (0.00346 \pm 0.00209) \cdot M$$
$$r = 0.9932; \quad F = 54.24; \quad s = 0.0750 \tag{336}$$

For 9 aromatic alcohols the equations were slightly modified:

$$R_M = (10.1652 \pm 0.8043) - (3.5323 \pm 0.3018) \cdot {}^0B - (0.3882 \pm 0.0313) \cdot {}^0\chi^v$$
$$r = 0.9843; \quad F = 93.32; \quad s = 0.0665 \tag{337}$$

$$R_M = (13.0513 \pm 0.7251) - (18.1623 \pm 1.4292) \cdot {}^1B - (1.4746 \pm 0.0562) \cdot \chi_{012}$$
$$+ (44.9454 \pm 4.9719) \cdot {}^3B$$
$$r = 0.9980; \quad F = 426.73; \quad s = 0.0258 \tag{338}$$

$$R_M = (11.4337 \pm 0.7236) - (0.6613 \pm 0.0440) \cdot \log P - (4.5126 \pm 0.2918) \cdot {}^0B$$
$$r = 0.9892; \quad F = 136.54; \quad s = 0.0553 \tag{339}$$

$$R_M = (12.9331 \pm 1.0723) - (0.8446 \pm 0.1129) \cdot \log P + (0.2492 \pm 0.1445) \cdot {}^3\chi^v$$
$$- (5.2047 \pm 0.4745) \cdot {}^0B$$
$$r = 0.9932; \quad F = 121.96; \quad s = 0.0479 \tag{340}$$

The correlations improved when other then topological indices were included in the calculations. This result indicates that the sterical characteristics of the analyte molecules are not sufficient to predict the retention behavior in adsorption TLC. The inclusion of other molecular parameters markedly enhanced the ratio of variance explained by the equations, which means that the retention of analytes in adsorption TLC depends considerably not only on the sterical, but also on the hydrophilic and hydrophobic parameters of analyte molecules.

A new topological index (I_{STI}) has been introduced to differentiate between the stereoisomers of menthol and thujol derivatives with hydroxyl groups in axial and equatorial positions.[230] Significant linear correlations were found between the retention parameters (R_f and R_M) of menthol and thujol derivatives and the new topological index. For the stereoisomers of menthol:

$$R_f = (0.05971 \pm 0.05506) + (1.07 \pm 0.11) \cdot 10^{-3} \cdot I_{STI}$$
$$r = 0.9900; \quad F = 98.41; \quad s = 0.031; \quad n = 4 \tag{341}$$

$$R_M = (0.8256 \pm 0.1129) - (2.0 \pm 0.2) \cdot 10^{-3} \cdot I_{STI}$$
$$r = 0.9882; \quad F = 83.42; \quad s = 0.0636; \quad n = 4 \tag{342}$$

$$R_f = (1.5582 \pm 0.0849) - (0.5532 \pm 0.0476) \cdot {}^0B_{STI}$$
$$r = 0.9927; \quad F = 135.18; \quad s = 0.0265; \quad n = 4 \tag{343}$$

$$R_M = -(1.9945 \pm 0.2174) + (1.0394 \pm 0.1219) \cdot {}^0B_{STI}$$
$$r = 0.9865; \quad F = 72.71; \quad s = 0.0680; \quad n = 4 \tag{344}$$

For the stereoisomers of thujol:

$$R_f = (0.1221 \pm 0.0337) + (7.04 \pm 0.66) \cdot 10^{-4} \cdot I_{STI}$$
$$r = 0.9912; \quad F = 112.63; \quad s = 0.0201; \quad n = 4 \tag{345}$$

$$R_M = (0.6751 \pm 0.0619) - (1.25 \pm 0.12) \cdot 10^{-3} \cdot I_{STI}$$
$$r = 0.9907; \quad F = 105.70; \quad s = 0.0371; \quad n = 4 \tag{346}$$

$$R_f = (1.0093 \pm 0.0834) - (0.2999 \pm 0.0450) \cdot {}^0B_{STI}$$
$$r = 0.9782; \quad F = 44048; \quad s = 0.0316; \quad n = 4 \tag{347}$$

$$R_M = -(0.9080 \pm 0.1445) + (0.5355 \pm 0.0779) \cdot {}^0B_{STI}$$
$$r = 0.9792; \quad F = 47.24; \quad s = 0.0548; \quad n = 4 \tag{348}$$

The good fit of the equations to the experimental data indicates that this new stereoisomoric topological index is a valuable tool for the prediction of the retention behavior of various stereoisomers in adsorption TLC.

C. Molecular Basis of Separation in Reversed-Phase Thin-Layer Chromatography

It is generally accepted that the retention of analytes in reversed-phase thin-layer chromatography (RPTLC) is mainly governed by their lipophilicity (hydrophobicity), that is RPTLC can be used for the determination of this important physicochemical parameter. However, the practice proved that the situation is much more complicated than was previously believed. The hydrophobic ligand bonded covalently or by adsorptive forces to the surface of the support never covers entirely the adsorptive centers on the support surface. They retain to some extent their original retention characteristics, which has to be taken into consideration. When the analyte contains one or more dissociable polar substructures the pH of the eluent modifies the apparent polarity of these groups, resulting in modified retention. Salts in the eluent may have a similar effect; they decrease the degree of dissociation of polar groups which increases retention. However, the dissociated ions of salts can adsorb on the active centers of support not covered by the hydrophobic ligand, modifying in this manner the overall retention capacity of the RPTLC system.

The model based on the differences in the strength of association between the analytes and the components of the eluent[196–198] has been slightly modified and applied for the description of the retention mechanism in RPTLC.[231] The model assumes that four different components are present in the eluent. In the case of water–acetonitrile (ACN) mixtures: the components are (1) pure acetonitrile, (2) pure water, (3) ACN–water associates with fixed stoichiometry held together with hydrogen bonds, and (4) ACN–water associates without fixed stoichiometry held together with dispersive forces. The time that an analyte molecule spends in the mobile phase (m) can be expressed as the sum of time spent in the different fractions of the mobile phase:

$$t_m = t_1 + t_2 + t_3 + t_4 \tag{349}$$

It has been proposed that the parameters of the above equation can be calculated by

$$t_1 = a \cdot x_1; \ t_2 = b \cdot x_2; \ t_3 = c \cdot (2.92 \cdot x_2 \cdot n'' + x_2); \ t_4 = d \cdot (x_1 + x_2) = d \tag{350}$$

where x_1 and x_2 are the volume fractions of ACN and water, respectively, and $n'' = 1/100^{x_2}$. The retention of any analyte molecule can be described by

$$R_f = R_1^* \cdot x_1 + R_2^* \cdot x_2 + R_3^* \cdot (2.92 \cdot x_2 \cdot n'' + x_2) + R_4 \tag{351}$$

where $R_2^* = R_2^{*0} \cdot \beta_{S2}$ and $R_3^* = R_3^{*0} \cdot \beta_{S3}$; β_{S2} and β_{S3} are the degrees of dissociation of the water clusters and of ACN–water associate with fixed stoichiometry. The degree of dissociations are related to

$$\beta_{S2} \approx \Delta K_2/(c_2 \cdot x_2) \text{ and } \beta_{S3} \approx \Delta K_3/(c_3 \cdot x_3) \tag{352}$$

where K_2 and K_3 are the dissociation constants of water clusters and ACN–water aggregates, respectively; c_2 and c_3 are the corresponding molar concentrations; and $x_3 = (2.92 \cdot x_2 \cdot n'' + x_2)$.

$$R_f = R_1^* \cdot x_1 + R_2^{*0} \cdot \Delta x_2 \cdot (K_2/c_2)^{1/2} + R_3^{*0} \cdot (2.92 \cdot x_2 \cdot n'' + x_2)^{1/2} \cdot (K_3/c_3)^{1/2} + R_4^* \tag{353}$$

$$R_f = A \cdot x_1 + B \cdot x_2 + C \cdot (2.92 \cdot x_2 \cdot n'' + x_2)^{1/2} + D \tag{354}$$

where $A = R_1^*$; $B = R_2^{*0} \cdot (K_2/c_2)^{1/2}$; $C = R_3^{*0} \cdot (K_3/c_3)^{1/2}$; and $D = R_4^*$. The fit of the equation discussed above was proved in the RPTLC chromatography of polycyclic aromatic hydrocarbons on RP-8 plates developed with ACN water. The parameters are listed in Table 76. It was stated that the model is suitable for the exact prediction of the RPTLC retention of these analytes. The retention parameters of other analytes determined under slightly different RPTLC conditions are compiled in Table 77.[232] It was concluded from the data that these polar analytes containing oxygen form hydrogen bonds with the organic modifier and this interaction accounts for their retention behavior.

The influence of the length of the hydrophobic alkyl chain on the retention of higher fatty alcohols and acids was studied in separate experiments using RP-2, RP-8, and RP-18

TABLE 76

Retention Parameters of Selected Polycyclic Aromatic Hydrocarbons on RP-8 Plates Developed with Acetonitrile–Water

Analyte	A	B	C	D
Naphthalene	2.209	1.349	−0.892	−1.198
Anthracene	−6.924	−10.631	4.087	8.045
Phenanthrene	1.666	0.805	−0.897	−0.634
Chrysene	−1.750	−3.945	1.239	2.726
Pyrene	−0.366	−2.353	0.796	1.238

From Kowalska, T. and Podgorny, A., *J. Planar Chromatogr.*, 4, 1991, 313–315. With permission.

TABLE 77

Parameters A, B, and C Measured by Reversed-Phase TLC on RP-2, with Methanol–Water Mixtures as Mobile Phase

Analyte	A	B	C
Phenol	3.438	1.519	−3.007
α-Naphthol	4.571	1.799	−4.229
β-Naphthol	4.447	1.681	−4.037
2-Hydroxybiphenyl	2.916	0.469	−2.194
4-Hydroxybiphenyl	3.761	1.061	−3.218
Benzophenone	2.291	0.229	−1.536
Acetophenone	1.444	0.192	−0.657
Methyl undecyl ketone	4.582	0.164	−3.836
Diphenyl ether	1.998	0.100	−1.226
Dibenzyl ether	2.387	0.066	−1.585

From Kowalska, T. et al., *J. Planar Chromatogr.*, 5, 1992, 192–196. With permission.

RPTLC plates.[233] Because the carbon loading of the support is different, the data may help in the elucidation of the role of uncovered polar silanol groups in the reversed-phase retention mechanism. The retention parameters for higher fatty alcohols and acids are compiled in Tables 78 and 79, respectively. The data in Tables 78 and 79 prove that both the character of the bonded hydrophobic ligand and the nature of the polar group in the analyte molecule exert a considerable influence on the retention. The supposed retention mechanism is presented in Figure 7. The orientation of the molecules of fatty acids at the interface between eluent and silica seems to favor intermolecular interactions both with methanol and water. Alcohols behave differently in the given RPTLC system. The experimental results concretely prove the so-called mixed retention mechanism: the apolar alkyl chains of analytes interact with the hydrophobic ligands and the polar –OH or –COOH groups bind to the free silanol groups by electrostatic forces (probably hydrogen bonding). It was further established that not only the length of the apolar alkyl chain, but also its density on the silica surface has a considerable impact on the retention and, moreover, on the retention mechanism.[234] A mixed retention mechanism has also been supposed in the case of heterocyclic amines on RP-2, RP-8, and RP-18 supports using methanol–water mixtures as eluents.[235]

To regulate the pH of the eluent for theoretical studies and to improve separation many RPTLC systems contain buffers. The model mentioned above has been extended for the description of the retention behavior also in RPTLC systems containing buffer. The validity of the model was proven using RP-18 stationary phase and acetonitrile–buffer eluents.[236]

Many efforts have been devoted to the elucidation of the relationship between the physicochemical parameters and structural characteristics of analytes and their retention behavior also in RPTLC.

The retention of some dansylated amino acids was determined in aqueous eluents containing LiCl, NaCl, KCl, RbCl, and CsCl at different concentrations.[237] The R_M values were extrapolated to zero salt concentration by

$$R_M = R_{M0} + b \cdot C \tag{355}$$

where R_M is the actual R_M value of a dansylated amino acid determined in the presence of C concentration of salt; R_{M0} is the R_M value extrapolated to zero concentration of salt in the eluent; b is the change in R_M value caused by a unit change of salt concentration (related to the strength of salting-out effect of salt). The parameters of relationship

TABLE 78

Parameters of A, B, and C for Higher Fatty Alcohols

Analyte	A	B	C
RP-2 Support			
Lauryl alcohol	8.594	1.781	−8.019
Myristyl alcohol	8.878	1.559	−8.268
Cetyl alcohol	16.989	4.266	−16.887
Stearyl alcohol	16.980	4.013	−16.854
RP-8 Support			
Lauryl alcohol	7.410	0.287	−6.651
Myristyl alcohol	8.509	0.282	−7.784
Cetyl alcohol	10.270	0.521	−9.609
Stearyl alcohol	10.792	0.363	−10.171
RP-18 Support			
Lauryl alcohol	1.816	−0.210	−1.243
Myristyl alcohol	1.948	−0.267	−1.460
Cetyl alcohol	1.931	−0.268	−1.529
Stearyl alcohol	1.648	−0.305	−1.336

From Kowalska, T. and Wikowska-Kita, B., *J. Planar Chromatogr.*, 5, 1992, 192–196. With permission.

TABLE 79

Parameters of A, B, and C for Higher Fatty Acids

Analyte	A	B	C
RP-2 Support			
Lauric acid	−1.072	−1.487	2.196
Myristic acid	−0.666	−1.757	1.850
Palmitic acid	−4.988	−3.375	6.384
Stearic acid	−2.294	−2.622	3.540
RP-8 Support			
Lauric acid	6.435	0.414	−5.665
Myristic acid	5.932	−0.034	−5.131
Palmitic acid	4.695	−0.661	−3.844
Stearic acid	−2.292	−2.665	3.299
RP-18 Support			
Lauric acid	6.000	0.346	−5.383
Myristic acid	5.738	0.192	−5.193
Palmitic acid	4.353	−0.062	−3.891
Stearic acid	3.050	−0.268	−2.659

From Kowalska, T. and Wikowska-Kita, B., *J. Planar Chromatogr.*, 5, 1992, 192–196. With permission.

between the effect of salts and their physicochemical characteristics are compiled in Table 80. The data in Table 80 indicate that the cation radii and the hydration energy of cations have significant impacts on the retention and the impacts are similar (compare the value of path coefficients). No explanation was presented for the nonlinear influence

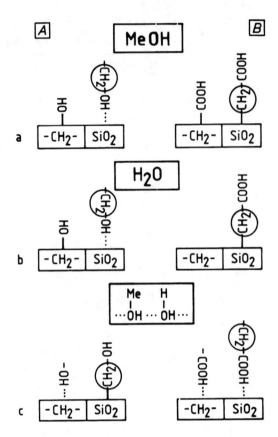

FIGURE 7
Preferential orientation of the single –CH$_2$– unit and the respective functional group of an alcohol (A) and an acid (B) on the solid–liquid interface between the stationary phase on the one hand, and pure methanol (a), pure water (b) or hydrogen-bonded methanol–water (c). (From Kowalska, T. and Witkowska-Kita, B., *J. Planar Chromatogr.*, 5, 1992, 192–196. With permission.)

of the physicochemical parameters of salts on the retention. The parameters of relationships between molecular characteristics and retention are compiled in Table 81. The data in Table 81 clearly show that the hydrophobicity of dansylated amino acids depends not only on the hydrophobicity of the free amino acid as expected, but also on the pK value of the α-amino group. The interaction between the acidic silanol groups on the support surface and the basic amino groups is probably of electrostatic character. This result proves again the occurrence of mixed retention mechanism in the RPTLC separation of more or less polar analytes.

The retention of 22 commercial nonionic surfactants was studied in RPTLC using methanol–water mixtures as eluents.[238] Linear correlations were calculated between the R_M values of surfactants and the concentration of methanol in the eluent. The slope (R_{M0}) and intercept values (b) were regarded as the best means of estimating the lipophilicity and specific hydrophobic surface area of surfactants. The hydrophobicity parameters were correlated with the structural characteristics of surfactants (Table 82). It can be concluded from the data in Table 82 that the hydrophilic ethyleneoxide chain exerts a stronger influence on the hydrophobicity and specific hydrophobic surface area of surfactants than the number of double bonds in the hydrophobic moiety (see path coefficients). The significant impact of the polar ethyleneoxide chains on the RPTLC retention can be explained by the supposition that they point toward the more hydrophilic mobile phase drawing away the surfactant molecule from the apolar support. It has been assumed that the double bond modifies the spatial arrangement of the hydrophobic alkyl chains, decreasing in this manner the contact surface between the hydrophobic moiety of surfactants and the apolar surface of support. This effect results in decreased retention capacity.

Molecular structure parameters combined with hydrophobicity have also been used in RPTLC for the description of the retention behavior of various analytes. Thus, the log k′

TABLE 80

Relationship between the Salting-Out Effect of Salts and Their Physicochemical Parameters. Results of Stepwise Regression Analysis

$$b = a + b_1 \cdot (x_1)^2 + b_2 \cdot \log x_1 + b_3 \cdot \log x_2$$

$$n = 64; \quad a = -349.5; \quad F = 23.02; \quad r^2 = 0.5350$$

Parameter	Independent variable		
	$(x_1)^2$	$\log x_1$	$\log x_2$
b	7.68	85.17	174.5
s_b	1.57	19.88	39.3
Path coefficient %	17.41	32.96	49.63

Note: x_1, cation radii; x_2, hydration energy of the cation.
From Cserháti, T. and Illès, Z., *Chromatographia*, 36, 1993, 302–306. With permission.

TABLE 81

Relationship between the Retention Characteristics of Dansylated Amino Acids and the Physicochemical Parameters of the Free Amino Acids. Results of Stepwise Regression Analysis (n = 12)

I. $R_{M0\text{(mean value)}} = a + b_1 \cdot x_1 + b_2 \cdot x_2$

II. $b_{\text{(mean value)}} = a + b_1 \cdot x_1$

Parameter	No. of equation	
	I	II
a	−3.22	6.05
b_1	0.41	−0.28
s_{b1}	0.18	0.11
Path coefficient %	44.31	—
b_2	−0.17	—
s_{b2}	0.06	—
Path coefficient %	55.69	—
r^2	0.7241	0.3814
F	11.81	6.16

Note: x_1 = pK value of the α-amino group; x_2, hydrophobicity parameter.
From Cserháti, T. and Illès, Z., *Chromatographia*, 36, 1993, 302–306. With permission.

values of 12 amino alcohols were determined on silanized silica, RP-8, RP-18, and RP-18 layers with concentrating zone using methanol:water ratio of 9:1 (v/v) as eluent.[239] No significant correlation was found between the log k' values determined on silanized silica and the molecular parameters included in the calculations (χ_R = molecular connectivity, χ_R^v = valence molecular connectivity, W = Wiener number, J = Balaban index, log P = hydrophobicity). Significant relationships were found between the physicochemical parameters listed above and the retention of amino alcohols both on RP-8 and RP-18 supports:

$$\log k'_{(RP-8)} = -(2.0317 \pm 0.3360) + (6.91 \pm 0.07) \cdot 10^{-2} \cdot W; \quad r = 0.9961 \quad (356)$$

TABLE 82
Relationships between the Structural Characteristics, Lipophilicity (R_{M0}), and Specific Hydrophobic Surface Area (b) of Nonionic Surfactants. Results of Stepwise Regression Analysis

I. $R_{M0} = a + b_1 \cdot n_e + b_2 \cdot C{=}C$

II. $b = a + b_1 \cdot n_e + b_2 \cdot C{=}C$

Parameter	No. of Equation	
	I	II
a	4.92	6.27
$-b_1 \cdot 10^{-2}$	2.38	3.94
$s_{b1} \cdot 10^{-3}$	5.37	6.30
Path coefficient %	55.86	57.76
b_2	0.55	0.84
s_{b2}	0.16	0.18
Path coefficient %	44.16	42.24
r^2	0.6752	0.7959
F	19.75	37.06

Note: n = 22; n_e, number of ethyleneoxide groups per molecule; C=C, number of double bonds in the hydrophobic moiety of surfactants; s_{b1} and s_{b2}, standard deviations of slope b_1 and b_2 values, respectively.

From Cserháti, T., *J. Biochem. Biophys. Methods*, 27, 1993, 133–142.

$$\log k'_{(RP\text{-}8)} = -(3.5512 \pm 0.3786) + (0.7594 \pm 0.0936) \cdot J; \quad r = 0.9681 \tag{357}$$

$$\log k'_{(RP\text{-}8)} = -2.3478 + 0.0832 \cdot J + 0.0854 \cdot J^2; \quad r = 0.9890 \tag{358}$$

$$\log k'_{(RP\text{-}8)} = -1.8403 - 0.0525 \cdot \chi_R - 0.1236 \cdot \chi_R^v + 0.0074 \cdot W + 0.1032 \cdot J \tag{359}$$

$$\log k'_{(RP\text{-}18)} = -(2.3787 \pm 0.2861) + (0.6051 \pm 0.0749) \cdot \chi_R^v; \quad r = 0.9671 \tag{360}$$

$$\log k'_{(RP\text{-}18)} = -(1.1956 \pm 0.3088) + (5.0 \pm 0.7) \cdot 10^{-3} \cdot W; \quad r = 0.9451 \tag{361}$$

$$\log k'_{(RP\text{-}18)} = -2.1924 + 0.4834 \cdot \chi_R^v + 0.0854 \cdot (\chi_R^v)^2; \quad r = 0.9876 \tag{362}$$

$$\log k'_{(RP\text{-}18)} = -2.1334 - 0.0786 \cdot \chi_R - 0.2980 \cdot \chi_R^v + 0.0017 \cdot W + 0.0483 \cdot J \tag{363}$$

The significant relationships between the retention of amino alcohols and their structural characteristics indicate the high predictive power of the calculations. They further prove that the sterical parameters have a considerable impact on the retention also in RPTLC.

Water-insoluble polymer beads prepared from β-cyclodextrin have also been used as TLC support. The retention of 14 3,5-dinitrobenzoic acid esters was determined using water–methanol mixtures as eluents.[240] The hydrophobicity value extrapolated to zero organic phase concentration (R_{M0}) was correlated with the following physicochemical parameters of analytes: π = Hansch–Fujita's substituent constant characterizing hydrophobicity; H-Ac and H-Do = indicator variables for proton acceptor and proton donor properties, respectively; M-RE = molar refractivity; F and R = Swain–Lupton's electronic parameters, characterizing the inductive and resonance effect, respectively; σ = Hammett's constant, characterizing the electron-withdrawing power of the substituent; Es = Taft's constant, characterizing steric effects of the substituent; B_1 and B_4 = Sterimol width parameters determined by distance of substituents at their maximum point perpendicular to attachment. Calculations found marked nonlinear relationships between the retention behavior and steric characteristics of 3,5-dinitrobenzoic acid esters (Figures 8 and 9). The nonlinearity of the relationships was tentatively explained by the well-defined dimensions of the cyclodextrin cavity. Only analytes corresponding to these dimensions can be easily accommodated in the cavity. In the case of larger or smaller analytes, the stability of the inclusion complex becomes lower, resulting in decreased retention. The significant role of steric parameters further indicates that accessible cavities are present on the polymer surface, influencing the retention of analytes liable to form inclusion complexes with β-cyclodextrin. The retention behavior of eight 4-cyanophenyl herbicides on the same β-cyclodextrin polymer support was determined using water–methanol eluents with HCl, NaOH, $MgCl_2$, and glycine as eluent additives.[241] The similarities and dissimilarities between the effect of eluent additives and physicochemical parameters were elucidated by principal component analysis. The results indicate that the retention behavior of herbicides markedly depends on their steric parameters: they form separate clusters according to the number of substituents on the two-dimensional nonlinear map of principal component variables (Figure 10). The hydrophobicity and steric effect of substituents form a loose cluster with the retention parameters determined in various eluent systems on the two-dimensional nonlinear map of principal component loadings (Figure 11). It was concluded from the data that the cyclodextrin support shows mixed retention characteristics. Analytes enter the cyclodextrin cavity depending on its dimensions; however, the binding forces accounting for the strength of interaction between the apolar inner wall of the cyclodextrin cavity and the analyte are of hydropbobic character.

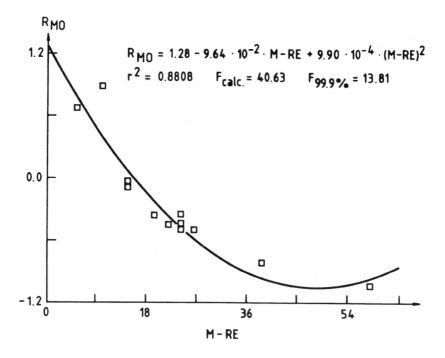

FIGURE 8
Relationship between the retention capacity of 3,5-dinitrobenzoic acid esters (R_{M0}) and their molar refractivity (M-RE). (From Cserháti, T., *Anal. Chim. Acta*, 292, 1994, 17–22. With permission.)

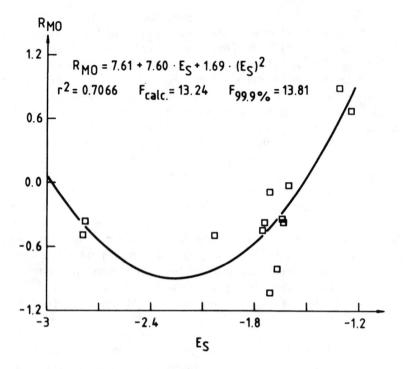

FIGURE 9
Relationship between the retention capacity of 3,5-dinitrobenzoic acid esters (R_{M0}) and the steric effect of substituents (E_s). (From Cserháti, T., *Anal. Chim. Acta*, 292, 1994, 17–22. With permission.)

Ion-pair RPTLC has not been frequently used in the chromatographic practice.[242,243] The theory of optimization in ion-pair RPTLC has been recently published.[244] It was supposed that the concentration of the ion pairing agent (in the case of sulfonated dyes: tetrabutylammonium bromide) on the surface of the apolar stationary phase can be described by the Langmuir adsorption isotherm. The dye anion (D^-) and the counterion (A^-) of the quaternary ammonium ion compete for the ion pairing agent both in the mobile and in the stationary phases. At the beginning of the chromatographic separation process the initial concentration of the quaternary ammonium ion in the eluent is $c_0(Q^+)$. Its concentration in the mobile phase in the first theoretical plate $c_0^m(Q^+)$ is the same:

$$c_0^m(Q^+) = c_0(Q^+) \tag{364}$$

The total concentration of the ion pairing agent in the i-th theoretical plate is

$$c_{s,i} = c_i^m(Q^+) + c_i^s(Q^+) \tag{365}$$

Assuming the Langmuir character of the adsorption, the relationship between the capacity factor, k, and the concentration of the ion pairing agent can be expressed by

$$k = \alpha/[1 + \beta \cdot c_i^m(Q^+)]^2 \tag{366}$$

$$c_{s,i} = k \cdot c_{s,i}/(1 + k) \tag{367}$$

$$c_i^m = c_{s,i} - c_i^s \tag{368}$$

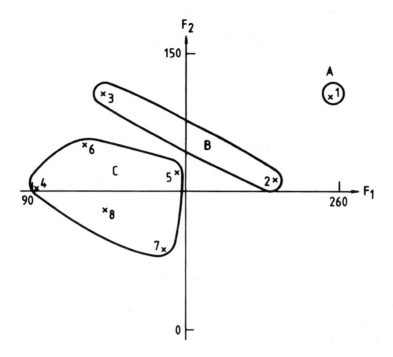

FIGURE 10
Distribution of 4-cyanophenyl herbicides according to their retention behavior. Two-dimensional nonlinear map of principal component variables. No. of iterations, 111; maximum error, 1.80×10^{-2}; A, unsubstituted 4-cyanophenol; B, monosubstituted 4-cyanophenols; C, disubstituted 4-cyanophenols. (From Cserháti, T. and Forgács, E., *J. Chromatogr. A*, 685, 1994, 295–302. With permission.)

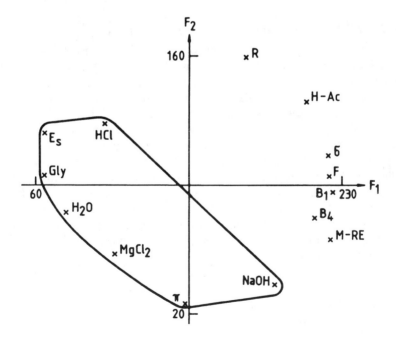

FIGURE 11
Relationship between the retention behavior and physicochemical parameters of 4-cyanophenyl derivatives. Two-dimensional nonlinear map of principal component loadings. No. of iterations, 79; maximum error, 1.62×10^{-2}. (From Cserháti, T. and Forgács, E., *J. Chromatogr. A*, 685, 1994, 295–302. With permission.)

When L is the stationary phase the competition can be described by

$$Q^+ + A^- + L \xrightleftharpoons{K_\alpha} [Q^+A^-L] \qquad (369)$$

$$Q^+ + D^- + L \xrightleftharpoons{K_\delta} [Q^+D^-L] \qquad (370)$$

$$K' = K_\delta \cdot c_i^m(Q^+)/[1 + K_\alpha \cdot c_i^m(Q^+)^2] \qquad (371)$$

The total concentration of the dye in the i-th theoretical plate is

$$c_{s,i} = c_i^m(Q^+D^-) + c_i^s(Q^+D^-L) \qquad (372)$$

$$c_i^s = c_{s,i}/(1 + K') \qquad (373)$$

$$c_i^m = c_{s,i}^s - c_i \qquad (374)$$

Because the molecular hydrophobicity governs the retention in RPTLC, the hydrophobicity can be easily calculated from the retention behavior of the analytes.[245] The hydrophobicity values are generally well correlated both with the experimental and calculated partition coefficient log P[246] and with the hydrophobicity values determined by reversed-phase HPLC.[247,248]

II. HIGH-PERFORMANCE LIQUID CHROMATOGRAPHY

A. Fundamentals

1. Adsorption High-Performance Liquid Chromatography

Liquid–solid chromatography was developed in 1906. Separation is carried out with a liquid mobile phase and a solid stationary phase that reversibly adsorbs solute molecules. The stationary phase is usually polar such as silica, porous glass, or alumina and the mobile phase is a relatively nonpolar organic solvent. Due to separation selectivity of adsorption HPLC, the compounds with different chemical structure are easily separated on silica and alumina. It can provide easy resolution of compounds with differing numbers of functional groups and provides maximum differentiation among isomeric mixtures. The retention of a solute requires displacement of an equivalent number of adsorbed solvent molecules. Polar functional groups are strongly attracted to the polar adsorbent surface, so that compounds with substituents of differing polarity are readily separated. Another feature of adsorption HPLC is the presence of discrete adsorption HPLC, the presence of discrete adsorption sites on the surface of the adsorbent. Optimum interaction between a solute molecule and adsorbent surface occurs when solute functional groups exactly overlap these adsorption sites.

Solvents in liquid chromatography have the following important properties: polarity, viscosity, boiling point, detector compatibility (UV adsorbance and cutoff) reactivity, miscibility, and safety. There are four major interactions between molecules of solvent and solute that are important in HPLC: dispersion, dipole, hydrogen bonding, and dielectricity. The larger the dispersion, dipole, hydrogen bonding, and dielectric interactions in combination, the stronger the attraction of solvent and solute molecules. The ability of a sample or solvent molecule to interact in all four of these forces is referred to as the polarity of the compound. Thus the polar solvents preferentially attract and dissolve polar solute molecules. Solvent strength increases with solvent polarity in adsorption HPLC. (The opposite is true in reversed-phase HPLC.)

Solvent polarity has been defined in various ways. Reichardt[249] provided a detailed summary of the empirical parameters of solvent polarity for 204 solvents. The most widely applied solute polarity parameter (P') is based on experimental solubility data reported by Rohrscheider[250] and Snyder.[251]

In general, a change in P' by two units causes roughly a tenfold change in the retention expressed by the capacity ratio k'.

$$k' = (t_R - t_0)/t_0 \tag{375}$$

where t_R and t_0 are retention time and dead time, respectively, as described by the following equation.

$$k'_2 / k'_1 = 10^{(P'1-P'2)/2} \tag{376}$$

where k'_2 and k'_1 refer to the capacity ratios obtained by the first and second mobile phases, respectively, and P'1 and P'2 refer to the solvent polarity (P') of the first and second mobile phases. Since P' is based on actual solubility data, and the sample solubility in the mobile phase and stationary phases determines the k' values, it can be expected that P' will provide a reasonably accurate measure of solvent strength in liquid–liquid partition chromatography. A better index of solvent strength in this case is afforded by the experimental adsorption solvent strength parameter ϵ^0. In adsorption HPLC the solvent strength ϵ^0 does not vary linearly with solvent composition. Instead, ϵ^0 increases rapidly for small additions of a stronger solvent B to a weaker solvent A, then asymptotically approaches the value of pure B. In practice it means that we have to be very careful about the polar impurities of the organic mobile phase used for the retention determination. A very small change in the water concentration can cause a dramatic change in the adsorption chromatographic retention of a solute. When the mobile phase contains a few percent water on purpose, the various trace amounts of water in the organic solvents will not cause too much of a deviation in the retention. Solvent selectivity is the other very important characteristic of the mobile phase. By changing the selectivity without changing the solvent strength two overlapping peaks can be separated. The greatest change in the mobile phase selectivity results when the importance of the various intermolecular interactions between solvent and sample molecules is markedly changed.

It is also important to check solvent miscibility, not only when preparing a mobile phase mixture, but also when the mobile phase is changed. It is also important to know what kind of mobile phase was used previously to avoid the formation of emulsion or precipitation in the column.

In adsorption liquid chromatography it is customary to add small amounts of water or a polar modifier such as methanol to the mobile phase. These polar compounds are preferentially adsorbed onto strong adsorption sites, leaving a more uniform population of weaker sites that serve to retain the sample molecules. This deactivation of the adsorbent leads to a number of improvements in subsequent separation. It causes less variation in sample retention from run to run, and we can observe higher column efficiency and reduced peak tailing in some cases. We must also consider that the equilibrium process in adsorption chromatography takes a minimum of 2 h or even more when we wish to equilibrate the column from a stronger mobile phase to a weaker one.

2. Reversed-Phase High-Performance Liquid Chromatography

The popularity of the technique rests with the reproducibility offered by use of the hydrocarbon bonded phase. Very efficient and high-speed separations can be achieved on chemically bonded hydrocarbon reversed-phase columns. The most popular RP columns are octyl and octadecyl silica, which are relatively stable with most aqueous eluents having a pH value less than 8. Both spherical and irregular silica particles are used as supports and n-octyl or n-octadecyl functions are the most popular organic ligands attached covalently to the support surface. The inner diameter (I.D.) of the analytical column is usually 4 to 5 mm and the length varies between 50 and 300 mm. They are operated at eluent flow rates ranging from 0.5 to 3 ml/min.

The maximum coverage (i.e., surface concentration) of organic materials reported for so-called monomolecular phases varies from 3.5 to 4 µmol/m$_2$, with coverage for many commercial octadecyl supports well below these values. The coverage percentage of the available silanol groups usually varies from 10 to 60%. The coverage or the presence of free silanol groups can be tested by measuring retention of some standard compounds (benzene, nitrobenzene, or some basic compounds). The extent of coverage affects the retention properties of both the stationary phase and the stability. At low coverage where a higher proportion of silanol is available, the stability in basic aqueous solution is lower than that for a higher coverage where the underlying siloxane bonds are protected by the hydrophobic ligand. In general, chromatographic retention increases with the degree of phase coverage. Selectivity of the support may be affected in this case of low coverage, due to a mixed retention mechanism (partition and adsorption on unreacted silanol groups). At higher coverage the change in selectivity is lower since solute retention is governed mostly by hydrophobic forces and even small polar molecules are unable to reach the free silanol groups.

In general, for reversed-phase packing of equal loading, the longer the chain length of the alkyl hydrocarbonaceous group, the greater the chromatographic retention when fixed conditions are used. The retention of polar and nonpolar samples was found to be influenced by the eluent, the length of the organic ligand, carbon content, pore size distribution of the silica support, and total porosity of the stationary phase. The authors described[252] that the rate of analysis increases with decreasing length of the alkyl chain.

For lower pressure the spherical particles are more suitable: for higher separation efficiency, 3- to 5-µm particle size stationary phases may be suggested.

Eluents used in reversed-phase chromatography with bonded nonpolar stationary phases are generally polar solvents or mixtures of polar solvents, such as acetonitrile or methanol with water. The most significant properties of solvents that may affect the chromatography are surface tension, dielectric constant, viscosity, UV adsorbance, and elutropic value. In general, the lower the polarity, the higher the strength of eluent in RPHPLC. Many samples contain ionogenic substances, which may undergo ionization in aqueous eluent and as a result severe peak broadening may occur. The effect is eliminated or reduced by use of buffer solutions. In practice, a buffer concentration between 10 to 100 mM seems to be sufficient in the pH region of peak buffering capacity.

3. Retention Determination

The capacity ratio (k') can be defined by the following equation in both adsorption and reversed-phase chromatography:

$$k' = t_R - t_0/t_0 \tag{377}$$

where the t_R is the retention time, which means the time passed from the injection to the appearance of the solute peak maximum, and t_0 is the dead time. The exact dead time is one of the most important parameters in the accurate measurement of k'. The problem starts with the definition of dead time, i.e., the retention time (or if multiplied by the flow rate, retention volume) of the unretained solute. The dead volume can be the elution volume of a solvent disturbance peak obtained by injecting an eluent component, or the elution volume of an unionized solute that gives the lowest retention volume, or the elution volume of an isotopically labeled component of the eluent or an isotopically labeled water molecule (only in RPHPLC), or the elution volume of a salt ion, or the volume of the liquid the column contains, or the extrapolated elution volume of a zero number of a homolog series. Wells and Clark,[253] Berendsen et al.,[254] and Knox and Kaliszan[255] provided reviews of various techniques for experimental determination of dead volume in HPLC.

The most accurate method is to inject a UV active salt (potassium iodide or sodium nitrate) in reversed-phase chromatography together with the solute for which the capacity ratio has to be determined. The only objection to the method may be the interaction of salt with the solute or the mobile phase additives. It may also happen that the dead time or dead volume vary with mobile phase composition. Especially when high concentrations of organic phase are applied in the mobile phase, the retention time of sodium nitrate or the potassium iodide may increase. In these cases the injection of the eluent component can provide a disturbance peak that can be used to calculate the capacity ratio.

In HPLC the retention of a solute is determined by the strength of solute-stationary phase and solute-mobile phase interactions; therefore, it is essential to reveal quantitatively how mobile phase composition affects solute retention. The dependence of the logarithm of the retention factor (log k') on the volume fraction of the organic component g, in the mobile phase can be described by the following equation:

$$\log k' = \text{slope } b + \log k'_0 \tag{378}$$

where the log k'_0 value (the intercept value of the straight line) refers to the extrapolated log k' value to neat water as mobile phase. The slope values shows the sensitivity of the change in the retention of a given solute caused by the change in organic phase concentration in the mobile phase. In practice the determination of these two parameters can be carried out by measuring the log k' values of the compounds by using a minimum of five organic phase concentrations in the mobile phase.

As we cannot expect linearity in a wide range of mobile phase compositions, the change of the organic phase concentration in 5% steps is advantageous. Then the log k' values obtained in five different concentrations will cover a concentration range of 25%. Compound retention should be within the range of 0 to 5 log k'. Retention in chromatography reflects solute distribution between the stationary and mobile phases as well as the temperature. Melander et al.[256] described a quantitative dependence of the log k' values on the temperature and the mobile phase composition:

$$\log k' = A_1 X (1 - T_C/T) + A_2/T + A_3 + A_4 f(X,T) \tag{379}$$

where X is volume fraction of the organic cosolvent in the hydroorganic mobile phase; T is absolute temperature; and T_C is the so-called compensation temperature derived by taking advantage of the enthalpy–entropy compensation. Linear relationships were found in RPHPLC between enthalpy and the entropy for reversible solute binding to hydrocarbonaceous stationary phases, as indicated by Melander et al.[256] The effect is called enthalpy–entropy compensation. Comparison of the compensation temperatures obtained from such data can be used to study the retention mechanism, whether they are identical or not in two different chromatographic systems.

4. Development of Column

During the last three decades the advancement of high-performance liquid chromatography (HPLC) column has paralleled, and sometimes even exceeded, the development of HPLC equipment. The rapid progress in HPLC techniques in the late 1960s was mainly due to the use of pellicular packings, a low-flow rate reciprocating pump and the flowthrough UV detector.

In the early 1970s microparticulate packings (particle size <10 μm) were predominant, which provided greater speed efficiency, and sample capacity benefits. Stable chemically bonded phases spurred the development of reversed-phase chromatography. The 1980s saw the introduction of biochromatography materials which provided better analytical and preparative problem-solving tools for life scientists.

The solid-phase extraction cartridges enabled more convenient sample cleanup. In the 1990s further refinements in column technology resulted in columns that provided large separation factors, more stable and durable reversed-phase columns, and special columns for solving specific chemical, environmental, and other speciality problems.

B. Direct (Normal)-Phase High-Performance Liquid Chromatography

1. Silica Supports

Currently more than 90% of column packings used in normal- and reversed-phase liquid chromatography are based on silica. Silica exists in various forms with the stoichiometric composition SiO_2. The atomic number of silicon is 14 and its atomic weight is 28.08. Silicon comprises 25.7% (w/w) of the earth's crust; it is the second most abundant element. Silicon is not found free in nature, but occurs mainly as oxide and silicates. Over 55% (w/w) of the earth's crust is made up of silicon oxide and silicates.

The silica used for liquid chromatography (LC) column packings is essentially porous and noncrystalline with the general formula $SiO_2 \cdot xH_2O$. Water is chemically bound in a nonstoichiometric amount, forming Si–OH. Silinol groups proved most useful for LC packing.[257]

Silanols are responsible for the polar character of silica packings used in normal phase LC to graft organic moieties. This treatment changes the silica surface and allows the development of bonded phases used in reversed-phase and other modes of LC.[258] The surface of silica forms is covered by hydroxylated silanol (Si–OH). Two hydroxyl groups on two vicinal silicon atoms are vicinal silanols (see Figure 12).

When borne by the same silicon atom, the two groups are termed geminal silanols and the other type of silanols can be isolated as free silanol groups. The surface of a crystalline form of silica is covered chiefly by isolated silanol groups. Crystalline forms of silica are found in nature. The unique amorphous silica source of biological origin is diatomaceous earth. All other amorphous silicas used in the industry are produced by synthetic methods. There are two main methods for obtaining pure silica: the liquid-phase route and the gas-phase route. Amorphous silica with a porous structure bears the above

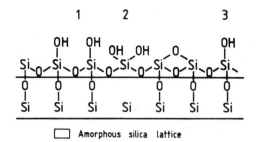

Figure 12. Three different types of hydroxyl groups on the amorphous silica surface: (1) vicinal silanols; (2) geminal silanols; (3) isolated silanols. (From Berthod, A., *J. Chromatogr.*, 549, 1991, 1–28. With permission.)

three types of silanol groups capable of hydrogen bonding with water molecules. The more elevated the surface silanol concentration, the greater the hydrophilic character of the surface. The role of silanols in LC selectivity is essential. It is very important to know the concentration of surface silanols for a given silica-based LC stationary phase.

There are three principal methods for the determination of surface hydroxyl groups: chemical, isotopic exchange, and spectroscopic methods.[259]

a. Chemical Methods

Water hydrogens are more acidic than silanol hydrogens, thus all physisorbed water must be removed by heating the silica sample at 150°C, under vacuum, for several hours. The most reliable estimate of surface silanol concentration was obtained by reaction of hydroxyl groups with methyllithium under nitrogen atmosphere. Provided that the sample was carefully dried, the solution reacted with hydroxyl groups as follows:

$$\text{Si–OH} + \text{LiCH}_3 = \text{Si–O}^- + \text{CH}_4 \qquad (380)$$

The amount of methane evolved was a direct measure of silanol surface concentration. If water molecules were present, methyllithium reacted with these first to produce lithium oxide (Li_2O) and methane; then it reacted with silanols. The methane was determined by gas chromatography.[260]

b. Isotopic Exchange Methods

Heavy water was used to determine surface silanol concentration by isotopic exchange

$$\text{Si–OH} + {}^2\text{H}_2\text{O} = \text{Si–O}\,{}^2\text{H} + \text{H}^2\text{HO} \qquad (381)$$

Surface silanol concentration was determined by weighing the dry sample before and after the exchange or by spectroscopic methods. Titrium-labeled water ^{3}HHO was preferably employed instead of heavy water, as this facilitates radioactive measurement of the exchanged sample.[261]

c. Spectroscopic Methods

Infrared (IR) spectroscopy was used to assess surface silanol concentration. Nuclear magnetic resonance (NMR) of ^{29}Si with cross-polarization and magic angle spinning (MAS) could also be applied to determine free and geminal silanol groups and silicon atoms with no silanols. Other methods such as thermogravimetric analysis, diffuse reflectance, Fourier transform infrared spectroscopy, or pH measurements also proved to be suitable for silanol determination.[262]

2. Chemical Properties of Silica

Most chemical properties of silica are due to surface silanols. Interactions of silica with water are silanol dependent. Hydroxyl acidity is responsible for the acid–base properties of silica. Silanols were used to graft various organic moieties on the silica surface and to obtain the wide variety of bonded silica essential for LC packings.

a. Acid–Base Properties

Silicic acid is a weak acid. The ionization constant of the reaction

$$Si(OH)_4 = Si(OH)_3O^- + H^+ \qquad (382)$$

was 1.6×10^{-10} at 20°C, which corresponds to a pK_a value of 9.8. The pK_a value of the second ionization

$$Si(OH)_3O^- = Si(OH)_2O_2^{2-} + H^+ \qquad (383)$$

was 11.7. The hydroxyl groups on the silica surface have also an acidic character. The pK_a value of the reaction

$$SiOH = SiO^- + H^+ \qquad (384)$$

was about 6.8, which was 3 pH units lower than the free silicic acid first ionization. There are several explanations for the higher mobility of silanol proton compared with that of silicic acid. The tendency for splitting off a proton from a particular silanol group was markedly promoted by its environment.[263]

The acidic character of surface silicas conferred some ion-exchange properties on porous silica. The hydronium ions (protons) of the silanol groups were exchangeable by cations of the solution in buffer or electronic solution. The ion-exchange properties of silica were highly dependent on the pH of the solution, the surface area, and the silanol concentration. Silica bore negative charges in solution at pH 7.[264] The rapid dissolution of silica at mobile phase pH values was higher than 10 and/or temperature higher than 40°C. A slow and insidious solubilization occurred from pH 8. This may cause small voids in the column packing, inducing changes in solute retention and leading to lower efficiency. This could be reduced with the use of a guard column or a precolumn installed between the pump and the injection valve to saturate the mobile phase in soluble silica. It is important to improve the pH stability of silica.

b. Silanol Acidity

Silanol groups are acidic, although there has been some controversy over the origin of the acidity and the variability of this acidity between different silanol groups on the same silica surface and similar silanol groups on silica surfaces with different histories. There have been several attempts to define silica acidity as a measurable parameter for the selection and characterization of silicas. One method reported is the measurement of the pH values of an aqueous suspension of silica.[265] Such a measurement may lead to surprising results, e.g., pH values ranging from 3 to 9.5. One problem in the synthesis of silica is the removal of traces of acid or alkali used in the synthesis and washing of the packing. Silicas that have very high pH values in the tests generally have high residual content of sodium or other metal ions present within the matrix. The material (Zorbax SIL) reported to have the lowest pH value involved extensive acid treatment for rehydroxylation and even though this is one of the more acidic silicas, it is highly probable that the pH measurement indicated the degree of acid removal after washing rather than a fundamental property of the silica. Because of the need for extremely careful pretreatment prior to pH measurement or titration of the silica, much of the measurement of acidity has relied upon physicochemical procedures such as IR, where the stretching frequency of the hydroxyl band is related to its acidity. Such infrared studies have suggested that the pK_a of the silanol groups is around 7.1.[266] This result must be treated, however, with caution since these authors remarked that although the average pK_a of the surface hydroxyls for a silica/alumina catalyst was found to be 7.1, there were a few groups showing pK_a values lower than –0.4. Thus, such measurements generally give average values and do not detect a small number of extremely active silanols.

c. *Water Adsorption and Desorption*

Silica hydroxyl groups are dissociated with water molecules by hydrogen bonding. This adsorbed water is tightly bound to the silica surface and it is removed by heating at 150°C under vacuum for several hours. Overheating of the silica sample led to dehydroxylation. Geminal silanols were reversibly dehydroxylated resulting in siloxane bridges. Between 200 and 800°C the temperature treatment under vacuum of a silica sample induces a monotonous decrease in the hydroxyl group concentration with siloxane bridge formation. If the calcination temperature does not exceed 600°C, the reactive siloxane group completely rehydroxylates when exposed to water.[267]

Complete dehydroxylation and sintering occur on heating above 1200°C. At this elevated temperature the amorphous character of silica slowly ceases. The hydrophilic surface is converted into a hydrophobic surface.[268]

The conversion of amorphous silica into crystalline form is dependent on the purity of silica and particularly on its sodium. At room temperature porous silica can adsorb water reversibly.

d. *Water Solubility*

Water solubility is very dependent on the crystalline or amorphous form of the sample and also on physicochemical properties, such as porosity, surface area, and particle size. The solubility of silica changes the effect of water temperature, pressure, and pH. The solubility of crystalline and glass form of silica is in the parts per million (ppm) range.

e. *Particle Size*

Silica packings are always obtained with particle size distribution, and 5 µm is generally considered as an average diameter. The average particle diameter can be defined as the average diameter in number d_n, the average particle diameter in surface d_s, or the average diameter in mass d_m, as follows:

$$d_n = \Sigma n_i d_i / N = \Sigma n_i d_i / \Sigma n_i \tag{385}$$

$$d_s = (1/S)\Sigma n_i d^3 \pi/4 = \Sigma n_i d_i^3 / \Sigma n_i d^2 \tag{386}$$

$$d_m = (4\pi p/3M)(\Sigma n_i d^4) = \Sigma n_i d^4 / \Sigma n_i d^4 / S n_i d^3 \tag{387}$$

where n_i denotes the number of particles with diameter d_i; N is the total number of particles in the studied sample; S stands for the surface area of the sample; M designates the sample mass; and p is the density of the particles.

Silica-based LC packings have a particle size ranging from 1 to 100 µm. Microscopy is the simplest method to determine particle size. Electronic counting scanners can give particle distribution, with few or no assumptions of the particle shape. Sedimentation is another widely used technique for particle size analysis.

The particle size is one of the most important parameters, greatly influencing the chromatography efficiency. The smaller the LC particle size, the higher the efficiency. The plate height (H) of a well-packed column can be as low as twice the mean particle diameter (100,000 plates per meter, $H = 10$ μm, $d^p = 5$ μm). Unfortunately, particle size also affects column permeability P (cm_2) in the following way:

$$P = (d_p^2/Y)[\epsilon_0^2/(1-\epsilon_0)^2] \tag{388}$$

where Y is a dimensionless shape factor ($Y = 180$ for spherical particles) and ϵ_0 is the interstitial volume or external porosity (cm^3). Permeability is a parameter of the column pressure drop (P).

$$P = uvL/P \tag{389}$$

where u is the linear mobile phase velocity (cm/s); v is the mobile phase viscosity (g/cms); and L is the column length (cm). If the particle diameter is divided by a factor of two, the column driving pressure has to be four times as high in order to obtain the same flow rate. Optimum particle size with regard to the analysis time, plate number, and pressure drop is approximately 2 to 4μm. Packings with average particle diameter of 5 μm are the most commonly used for 10- to 25-cm analytical columns with I.D. of 4 to 4.6 mm. Packings of 3 μm are used to obtain high efficiency with short columns; 10 μm, 20 μm, or larger particle size packings are used when the pressure is an important factor, such as in preparative chromatography.

f. Surface Area

The specific surface area of a porous solid material is equal to the sum of its internal and external surface areas. The external surface area corresponds to the geometric surface of particles per gram of sample. For spherical particles of equal size d the external surface area is obtained by

$$A = 6/d\sigma \tag{390}$$

where σ is the particle density, defined as the ratio of particle mass to particle volume. For nonporous particles σ is the solid silica density 2.2 g/cm^3. The external surface area is inversely proportional to the particle diameter. A monodisperse nonporous silica with particle size of 3 μm has a surface area of 0.91 m^2/g. Polydisperse nonporous silica with an average particle diameter of 3 μm has a surface area of 0.74 m^2/g. Surface area is measured by the routine BET method.[269] The complete theory of gas adsorption and specific surface area determination is described in the book by Gregg and Sing.[270] The surface area of a silica sample is dependent on the porosity, which forms the internal surface area. The specific surface area of silica used in high-performance liquid chromatography ranges from 10 to about 500 m^2/g depending on the average pore diameter.

g. Porosity

A pore is a hole, a cavity, or a channel connected to the surface of solid. The spaces or interstices between particles are voids rather than pores. A cavity that does not communicate with the surface is called a closed pore or an internal void and will not contribute to porosity or specific surface area. The pore system and its characteristics are the topic of numerous papers.[271] The pore shapes shown have wide variety. The pore size covers a range of several orders of magnitude. A feature of special interest is the width p_d of the pores, which is the diameter of an equivalent cylindrical pore. According to IUPAC classification there are micropores with $p_d < 2$ nm, mesopores with $2 < p_d < 50$ nm, and macropores with $p_d > 50$ nm. The porosity of a sample ϵ_p is defined as the total pore volume V_p divided by the volume of the sample V_T:

$$V_T = V_p + V_s \tag{391}$$

where V_s is the volume of solid silica, assuming that there are no closed pores. Pore volume and pore size distribution are estimated by gas adsorption–desorption isotherms.

The adsorption–desorption isotherms of gas onto solids can be divided in six different types. Types I, IV, and V isotherms show hysteresis; the desorption branch of the curve does not coincide with the adsorption branch. Both branches (desorption and adsorption) are used in pore shape and volume determination. Types I, IV, and V isotherms correspond to microporous, mesoporous, and polydisperse porosity, respectively. Types II and III correspond to nonporous samples with strong interaction and non- or macroporous samples with weak interaction, respectively. The type VI isotherm corresponds to nonporous material.[270,271]

h. Relationship between Surface Area and Porosity

The specific surface area of silica has an important impact on solute retention and selectivity. The higher the specific surface area of silica, the higher the solute–stationary phase exchange area. Solute retention is roughly proportional to the specific surface area when all other parameters are constant (same kind of stationary phase, same mobile phase, same column length, same flow-rate, etc.)

Pore size, pore volume, and specific surface area are interrelated. Assuming that there are no closed pores, the specific pore volume, V_p (cm^3/g) to p_a, the apparent density of the particle is defined as the ratio of the mass of the porous particle to its total volume (cm^3/g):

$$p_a = p/(1 + V_p p) = 2.2/(1 + 2.2 V_p) \tag{392}$$

(2.2 g/cm^3 is the density of solid silica). If we assume a monodisperse distribution of perfectly cylindrical pores with diameter p_d, the internal surface area due to the pores is expressed by

$$A_p = 4V_p/p_d \tag{393}$$

where A_p is the internal surface area; V_p is the specific pore volume. The specific surface area is the sum of the external surface area ($A = 6/d\sigma$) with the corrected density ($p_a = p/(1 + V_p p) = 2.2/(1 + 2.2 V_p)$) and the internal surface area ($A_p = 4V_p/p_d$):

$$A_s = 6(1 + V_p p)/d\sigma + (4V_p/p_d) \tag{394}$$

Pore diameter on the surface area and the relative weight of the internal and external surface area are of great importance. For most samples more than 97% of the specific surface area is internal surface due to pores. The particle diameter has little importance with regard to the surface area, which depends mainly on the pore size and volume (Table 83).

i. Surface Silanols and Polarity

Silanol groups are responsible for the polarity of the silica surface. A silica completely dehydroxylated by prolonged heating at 1300°C is hydrophobic. The silanol concentration

TABLE 83

Effect of Pore Volume and Pore Size on the Specific Surface Area

Particle diameter (μm)	Pore volume (cm³/g)	Pore diameter (nm)	Surface area Total (m²/g)	Internal (% of total surface)
100	1.0	50	80.1	99.9
		10	400.1	99.9
		5	800.1	100.0
	0.5	50	40.1	99.8
		10	200.1	99.9
		5	400.1	100.0
	0.1	50	8.0	99.6
		10	40.0	99.9
		5	80.0	99.9
3	1.0	50	82.9	96.5
		10	402.9	99.3
		5	802.9	99.6
	0.5	50	41.9	95.4
		10	201.9	99.1
		5	401.9	99.5
	0.1	50	9.1	87.8
		10	41.1	97.3
		5	81.1	98.6

From Berthod, A., *J. Chromatogr.*, 549, 1991, 1–28. With permission.

of a fully hydroxylated silica is about 9 μmol/m². The pretreatment temperature under vacuum may decrease the silanol concentration and adjust the hydrophilic character of a sample.[272] The selectivity of thermally treated silica used in normal phase LC greatly depends on the water content of the apolar mobile phase.[273]

Reproducible results with activated silicas and apolar solvents such as *n*-hexane are only obtained when the water content of the silica and the mobile phase is controlled and adjusted. Water molecules, always present in trace amounts in an apolar solvent, adsorb on most polar silanol sites, decreasing the retention of polar solutes and modifying the solute–stationary phase exchanges. Peak shape and efficiency are also affected by the adsorption of water on silica. Methanol and other polar solvents can be added to the apolar mobile phase to adjust the selectivity.

Acid silanols have great affinity for basic groups. The low efficiencies and peak tailings obtained with the amino-containing solutes are due to –SiOH...NH– acid–base interaction. In order to reduce the residual silanol concentration of bonded silica, the end-capping treatment was used. The small trimethylchlorosilane molecule was used to react with silanols that were not accessible to the large octadecyl-containing reagent. An end-capped silica still bears unreacted silanols.

The modern trend in manufacturing base-deactivated silica packings is to increase the ligand density.[274]

j. Carbon Content and Bonding Density

Two important parameters for the characterization of bonded silica are carbon content, in grams of C per 100 g of packing, and surface concentration or bonding density, in micromoles of bonding moiety per square meter of initial silica surface area.[274] Although bonding density is almost never given by manufacturers, it is one of the most important parameters for a bonded silica-based stationary phase. High bonding density implies a low residual silanol concentration; this phase has better resistance to elevated pH mobile phases and higher capability to separate amino-containing compounds.

k. Chemical and Physical Chromatographic Requirements

The safe pH range for the mobile phase is 2 to 9. Below pH 2, Si-C bonds of derivatized LC packings can be split, above pH 9, silica is slowly solubilized into silicate. The working pH range increases when bonding density and/or silica purity is high. A small addition of sodium silicate to the mobile phase greatly reduces the dissolution of silica at elevated pH.[275] Zirconia cladding enhances the resistance of silica to alkaline solutions.

The stationary phases must have some mechanical strength to withstand the column back-pressure, in the range 10 to 200 bar or 140 to 2900 psi. Silica pore size may be a critical parameter in bulky solute analysis. For example, large-pore silica supports are required for biopolymer separation, i.e., for proteins and peptides.[276] In these analyses, 30 and 50 nm pore size packings are commonly used to avoid internal interactions such as restricted and slow diffusion of biopolymers. The large molecules can be trapped inside medium sized pores, inducing peak tails. A narrow pore size distribution is also required in size-exclusion chromatography to obtain sufficient efficiency.

l. Application of Normal-Phase Chromatography

Normal-phase (NP) chromatography is a powerful complement to the more popular reversed-phase HPLC method for separating nonionic compounds. About one fifth of all HPLC separations are now performed by NPHPLC. The advantages of this method is that it can be performed with totally organic mobile phases of high solute solubility, which is important in preparative applications. NPHPLC is also capable of extensive selectivity changes with the use of various mobile phase constituents. NPHPLC methods are especially useful for the separation of difficult mixtures containing positional isomers.[277]

Current NPHPLC applications commonly use columns containing polar bonded phases (cyano-, diol-, etc.). Columns containing unmodified silicas are less popular, because of problems in maintaining a constant surface activity for repeatable separations. However, retention with bonded-phase NPHPLC columns is generally more consistent, because of insensitivity to small concentrations of water in the mobile phase of samples. Chromatographic reproducibility of bonded-phase columns for NPHPLC is generally viewed as superior to the column of unmodified silica.[278] The polar bonded-phase columns are less retentive in NP-HPLC than unmodified silica, often permitting the elution of highly polar compounds that are eluted with difficulty from the more retentive unmodified silica.

A common problem with unmodified silica columns with totally organic solvents is that polar compounds often show broad, tailing peaks, especially for basic components. This condition can exist even if the mobile phase contains basic additives designed to minimize this effect. Such undesirable characteristics complicate the design of rugged and reproducible quantitative methods. As a result, unmodified silica columns are not widely accepted for quantitative analysis.

Unmodified silica columns present a special problem that must be solved for reproducible separations: the activity of the adsorbent surface must be constant. The most effective procedure to stabilize this surface activity is to control the level of water adsorbed to the silica surface. This is often accomplished by fixing the concentration of water in the mobile phase, a procedure viewed as awkward by many users. Thus polar, usually protic mobile phase modifiers such as methanol and propanol routinely are used instead of water to control surface activity. The fundamental problem remains, however, that commercial unmodified chromatographic silica often exhibits an inhomogeneous adsorbing surface. This fact can result in broad, poorly shaped peaks when used for separating many polar solutes and basic drugs. Previous studies have shown that chromatographic silicas can be arbitrarily classified into types to define their potential utility as supports in reversed-phase chromatography.[262,279]

So-called type A silicas are generally more acidic and less purified. These materials apparently have less energy-homogeneous surfaces and are likely to exhibit tailing, misshapen peaks with more polar and basic compounds. Type B silicas are more highly purified and less acidic and are often preferred for many RPHPLC applications.

Kirkland et al.[280] determined the characteristics and chromatographic properties of a new porous silica microsphere, Zorbax RX-SIL. The chromatographic characteristics of this new type B silica were compared with conventional type A silica (Zorbax-SIL) to determine the type and level of mobile phase modifiers needed for good results. The effects of sample type and sample loading on retention column efficiency and peak shape were also measured.

Absolute differences in retention for the two silicas were found to be functions of differences in surface areas. It was assumed that the heterogeneous surface of type A silica led to isotherm nonlinearity and the related phenomena of large plate height and asymmetric tailing peaks. The effect of sample loading on column efficiency was similar for the two silica types. Sample loadability was essentially equivalent for the two silicas, with small variations dependent on differences in packing surface areas. As mentioned above, many users consider the necessity of maintaining a constant amount of water modifier in the organic mobile phase as experimentally awkward. Because of the convenience, other highly polar, organic-miscible modifiers such as methanol and propanol are often used in normal-phase separations to control and maintain the activity of the silica adsorbent.

These polar organic compounds strongly bind to the silica surface, although they do not bind as tightly as water. For the convenience of protic-solvent modifiers, a study was made with methanol and propanol to determine their effect on type A and B silicas. Similar results were found for the two silica types; both showed sharply increasing k' (capacity factors) vs. methanol characteristics at low methanol concentrations. The low absolute k' differences were due to different packing surface areas. These results suggest that methanol-modified silica is less homogeneous than water-modified silica. The new type B silica showed a constant plate height throughout the entire 0 to 0.4% methanol concentration range studied. These results again suggest superior surface-energy homogeneity of the new type B silica. These studies confirm that deactivation with water usually gives the best chromatographic results with all silica types. The surface of water-deactivated silicas appear most homogeneous, producing more efficient columns and superior peak shapes. Still, methanol and 2-propanol are effective and more convenient deactivating agents when used at appropriate levels. In some systems, however, these modifiers may cause misshapen peaks for some solutes, particularly when used at low concentrations.

m. Silanophilic Interaction

The silanophilic theory was developed by Horváth and co-workers.[281] Irregular retention behavior of crown ethers in reversed-phase chromatography with silica-bonded hydrocarbonaceous stationary phases is interpreted by using the dual binding model. It has been assumed that retention is caused not only by the usual solvophobic interactions, but also by silanophilic interactions between the eluate and the accessible silanol groups at the surface of various alkyl-silica bonded phases. Such behavior has also been documented for peptides.[282]

Both normal- and reversed-phase behavior have been observed for a wide range of species when separated under reversed-phase conditions. At low concentrations of the organic solvent system, the behavior is predominantly reversed-phase, but at higher concentrations the solutes show increasing normal-phase behavior. This results in typical

U-shaped plots of log k' vs. organic solvent concentration. In the absence of silanophilic interactions, these plots are linear with a negative slope.

n. Reduction of the Effects of Silanol Groups

The reaction of silica with long-chain alkylsilyl halides does not cover all the available silanol groups. The effects of the silanol groups can thus be eliminated by reacting them with some other smaller reagent, which covers as many of the remaining groups as possible. The usual procedure is to replace the hydrogen atom of silanol with a trimethylsilyl group (see Figure 13).[283]

o. Ionic Interactions

In the course of the reversed-phase chromatography of basic compounds peak shapes were found to be poor and efficiencies low. Some of the difficulties in the course of chromatography were established by Jane,[284] who employed the method of using bare silica rather than reversed-phase packings with methanol mobile phase containing a small percentage of water buffered to high pH. In the development of reversed-phase analytical methods for quaternary ammonium compounds used as animal feed additives, it has been observed that the retention of the components was strongly influenced by pH and buffer concentration. Surprisingly, the substitution of a bare silica column gave an almost identical separation. The result prompted a study of the chromatography of basic compounds on silica[285] using methanol–water mobile phases buffered to less extreme pH values. It can be concluded from the above studies that in contrast to the silanophilic effects noted for crown ethers, the principle mechanism of retention was ion exchange chromatography on the acidic silanol groups. By studying the separations of several probes of different basicity using packings of different surface coverages of bonded phase groups, retention mechanisms were observed comprising a number of competing interactions, which included hydrophobic interactions with siloxane bridges and silanophilic interactions in addition to ion exchange. A method for the assessment of the ion exchange character of retention was described by Stout et al.[286]

It has been shown that the predominant retention mechanism of basic, ionogenic solutes on unbonded silica in buffered aqueous organic solvents was one of ion exchange. Under such a retention mechanism, the plots of capacity factor against the inverse of competing ion concentration are linear. If ion exchange is the sole mechanism, such a plot will pass through the origin. When the separation mechanisms that are not influenced by ionic species are present, there is an intercept, corresponding to the retention that should be observed at an infinite competing ion concentration. Such an intercept is typical where silanophilic or reversed-phase retention mechanisms occur.

The slope of the capacity factor of an ionic solute vs. the inverse of ionic strength was shown to be directly related to the ion exchange distribution coefficient. For silicas of similar surface area, the slope of this plot acted as an indicator of the ion exchange character of the material. It has been shown that silicas with properties favorable for the reversed-phase chromatography of basic compounds had low values of the slope, while those of silicas unfavorable for the chromatography of bases had high slope values.

Different authors have shown that physicochemical properties, e.g., pore volume, mean pore diameter, specific surface area, concentration of accessible groups, and the structure of the siliceous support matrix, significantly affect the efficiency of the chemical process.[287-289] Moreover, purity of the support surface is one of the major factors influencing structural and physicochemical properties of the basic silica materials used in the preparation of packings with high coverage density of chemically bonded phases.[290,291] The effect of the surface purity of silica gel supports on the coverage density of chemically bonded phases was studied by Buszewski.[292] The silicas were washed with 20% v/v HCl before

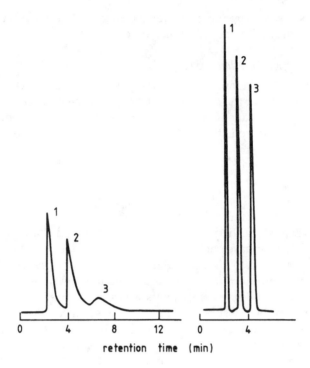

FIGURE 13
Illustration of the beneficial effects of end-capping. Left, uncapped C_{18}, prepared from a silica containing moderate metal concentrations; right, separation performed on the same material after end-capping. (From Ohtsu, Y. et al., *J. Chromatogr.*, 481, 1989, 147–157. With permission.)

chemical modification. Different physicochemical methods — porosimetry, pH measurement, secondary ion mass spectrometry (SIMS), CP/MAS NMR and chromatography — were used for the characterization of these materials. Application of SIMS shows the presence of eight different elements, forming strong specific adsorption centers on the surface of unmodified silica gels. Purification of the surface of silica gels with 20% HCl solution caused partial removal of these elements and a decrease in concentration. Surface chemical modification with C_{18} ligands of adsorbent after washing caused approximately a 15% increase in coverage density values, which led to a decrease in retention times as well as an increase in the peak symmetry of substances by HPLC methodology.

It is very important to know the distribution, organization, and orientation of the alkyl chain ligands on the silica surface. Most silicas applied in chromatography are not pure. Typical levels of impurities found in silica gel are shown in Table 84. The first four rows are from silica samples prepared by Ohtsu et al.[283] Silica I was prepared from an impure sodium silicate. Silica II was obtained from silica I after extensive acid washing. Silica III was prepared from carefully purified sodium silicate, and silica IV was gained by washing the material with acid. The other data were derived from impurity measurements of commercial silicas.[293] Develosil (a silica prepared from a sodium silicate derived sol), Kromasil (prepared from tetraethylorthosilicate (TEOS), Nucleosil, LiChrospher Zorbax SIL, and Zorbax RX SIL.[280]

The effect of the metal ions may be seen in Figure 14. The separation of chelating compounds purpurin and quinizarin together with amylbenzene is shown in the figure. The Kromsil packing clearly gave a result superior to that of Develosil when bonded in an analogous manner. This may be directly related to the difference in the metal content of the media.[283] The material with high level metal concentrations gave very poor peaks

TABLE 84
Typical Impurity Levels in Silicas from Various Sources

Silica	Metal Concentration (µg/g)							
	Na	Mg	Al	K	Ca	Ti	Fe	Zr
I	308	11	264	42	16	230	154	11
II	0	8	188	0	13	162	33	13
III	66	0.1	8	0	1	10	15	3
IV	0	0	6	0	0	7	1	3
Develosil	25	51	54	0	215	38	0	5
Kromasil	25	0	0	0	0	2	0	0
Nucleosil	56	—	0	—	130	57	76	0
LiChrospher	2900	—	1100	—	0	235	445	0
Zorbax PSM60	105	—	0	—	0	41	68	<25
ZorbaxRxSIL	10	<4	<10	<3	4	—	—	—

From Ohtsu, Y. et al., *J. Chromatogr.*, 481, 1989, 147–157. With permission.

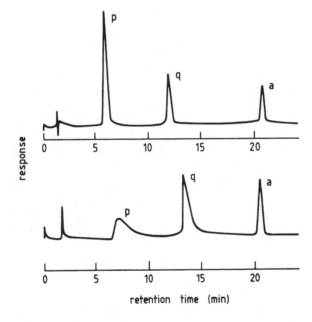

FIGURE 14
Comparison of the chromatography of chelating species on silicas with different metal constants. Top, Kromasil C_{18}; bottom, Develosil C_{18}. Peaks: p, purpurin; q, quinazarin; a, amylbenzene. Mobile phase: 75% methanol–0.05 M phosphate buffer pH = 3. (From Ohtsu, Y. et al., *J. Chromatogr.*, 481, 1989, 147–157. With permission.)

in this experiment, while purified silica gave acceptable performance. Other researchers also noted the effects of metal ions that are incorporated in the silica matrix. It has been calculated that aluminum present in the structure confers high acidity upon adjacent silanol groups[294] while Sadek and co-workers[295] have shown that the removal of metal impurities from silicas leads to a fundamentally different retention behavior of hydrogen bond acceptor solutes, which was attributed to a lower number of metal hydroxides and the reduced influence of metal ions upon surface silanol groups. These data, together with observations on the gas chromatographic adsorption of amines, led Nawrocki[296] to the conclusion that the metals play an all important role in enhancing the acidity of surface silanol groups in chromatographic silicas.

p. Reversed-Phase Character of Silica

Besides polar silanol groups, silica possesses nonpolar siloxane bonds. When these are at the surface, they are expected to reduce polarity on the surface. This reduction has been observed for heat treated silicas. It should also be possible, under appropriate conditions, to observe retention of nonpolar molecules on these silanol bridges. Plots of logarithm capacity factor for both propylbenzene[297] and toluene[298] against methanol concentration were linear, as expected for a reversed-phase mechanism. The retention was, however, small compared with that found for an alkyl bonded-phase packing, although the contribution of reversed-phase mechanism to the separations of basic compounds on bare silica with predominantly aqueous mobile phases was shown to be significant. The reversed-phase component of the retention mechanism was a function of the degree of silica hydroxylation with the more highly rehydroxylated silicas showing a significantly lower reversed-phase character. These facts confirm the assumption that the siloxane groups are the site of reversed-phase interactions on bare silica.

C. Reversed-Phase Chromatography

Reversed-phase chromatography has become the dominant branch of high-performance liquid chromatography over the last few years. The name *reversed-phase chromatography* was a rational choice at a time when chromatography was practiced almost exclusively by using polar stationary phase and a nonpolar eluent. However, today an estimated 80 to 90% of chromatographic systems used in HPLC work consist of nonpolar stationary phases and polar eluents. The popularity of the technique rests with the reproducibility offered by the use of hydrocarbon bonded phases. The aqueous eluents have high optical transparency at low UV wavelength and are cheaper, less toxic, and less flammable.

The first attempt to bind an organic moiety to the surface of silica gel was made by Halász and Sebestian in 1969, who attached aliphatic hydrocarbon chains to the silica gel surface by means of silicon–oxygen–carbon linkage. The original synthesis by Halász and Sebestian involved refluxing the silica gel with an aliphatic alcohol, but, unfortunately, the silicon–oxygen–carbon bonds were very weak and the bonded hydrocarbon chain rapidly hydrolyzed from the surface, generating the original silica gel. Nevertheless, the material was sufficiently stable to allow Halász to identify the highly desirable chromatographic properties of the bonded phase. Gilpin and Burke[299] described in 1973 the use of chlorosilanes as bonding reagents. When the hydroxyl group of silica gel reacted with a chlorosilane, hydrogen chloride was released and the organic moiety was attached by means of the silicon–oxygen–silicon bond. The silicon–oxygen–silicon bond was much stronger than the silicon-oxygen-carbon bond and such bonded phases could be used satisfactorily in a liquid chromatography column over long periods of time, provided the extremes of pH were avoided.

This type of bond was the basis for the synthesis of the vast majority of contemporary bonded phases. There are basically three types of bonded phases: the "brush" phase, the "bulk" phase, and the oligomeric phase.

The brush phase is created by using monofunctional silanes such as dimethyloctylchlorosilane that react directly with surface silanol groups with the elimination of hydrogen chloride. As a result the surface is covered with dimethyloctyl chains like bristles on a brush, hence the term *brush phase* evolved. The brush phase, synthesized under carefully controlled conditions, is the most reproducible, and, consequently, it is the most commonly used phase in LC analysis.

When bifunctional silanes such as methyl-octyldichlorosilan were used in the synthesis it was possible to produce an oligomeric phase. Dichlorosilane reacted with a silanol group producing a methyloctylmonochlorosyl group on the surface with the evolution of hydrogen chloride. When this monochlorosilane reacted with water more hydrogen chloride was generated and the bonded moiety became a methyloctyl-monohydroxysilyl group. The bonded silica containing the hydroxyl groups could then be again treated sequentially with the dichlorosilane and then water, each time building another methyloctylsilyl group onto the surface. In this way a layer of hydrocarbon chains can be laid down on the silica surface as an oligomer, producing a very stable type of reversed phase. Employing a fluidized bed technique of synthesis[300] and their properties was reported.[301]

Trifunctional silanes such as octyltrichlorosilane could produce the third type of bonded phase, the bulk phase. When the surface of silica gel was saturated with water and treated with octyltrichlorosilane, the reaction occurred between both hydroxyl groups and adsorbed water. The water caused the formation of octylsilyl polymer, which is cross-linked, and consequently the stationary phase assumed a multilayer character. The synthesis could also be accomplished by applying a procedure similar to that used in the preparation of the oligomeric phases, that is, by employing a sequence of two-stage reactions involving first treatment with water and then with trichlorosilane. In this way the polymeric layer could be increased to whatever thickness was desired. After the last stage the product was treated with water and finally end capped. The multilayer character of this type of bonded phase evoked the term *bulk* or *polymeric bonded phase*.

Akopo et al.[301] carried out some retention measurements on samples of bulk and brush phases using methanol–water mixtures as mobile phases. It has been established that at very low methanol concentrations the bulk phase performed in the expected manner and the retention of a solute decreased with the increase of methanol concentration. However, solute retention on the brush phase first increased with methanol concentration until a maximum was reached. At higher concentrations solute retention started to decrease and subsequently gradually became reduced as methanol concentration continued to increase. Examples of the curves relating solute retention to solvent concentration for the two phases are shown in Figure 15. It can be seen that the retention volume of ethanol decreases continuously from 0 to 10% (w/w) of methanol on the bulk phase ODS-3. In contrast to this, in the brush phase the retention volume of ethanol reaches a maximum at about 2.5% (w/w) methanol in the mobile phase and subsequently falls; the graph of retention volume against methanol concentration finally becomes parallel to the curve of the bulk phase.[302]

Other chromatographers described the same phenomenon[303] and suggested that it was due to dispersive forces between the hydrocarbon chains themselves, which are greater than the dispersive forces between hydrocarbon chains and the mobile phase. As methanol concentration increased the dispersive interactions of the hydrocarbon with the mobile phase eventually became sufficiently large to allow chains to disengage from one other and, consequently, to increase the effective surface area of the stationary phase.

The behavior of the oligomeric phase when in contact with water has apparently not been studied so far. However, as the polymer is linear and not cross-linked, it is likely to have properties similar to those of the brush phase. However, the oligomeric phase differs in one important aspect from the brush and bulk phases: in its stability to mobile phases with very low pH values.

1. *Solvent–Stationary Phase Interactions In Reversed-Phase Liquid Chromatography*

The mechanism of interaction between solvent molecules and a reversed-phase surface is similar to the complementary interactions of solvent molecules with a silica gel surface. A layer of solvent is built up on the surface by absorption. However, the interactive forces between the solvent and reversed phase are dispersive in nature, as opposed to those with silica gel, which are mainly polar.

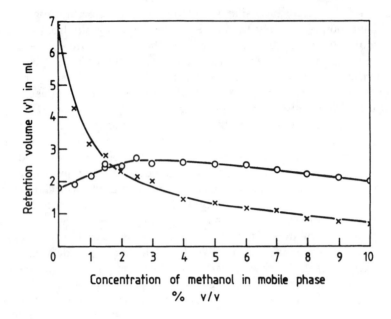

FIGURE 15
Graph of retention volume of ethanol (V') against concentration of methanol in the mobile phase. (O) RP-18; (x) ODS-3. (From Scott, R. P., *J. Chromatogr.*, 656, 1993, 51–68. With permission.)

The adsorption isotherm can be described by the Langmuir equation. However, since the interactions with reversed phase are almost exclusively dispersive and are not a mixture of dispersive and polar interactions as in the case of silica, a detailed study of the adsorption isotherm can provide a more exact understanding of the surface than can be attained by silica gel.

a. Molecular Interactions and Retention in Liquid Chromatography

The first major contribution to the molecular interaction theory of solutes was discovered in gas chromatography where retention is solely controlled by interactions in the stationary phase. Laub et al.[304,305] studied the effect of mixed phases on solute retention and arrived at the surprising conclusion that the corrected retention volume of a solute was linearly related to the volume fraction of either one of the two phases.

$$V'_{ab} = \alpha V'_a + (1 - \alpha) V'_b \qquad (395)$$

where V'_{ab} is the corrected retention volume of a solute on the mixture of phases; V'_a is the corrected retention volume of the solute in phase a; V'_b is the corrected retention volume of the solute in phase b; and α is the volume fraction of phase a. Rearranging this equation, we have:

$$V'_{ab} = \alpha (V'_a - V'_b) + V'_b \qquad (396)$$

Laub and Purnell[304] confirmed the above relationship in a number of interesting ways. They showed that the following alternative chromatography systems all provided the same value for the corrected retention volume of a substance. A given α fraction of phase a could be mixed with a fraction $(1 - \alpha)$ of phase b, and coated on a support and packed in a column. Also, the two fractions could individually be coated on some support and the coated supports mixed and packed in a column. Finally, each fraction could be coated on a support and packed into separate columns and the columns joined in series. It has been demonstrated that all three columns gave exactly the same corrected retention volume for a given solute. The effect of a stationary phase volume was determined. More importantly, it shows that for the distribution systems examined, the distribution coefficients but not their logarithms can be summed. It is now of interest to see if the same relationship can be obtained in liquid chromatography.

The same experiments cannot be directly carried out in liquid chromatography, for, as has already been shown, any modification in volume fraction of one component of the mobile phase will simultaneously change both the nature of the stationary phase surface as well as the interactions in the mobile phase. Katz et al.[306] avoided the problem by employing a liquid–liquid distribution system using water and a series of immiscible solvent mixtures and by measuring absolute distribution coefficients as opposed to retention volumes. The distribution coefficient was determined of n-pentanol between water and mixtures of n-heptane and chloroheptane, n-heptane and toluene, and n-heptane and heptyl acetate. The two-phase system was thermostated at 25°C and after equilibrium had been established the concentration of the solute in two phases was determined by GC analysis. The results are shown in Figure 16. It can be seen that the same linear relationships between solvent composition and the distribution coefficient was obtained for all three solvent mixtures simulating the results that Purnell and Laub [304] obtained in their GC experiments. The results obtained can be described by exactly the same equation and thus, with knowledge of the distribution coefficient of pentanol between water and any of the pure solvents, the distribution coefficient can be calculated for any binary mixture of those solvents and water. Again, the results demonstrate that in the distribution systems examined, the actual distribution coefficients but not their logarithms can be summed.

As can be seen, the concentration of any solvent in the mixture controls the probability of interaction between the solute and solvent. Katz et al.[306] further measured the distribution coefficient of n-pentanol between mixtures of n-heptane and chloroheptane and pure water. The results obtained are shown in Figure 17. The linear relationship between the distribution coefficient and the volume fraction of the respective solvent is again clearly demonstrated. As can be seen, the distribution coefficient of n-pentanol between water and carbon tetrachloride is about 2.2 and an equivalent value for the distribution coefficient of n-pentanol was obtained between water and the mixture of chloroheptane–n-heptane (82:18 v/v). Katz et al. repeated the experiment with toluene, using this time a chloroheptane–n-heptane (82:18 v/v) mixture instead of carbon tetrachloride. The chloroheptane–heptane/n-heptane mixture showed in an identical behavior to carbon tetrachloride and all points were on the same straight line as that produced with a mixture of carbon tetrachloride and toluene. These experiments were similar to those with normal-phase chromatography using pure water instead of silica gel, except for the water phase, which was not modified by the solvents in the way a silica surface would have been.

b. Stationary Phase Effects in Reversed-Phase Liquid Chromatography

A retention mechanism universally operative in reversed-phase liquid chromatography is hydrophobic interaction between a solute and a stationary phase in the presence of an aqueous mobile phase. The simplest model for the retention mechanism involves the intermolecular association between the hydrophobic moieties, one in the solute and the other in the stationary phase. A similar interpretation has been provided by taking into account the solvation of these components.[307,308]

It is possible to explain most observations in reversed-phase liquid chromatography based on these mechanisms, especially when the results were obtained for compounds with a similar skeleton under a limited range of conditions. Good correlation has been

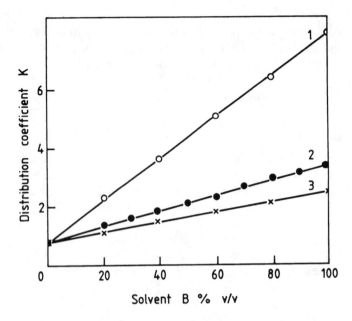

FIGURE 16
Graphs showing the distribution coefficient of *n*-pentanol between water and a binary solvent mixture plotted against solvent composition. Solvent A, *n*-heptane. Solvent B: 1, heptyl acetate; 2, toluene; 3, heptyl. (From Scott, R. P., *J. Chromatogr.*, 656, 1993, 51–68. With permission.)

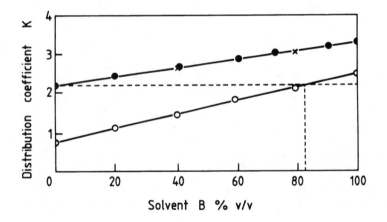

FIGURE 17
Graphs showing the distribution coefficient of *n*-pentanol between water and a binary solvent mixture plotted against solvent composition. (○) Solvent A, carbon tetrachloride; solvent B, toluene, (●) solvent A, *n*-heptane, solvent B, heptyl chloride; (x) solvent A, heptyl chloride–*n*-heptane (82:18, v/v); solvent B, toluene. (From Scott, R. P., *J. Chromatogr.*, 656, 1993, 51–68. With permission.))

observed between log k' values in RPLC under such conditions and log P values in 1-octanol–water two-phase systems.[309]

As is well known, a methylene group in a molecular structure contributes to the relative increase in k' values as in the distribution coefficients of liquid–liquid partitioning. The free-energy changes associated with transfer of one methylene group from water to the organic phase is 3421 to 3682 J/mol in liquid–liquid partitioning[310] and about 3389 J/mol in RPLC with a C_{18} phase and water.[311] These results imply the partitioning mechanism in a primitive sense to be operative between an aqueous mobile phase and stationary phase alkyl groups, the latter not contributing to the selectivity but merely controlling k' values via phase ratios. However, there are numerous results that clearly indicate the contribution of additional effects of the stationary phase to solute retention and selectivity. Here secondary retention processes caused by the participation of silanols and metal impurities on the silica support are omitted. The effect of stationary phase (alkyl groups and solvent molecules existing in the stationary phase under elution conditions) on retention is not very substantial with the alkyl type bonded phases, but is of great significance for at least two reasons. First, only a minor change in selectivity is needed to effect separation owing to the high efficiency of reversed-phase HPLC. Second, hydrophobic interaction alone is inadequate as a means of varying the selectivity — changes in the composition of the alkyl-bonded phase and in the organic solvent can provide additional selectivity effects. Although several mechanisms have been proposed to account for the retention process in reversed-phase liquid chromatography, most of these were based on or could be applied to the retention behavior of a limited range of solute and mobile phase compositions. It is desirable to have a uniform understanding of the retention process to explain each retention behavior, including that caused by the stationary phase effects. Important information can be gained from the stationary phase effects in terms of the retention mechanism, which points to an active role of the stationary phase. The participation of not only bonded moieties, but also organic solvents residing in the stationary phase is important to effect separation in practice. In the case of an electron donor or acceptor bonded silica phase, the stationary phase effect may predominate in the retention and separation.

2. Alkyl-Bonded Silica Phase

a. *Effect of Alkyl Chain Length of Bonded Phase on Selectivity*

A simple model of reversed-phase retention as a means of predicting the effect of alkyl chain length on retention is presented. We assume that the primary factors in determining retention are

1. unfavorable interaction between the hydrophobic portion of a solute and aqueous solvent, and
2. association of a solute with the individual alkyl chains so as to reduce the hydrophobic surface area, referred to as hydrophobic or solvophobic interactions.[312]

Here absolute retention is dependent on the phase ratio determined by alkyl chain length and the surface density, but the selectivity will not be affected by solute structures in terms of shape and polar properties.

Early studies dealt with the linear relationships between k' values and surface coverage and the alkyl chain length of the stationary phase or phase ratios, assuming that the retention is determined by the area of contact between the hydrophobic moiety of the solute and the stationary phase. This argument, however, does not take into account how the alkyl groups are arranged and solvated in the stationary phase. The k' values are not necessarily proportional to the phase ratios determined by the length of alkyl chains and surface coverages. This means that the selectivities based on steric and polar characteristics of solutes were clearly affected by these factors beyond the predictions based on the ratios.

Table 85 shows the k' values obtained for several polar and nonpolar compounds on C_{18} and C_8 stationary phases. The more hydrophobic solutes were eluted with 80% methanol, whereas the more hydrophilic solutes were eluted with 20% methanol. Any difference in selectivity between the two stationary phases in the same mobile phase must be attributed to the difference in the alkyl chain length of the stationary phase. C_{18} stationary phase showed up to two to three times longer retentions than the C_8 stationary phase in 80%

TABLE 85

Retention of Polar and Nonpolar Compounds on Silica C_{18} and C_8 Stationary Phases

Mobile Phase	Solute	k' C_{18}	C_8
80% methanol	Pyrene	9.0	2.9
	n-C_6H_{14}	6.7	3.3
	n-$C_{12}H_{25}OH$	9.4	5.3
20% methanol	n-C_3H^7OH	1.2	1.1
	n-C_4H_9OH	4.2	3.7
	trans-1,4-Cyclohexanedimethanol	4.2	3.3
25% CH_3CN in 0.2 M H_3PO_4	Somatostatin	3.6	3.3

From Scott, R. P., *J. Chromatogr.*, 656, 1993, 51–68. With permission.

methanol. However, in 20% methanol eluent the C_8 stationary phase, although having a much lower carbon content (about 11%) than C_{18} (about 20%), showed comparable retention times for the polar compounds. The retention times with 20% methanol are given together with k' values in Table 85 to emphasize the preferential retention of more hydrophilic solutes on C_8. These results indicate the preferential retention of polar, hydropohilic solutes by the C_8 phase compared with the C_{18} phase.

In contrast to the data shown in Table 85, the C_{18} stationary phase preferentially retains rigid, planar polynuclear aromatic hydrocarbons (PAHs), n-alkanes, or nonplanar aromatic compounds with rotational freedom of phenyl groups (such as o-terphenyl and triphenylmethane) relative to the C_{18} stationary phase. These simple examples show that the mechanism assuming the retention to be determined by the sole contribution of the mobile phase effect (solvophobic effect), or by the quantity of hydrophobic groups in the stationary phase ratios, cannot explain each result.

b. *Steric Selectivity of Alkyl-Bonded Stationary Phase*

A C_{18} stationary phase is known to be relatively tangled especially for mobile phases with low organic solvent contents. As organic solvent content increases, the alkyl chains sorb organic solvents, swell, and become more ordered in their conformation. As a result, PAHs can more readily penetrate the stationary phase and their retention relative to the more bulky or flexible solutes increases.[313]

The steric discrimination of planar from nonplanar compounds was also given with the polymeric C_{18} phase in comparison to the monomeric C_{18} phase.[314-317] It has also been demonstrated that the steric selectivity and the difference based on the chain length are affected by the mobile phase organic solvent.[318,319] The authors established that the three stationary phases (C_1, C_8, and C_{18}) gave different selectivities between aromatic and saturated compounds in methanol and acetonitrile. The results preclude the possibility that the difference in selectivity was caused by the difference between aromatic and saturated compounds in the mobile phase interaction, as no such difference was observed with the C_1 phase. The difference could not be due to the steric effect either, as planarity of the solutes did not influence the results. It can only be explained by the difference in the solvation of solutes in stationary phase between C_1 and the longer alkyl-bonded phases. It should be emphasized that the effect of solute–solvent interaction on the C_{18} phase was found to be much greater than on the C_1 phase, and even larger with higher organic solvent contents. These results imply that the hydrocarbon solute molecules in the stationary phase can realize the orderliness of bonded chains determined by the chain length, surface density, and mobile phase composition, and simultaneously associate with the solvent molecules in the C_{18} phase. The effects are much lower with C_8 and negligible in the C_1 phase. The results suggest a mechanism based on partitioning of solutes between the mobile phase and the effective stationary phase, namely alkyl chains associated with solvent molecules.

However, the effective stationary phase should not be taken as a simple mixed solvent, as C_1 or C_8 and C_{18} phases showed considerable differences in their response to the change in the mobile phase. Insensitivity of the C_8 phase toward the change in mobile phase implies that the chain overlap is much less than with the C_{18} phase in this range of mobile phase, as would be the case with the C_1 phase.

The effect of organic solvent in stationary phase on polar group selectivity has been studied.[319] The results indicated a notable difference in polar group selectivity between tetrahydrofurane (THF)–water and methanol–water systems, which is higher than that between acetonitrile–water and methanol–water. Thus THF and methanol would constitute an interesting set-up pair to be used with water in a ternary mobile phase for controlling the separation of substances with different functional groups.

The k' values for these benzene derivatives in 50% methanol showed good correlation with log P values, but not with the k' values in THF–water. The retention in THF vs. methanol decreases in the order: phenols = nitro compounds > hydrocarbons, chlorobenzenes > esters = alcohols. Alcohols, especially alkanols, were preferentially retained in methanol–water compared with THF or acetonitrile systems. The difference between phenols and alcohols can be explained by the difference in their ability to stabilize the partial negative charge upon hydrogen bonding as indicated by their pK_a values.

Figure 18 shows the separation of four substances with different functional groups with methanol and THF mobile phases.[320] It can be seen that the elution order is exactly reversed with the two systems. This example clearly illustrates the significant role polar group selectivity can play with the change of the organic modifier. It has also been shown that the change in selectivity can be realized with the addition of small amounts of interacting solvents to a mobile phase. The result shows that the compounds with acidic functions, phenolic OH or π-acidic benzene ring, are preferentially retained in mobile phases containing THF which can serve as a base, and those with basic dipolar carbonyl groups are disfavored in mobile phases containing basic dipolar THF or acetonitrile. The results indicate the contribution of solute–solvent interactions in the stationary phase to the retention and selectivity. Such a contribution would be required to stabilize polar functionality in the nonpolar stationary phase. As is well known, stationary phases with medium alkyl chain C_4 or a half-coverage C_{18}-phase show maximum retention for peptides,[321] especially for those with high molecular masses. This is understandable, since the presence of hydrophilic or ionic species in the hydrophobic environment is energetically unfavorable.

The stationary phase effects are compatible with a mechanism based on the partitioning of solutes between the mobile phase and the effective stationary phase, anchored alkyl chains associated with solvent molecules. Simple solvated alkyl chains or simple mixed solvents, however, do not give an adequate description of the alkyl-bonded silica stationary phase. The same mobile phase composition results in a difference in the selectivity for different stationary phases. The difference in alkyl chain length or in surface coverage can be envisaged to produce the difference in steric requirement and hydrophilic–hydrophobic properties, which in turn determine the chromatographic properties. In this sense, the properties of the C_{18} phase are primarily determined by the extent of surface coverage with alkyl groups.

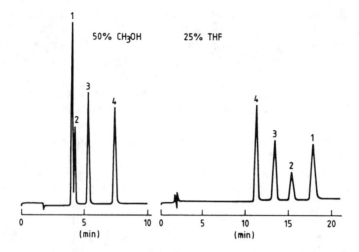

FIGURE 18
Chromatograms illustrating the difference in functional group selectivity caused by organic solvents with C_{18} phase. Peaks: 1, p-nitrophenol; 2, p-dinitrobenzene; 3, nitrobenzene; 4, methyl benzoate. (From Tanaka, N. et al., *J. Chromatogr.*, 656, 1993, 265–287. With permission.)

3. Polymer-Based Packing Materials in Reversed-Phase Liquid Chromatography

a. Selectivity of Polymer-Based Packings

Several high-efficiency polymer gel packings are available, and according to reports their selectivities are difficult to understand, or are considerably different from those of silica-based phases. The pore structures of polymer gels are also different from those of silica.[322,323]

Molecular mass-elution volume curves obtained for polymer gels are always associated with a second plateau in a molecular mass range below 500, corresponding to the micropores, whereas no such compounds are present in ordinary silica particles. Thus in the biporous structure of cross-linked polymer gels, macroporous particles are composed of microporous materials. Lightly cross-linked polymer gel chains on the surface of solid cores of polymer gels might be responsible for the microporosity. The micropores play a major role in determining the selectivity of the size-exclusion effect, which represents the characteristics of selectivity of all polymer gels.[324]

b. Steric Selectivity of Polymer Gels

The characteristics of steric selectivity of polymer gels are in fact the result of the size-exclusion effect of the micropores, and that of the micropore structure, and hence, the selectivity between bulky, flexible and rigid, compact solutes can be controlled by the choice of the diluents in suspension polymerization or in multistep swelling polymerization.

The difference in selectivity of two polymer gel was investigated by Hosoya et al.[325] The authors described the polymer gel packings prepared from methyl methahacrylate and ethylene dimethacrylate in two diluents, cyclohexanol and 2-octanol. Despite the similarity in meso- and macropore sizes, the two gels showed different selectivities for hydrocarbons with different planarity and size. This is due to the smaller micropores of the polymer gel prepared in 2-octanol. Similar results were obtained with aliphatic compounds. Figure 19 shows typical plots indicating the difference in selectivity terms of the shape of solute molecules between alkyl-bonded silica and polymer gel packings. Shodex DE-613 poly(alkyl methacrylate) gel, having short alkyl groups, showed a clear preference for aromatic compounds having more than one phenyl group compared with alkylben-

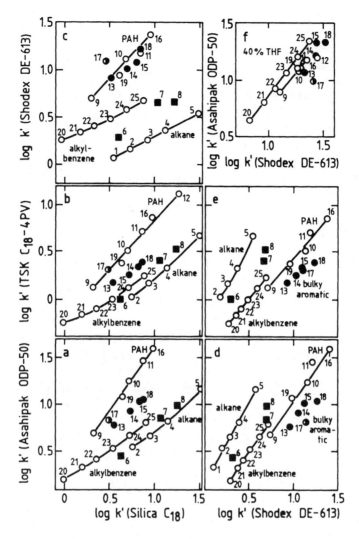

FIGURE 19
Comparison of retention selectivity between silica C_{18} and polymer gel columns. Mobile phase: 80% methanol unless indicated otherwise. (●) Aromatic compounds with phenyl groups having rotational freedom, or bulky compounds. (■) cycloalkanes: 1, pentane; 2, hexane; 3, heptane; 4, octane; 5, decane; 6, cyclohexane; 7, adamantane; 8, *trans*-decalin; 9, naphthalene; 10, anthracene; 11, pyrene; 12, benz[α]pyrene; 13, diphenylmethane; 14, 1,2-diphenylethane; 15, o-terphenyl; 16, triphenyl; 17, tritycene; 18, triphenylmethane; 19, fluorine; 20, benzene; 21, toluene; 22, ethylbenzene; 23, propylbenzene; 24, butylbenzene; 25, amylbenzene. (From Tanaka, N. et al., *J. Chromatogr.*, 475, 1989, 195–208. With permission.)

zenes, and for cycloalkanes compared with linear alkanes. Esterified Asahipak ODP-50 poly(vinyl alcohol) gel (PVA) showed a clear preference for planar PAHs compared with bulky aromatic compounds.[326] The general preference by polymer gels decreases in the order PAHs > polyphenylalkanes (Ar-Ar) > alkylbenzenes > cycloalkanes > linear alkanes. Such a preference was observed for all polymer-based stationary phases, regardless of the alkyl chain length or the size of the macropores. Shodex DE-613 with short alkyl-bonded polymer gel showed higher selectivity for different types of hydrocarbons than Asahipak ODP-50 and TSK C_{18}-4PW. C_{18}-bonded types for the effect of polymer network structure is predominant with Shodex DE-613 owing to the smaller extent of hydrophobic interaction. Polymer gels with alkyl backbones, DE-613 and TSK C_{18}-4PW, showed preferential retention of aromatic compounds in spite of their saturated structures, possibly owing to dipole–π interactions involving ester groups. The preference toward rigid, compact compounds over bulky, flexible compounds was also seen with saturated compounds having no functional groups such as cyclohexane and adamantane, which are expected to undergo minimum specific interactions with the stationary phase except hydrophobic and steric interactions. Therefore the shape selectivity associated with polymer gels can be attributed, at least in part, to the structural matching between solute and the rigid polymer network structure, or the micropores. The C_{18} type polymer packings, TSK C_{18}-4PW, alkylated poly(hydroxylalkyl acrylate or methacrylate), and Asahipak ODP-50 esterified poly(vinyl alcohol), showed greater retention of alkyl compounds than did DE-613, having short alkyl groups. The preference for compact solutes compared with bulky solutes increases in the order silica C_{18} < DE-613 < TSK C_{18}-4PW < ODP-50 < PLRP -S 300. The results indicate a greater population of the relatively large micropores of the DE-613 than ODP-50 or PLRP-S 300, probably due to the type of diluent used in preparation.

c. *Effect of Mobile Phase on Steric Selectivity*

The difference in the effect of solute structure on column efficiency can be explained on the basis of the difference in the structures of the polymer gels. DE-613 and TSK C_{18}-4PW are alkyl type gels containing no aromatic functionality, whereas PLRP-S 300 and ODP-50 contain aromatic groups in monomer and/or cross-linking reagents. These functionalities can provide interactions with aromatic compounds, especially when the solutes are planar. The fact that the retention of bulky compounds on ODP-50 and PLRP-S 300 is relatively weaker than those of PAHs compared with DE-613 and TSK C_{18}-4PW in similar mobile phases indicates that the former possess tighter network structures than the latter. With better solvation in THF and more so in acetonitrile with a closer solubility parameter to the PVA backbone, only PAH showed slightly higher h (reduced plate height) values, whereas excellent efficiencies were observed for the other solutes with ODP.

d. *The Effect of Biporous Structure on Chromatographic Properties*

Polymer-based packings showed excellent performance for polypeptides including bovine serum albumin (BSA) owing to the favorable macropore structure. Figure 20 shows the chromatograms obtained with four polymer gel packings for polypeptides (molecular mass 12,000 to 66,000) and a small molecule, 1-naphthalenemethanol.[327]

Although the retention times for polypeptides are comparable on all packings, those for 1-napthalenemethanol are considerably different. Whereas the retention and separation of large molecules indicate an abundance of macropores, the retention of a small molecule indicates an abundance of micropores in each packing. As can be seen in Figure 20, large molecules such as polypeptides are chromatographed by using macropores, and the small molecules are chromatographed by using micropores of polymer gels. Such difference cannot be seen with silica-based phases.

e. *The Effect of Polymer Structure on Polar Group Selectivity*

In addition to size-exclusion effect of macropores, some polymer gels possess another factor, the presence of dipolar carboxylate groups in methacrylate and esterified poly(vinyl alcohols) gels, to show unique selectivity. These polymer gels in methanol show a somewhat similar selectivity to cyanoalkyl-bonded phases on silica or to alkyl-bonded phases in dipolar solvents. Preferential retention of aromatic compounds, especially π-acids, is notable. A stronger mobile phase for aromatic compounds with about 10% more methanol, and a weaker mobile phase for aliphatic compounds with about 10% less methanol, are recommended with polymer gels in comparison with an ordinary silica C_{18} phase. The results obtained with polymer gels suggest the applicability of these packings in RPLC in wider mobile phase conditions than for silica C_{18}, although the full scope of application

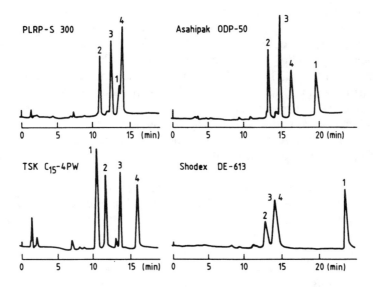

FIGURE 20
Elution of polypeptides with polymer gel columns. Peaks: 1, naphthalenemethanol; 2, cytochrome c (molecular mass 12,000); 3, lysozyme (molecular mass 14,000); 4, bovine serum albumin (molecular mass 66 000). Linear gradient of acetonitrile from 20% to 60% in 20 min in the presence of 0.1% trifluoroacetic acid. (From Tanaka, N. et al., *J. Chromatogr.*, 535, 1990, 13–31. With permission.)

is yet to be explored. A basic understanding of polymer gel structures and resultant selectivity will help with the selection of separation conditions.[328]

Silica and polymer-based packings are frequently compared and should complement each other in the separation of small molecules, because silica-based packings are not as chemically stable as polymer gels, and polymer gels are generally less efficient. One area of RPLC where silica- and polymer-based packings can be in serious competition is the separation of polypeptides, where micropores play no role, compared to packings on wide-pore silicas and those based on polymer gels. Silica C_{18} phases prepared from wide-pore silicas showed poorer performance than short alkyl-bonded phases owing to smaller pore sizes, which are much less stable than C_{18} in trifluoroacetic acid (TFA) solution. Polymer gels provided equal or better recoveries and performance and much greater stability in TFA solution compared to C_{18} phases. The easy control of pore size in both macro- and micropore size ranges is an additional advantage of polymer-based packings.[329]

4. Correlation of Retention and Selectivity of Separation in Reversed-Phase High-Performance Liquid Chromatography with Interaction Indices and with Lipophilic and Polar Structural Indices

Various models have been proposed to describe the retention in reversed-phase systems, for example, a competitive adsorption model with a modified scale of solvent strength,[330,331] a model based on probability of interactions of the solute in stationary and the mobile phases,[332,333] modified models of the distribution of a solute between two liquid phases such as the model using the Hildebrand solubility parameter theory,[334,335] or the model supported by the concept of molecular connectivity,[336] and rigorous models based on the solvophobic theory,[312,337] or on the molecular statistical theory. Unfortunately, sophisticated models introduce a number of physicochemical constants that are often not known or difficult and time-consuming to determine, so that such models are not very suitable for rapid prediction of the retention data. The most characteristic feature of reversed-phase chromatography is lower polarity of the stationary phase in comparison with the mobile phase. Theoretically, alkyl phases chemically bonded on a suitable support such as octadecylsilica or octylsilica materials commonly used in contemporary practice of reversed-phase HPLC should behave as almost ideally nonpolar materials with properties of long-chain aliphatic hydrocarbons. The bonded alkyl chains differ from the free molecules of hydrocarbons in liquid phase in their limited mobility. Futhermore, for steric reasons it is virtually impossible to modify all available silanol groups on the surface of silica gel by chemical reaction with the silanization reagent and the unreacted silanol groups may affect the retention by specific interactions, especially with basic solutes. Finally, organic solvents used as the components of the mobile phases in reversed-phase systems are preferentially adsorbed by the stationary phase and can significantly modify its properties.[338,339]

The solvophobic theory emphasizes the importance of mobile phase interactions in the control of the retention mechanism. The solvophobic interactions are considered as the driving force of the formation of associates of the solutes with the nonpolar stationary phase. The replacement of weaker interactions between the moderately polar solute and polar mobile phase by mutual interactions between the strongly polar molecules of the mobile phase in the space element of the mobile phase occupied by a solute molecule before transition results in an overall energy decrease in the system, which is the driving force of retention in the absence of strong interactions between the solute and the stationary phase. The theoretical background of this retention model is useful as the starting point in the derivation of a simplified semiempirical description of reversed-phase systems, making it possible to characterize and predict the retention and selectivity within a series of compounds to be separated.[340] Due to simplifications concerning the effects of the stationary phase accepted in the derivation, this approach can be applied to relative rather than absolute predictions of retention and selectivity and a suitable set of standard reference compounds is necessary for the calibration of the retention (or selectivity) scale.

a. Structural Correlations of Lipophilic and Polar Indices

In accordance with expected potentials for characterizing the lipophilicities of solutes by the indices n_{ce} or Δn_c, good agreement was found of the relative indices Δn_c with Hansch and Leo hydrophobic constants π (correlation coefficient was 0.991),[341] but no correlation could be detected between n_{ce} or Δn_c and Snyder's polarity indices P'.[342]

It has been shown that the relative polar index Δq is directly proportional to the difference between the interaction indices of the solute and the reference standard compound after subtraction of the lipophilic contributions from the alkyl substituents, $I_{0X,i}$ and $I_{0X,s}$, respectively:

$$\Delta q = (c_M (I_{H_2O} - I_{org} /2.3 \, RT)(c_X V_{0X,i} I_{0X,i} - c_X V_{0X,i} I_{0X,s}) = \text{constant}(I_{0X,i} - I_{0X,s}) \quad (397)$$

where $V_{0X,i} = V_{0X,s}$ is the molar volume of the zeroth member of the calibration homologous series. Therefore, Δq is expected to be directly proportional to the polarity of the functional groups in the solute. Good correlations of the relative indices q with Snyder's polarity indices P' was found experimentally (correlation coefficient = 0.988),[341] which seems to confirm that the indices q_i or Δq characterize the polarity of a solute. The relative polar indices Δq were found to increase with the polarities of the functional groups approximately in the order n-alkanes < polycyclic aromatic hydrocarbons < n-alkylbenzenes < benzene, styrene, biphenyl < halogenated benzenes < dialkyl ethers < alkyl aryl ethers, diaryl ethers < aromatic nitriles < aromatic ketones and aldehydes < aromatic amines < aromatic alcohols < phenols, alkylphenols < chlorophenols. For a given class of compounds, the Δq values were found within a relatively narrow range. The dependence of the relative polar indices q in acetonitrile–water on Δq in methanol–water mobile phases

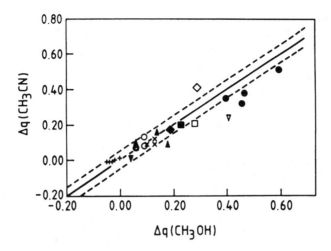

FIGURE 21
Correlation between q_i in acetonitrile–water, $q(CH_3CN)$, and methanol–water $q(CH_3OH)$, mobile phases on a Silasorb C_8 column from compounds from different classes. (+) alkylbenzenes; (▼) styrene; (△) halobenzenes; (○) alkyl aryl ethers; (×) esters of carboxylic acids; (▲) alkyl aryl ketones; (□) benzaldehyde; (■) benzonitrile; (◇) aniline; (▽) benzyl alcohol; (◆) nitrobenzene; (●) phenols. (From Jandera, P., *J. Chromatogr.*, 656, 1993, 437–467. With permission.)

is shown in Figure 21. The most apparent feature of this plot is a regular increase in Δq in the two sets of indices with increasing polarity of the sample solutes. The compounds with a common functional group are found in more or less limited regions in this graph. Some of these regions overlap and the areas in which the data for one class of compounds are spread increase with the increasing polarity of the compounds. In the absence of specific polar interactions, Δq in a mobile phase containing various organic solvents should be the same and should appear on the straight line with slope = 1. As can be seen in Figure 21, the experimental data for most compounds are spaced within the interval ±0.5 Δq units on both sides from this line (dashed limits). Because of selective polar interactions, the differences in q values are found for mobile phases containing different organic solvents, in that the indices are shifted in the direction of stronger selective interactions, either to the methanol region below the line or to the acetonitrile region above the line. In Figure 21, this behavior is observed for the compounds that form hydrogen bonds with –OH groups in water and alcohols, such as phenols, ketones, and aldehydes.[343] The points for these compounds are shifted downward toward the interval limited by the dashed lines to the methanol region. The data point for aniline is shifted in the opposite direction toward the acetonitrile area. The position of the data in the Δq–Δq diagrams can be used to estimate the polarities and the presence of specific functional groups in simple organic compounds from their reversed-phase retention data. Good correlation was found also between the polar indices Δq and the number of substituents in methylbenzenes and in chlorobenzenes, where the contribution of the ortho effect is almost negligible. The correlation was not as good for chloroanilines, where the ortho effect seems to influence the values of Δq more significantly. The correlation was very poor for Δq values of chlorophenols, probably because of interaction between the chloro substituents and the phenolic group. The polar Δq indices are only slightly or not at all influenced by alkyl, halo, or methoxy substituents, and the q_i indices of all the phenylurea herbicides studied are close to one another, with the exception of hydroxymerthoxyuron with a phenolic hydroxyl group. The contributions of the alkyl substituents to the Δn_c indices of substituted triazines are lower than the values for the substituents on the benzene ring and the Δq indices of the triazine herbicides are close to one another. The differences between the experimental Δn_c and the values calculated from the additive contributions is 0.13 or less and the difference between the experimental and calculated Δq indices is less than 0.06. The experimental data support the idea of the additives of the contributions to the Δn_c and Δq indices in a limited number of the classes of compounds studied so far, and more experimental data would be necessary to verify if the additivity rules for these indices are valid in general.

The interaction indices can be used for the prediction of retention and selectivity in reversed-phase systems with binary or ternary mobile phases, but the precision of prediction over a wide range of mobile phase composition is improved by introducing two structural indices for each solute, which is characteristic of the lipophilicity of the hydrocarbon moiety of the molecule and the polarity of functional groups(s). The structural contributions to these indices have proved to be additive in the classes of compounds studied so far. Combination of the two indices of the solutes controls the selectivity of their separation in binary and ternary mobile phases. Based on the lipophilic and polar indices, conditions for optimum isocratic or gradient separation can be predicted.

The prediction of retention on one column from the indices determined on another column is possible in principle, but the columns must possess the same length as the bonded alkyl chains. Limitations to this approach may originate

- in a limited precision of experimental data such as those caused by nonlinearities of the log k' vs. concentration of the organic solvent in aqueous-organic mobile phase, or
- in possible specific interactions with the stationary phase such as interactions between unreacted silanol groups in alkyl-bonded silicas and polar solutes, mainly basic compounds.[343]

5. Quantitative Structure–Retention Relationships in Reversed-Phase High-Performance Liquid Chromatography

Quantitative structure–retention relationships (QSRRs) are some of the most extensively studied manifestations of linear free-energy relationships.[344] Linear free-energy relationships are extrathermodynamic relationships, and are not necessarily due to thermodynamics. Extrathermodynamic approaches combine detailed models of processes with certain concepts of thermodynamics. The thermodynamic properties of a given system are bulk properties reflecting just the net interactive effects in that system. The magnitude of thermodynamic parameters represents the combination of individual interactions that may take place at the molecular level. Thus, from chemical aspects classical thermodynamics is inadequate.[345]

The presence of linear free-energy relationships suggests a real connection between some correlated quantities, and that the nature of this connection can subsequently be identified. It is also assumed that the correlations between specific quantities can be attributed to some physicochemical relationships. Statistically derived correlations encourage attempts to determine also the relationships in the background. Over a period of 15 years QSRR studies have been conducted with the following aims in view:

- prediction of retention for a new solute
- identification of the most informative structural descriptions
- elucidation of the molecular mechanism of separation operating in a given chromatographic system
- evaluation of complex physicochemical properties of solutes (other than chromatographic)
- estimation of relative biological activities within a set of solute xenobiotic compounds

Chromatographic retention data must be some function of the temperature, chemical structure of solute, stationary phase, and mobile phase, with all their mutual interactions. However, there is no general, strict, unequivocally verifiable canonical equation relating retention to the four main chromatographic variables, that is, the temperature, the structure of the solute, the stationary phase, and the mobile phase. Even if the stationary and mobile phases applied in a given chromatographic system remain constant, precise, quantitative description of the retention of a series of solutes may involve difficulties, the more so the more diverse the solutes considered, although the problem is by no means trivial for homologues. The plots of log k' vs. carbon number of a homologue are usually linear, but only for some limited range of aliphatic chain length.

In general, relationships between empirical or theoretically calculated parameters of molecular structure require statistical evaluation in order to check the significance of the resulting correlations. Basically, the research strategy now applied in QSRR studies was transferred from studies initiated in the 1960s on quantitative structure (biological) activity relationships (QSARs).[346] The advantage of QSRRs over other quantitative structure–property relationship studies is that chromatography can readily yield a great amount of relatively precise and reproducible data. In a chromatographic process all conditions may be kept constant. Thus, solute structure becomes the single independent variable in the system.

The first reported QSRRs were derived by multiple regression analysis of retention data against a set of structural descriptions. Another early approach to QSRRs was based on the assumption of additive substituent effects on retention analogously to the new nonparametric method of correlation analysis applied by Free and Wilson in medical chemistry.[347] One can attempt to derive QSRRs making no mention of any existing chromatographic theory. A typical strategy is to generate a multitude of solute descriptions that are next regressed against retention data. Observing all the statistical rules, one selects the minimum number of descriptors needed to produce an equation yielding the calculated retention data in satisfactory agreement with the observed values. The number of descriptors that can be assigned to an individual solute is virtually unlimited. The most commonly used structural descriptors in QSRRs are given in Table 86.

However, also numerous rare, sometimes ad hoc designed solute descriptors are reported. It is often difficult to assign any physical sense to such parameters and to interpret QSRR equations consisting of terms produced by various transformations and combinations of such descriptors, their square roots, cubes, reciprocals, or products. If QSRRs result from the analysis of tens or hundreds of descriptors, then most likely several equations with similar predictive abilities but consisting of different sets of variables can be derived. QSRRs that are not interpretable in physical terms are not very informative regarding the mechanism of retention. A more promising QSRR is to start from the existing theories of chromatographic separations and to attempt to quantify the abilities of solutes to take part in the postulated intermolecular interactions.[348] These fundamental intermolecular interactions involving solute molecules, molecules forming mobile phase, and molecules of stationary phase are as follows:

- dipole–dipole (Keesom)
- dipole–induced dipole (Debye)
- instantaneous dipole–induced dipole (London)
- hydrogen bonding
- electron pair donor–electron pair acceptor
- solvophobic interactions

The potential energy (E) of the first three types is approximated by

TABLE 86
Structural Descriptors in QSRR

Bulkiness-Related (Nonspecific) Parameters

Molecular mass
Refractivity
Molecular volume
Total energy
Solvent-accessible area
Taft constant

Geometry-Related (Shape) Parameters

Moments of inertia
Length-to-breadth ratio
Sterimol width parameters
Angle strain energy

Physicochemical Parameters

Hydrophobic constant (π)
Hammett constant
Solubility parameters
Boiling points
Solvatochromic parameters

Polarity-Related (Electronic) Parameters

Swain–Lupton's constants
Dipole moments
Atomic excess charges
Orbital energies
Superdelocalizabilities
Partially charged surfaces

Molecular Graph-Derived (Topological) Parameters

Adjacency matrix indices
Distance matrix indices
Information content indices

$$E = -W^2 \epsilon^{-1} r^{-6} [2\mu_1^2 \mu_2^2 / 3kT + \alpha_2 \mu_{12} + \alpha_1 \mu_{22} + 3I_1 I_2 \alpha_1 \alpha_2 / 2(I_1 + I_2)] \tag{398}$$

where W and k are constants; ϵ is relative electric permittivity of the medium; r is the distance between interacting molecules; T is the absolute temperature; and μ, α, and I are the dipole moments, polarizabilities, and ionization potentials, respectively, of the interacting molecules. Equation 398 substantiates the assumption that, within a set of solutes of similar hydrogen-bonding and charge-transfer properties, chromatographed under identical conditions, the retention parameters can be approximated by a combination of polarizabilities, ionization potentials, and squares of dipole moments. In pre-QSRR studies, attempts were made to select solutes either with similar dipole moments and varying polarizability[332] or with similar polarizability and varying dipole moments and to relate retention to the variable.

Tijssen and his co-workers[349] considered three types of interactions: dispersion, orientation, and so-called acid-base interactions. The ability of an individual compound to take part in the respective interactions is reflected by its specific partial solubility parameter. The problem encountered when testing the predictive potency of the approach was to determine precisely the solubility parameters. Similarly, Horváth et al.'s[312] Martine and

Boehmn's[350] molecular statistical theory and several other early theoretical approaches to RPHPLC required a knowledge of a number of physicochemical parameters that were not available for individual solutes. The individual solute properties affecting retention were identified, which, in turn, suggested the choice of the most informative structural descriptions for QSRRs. Carr and his co-workers,[351,352] in studies on the nature of RPHPLC separations, proposed an approach based on the solvotochromic comparison method and linear solvation energy relationships (LSERs). The chemical and physical characteristics of the solute that determines the retention can be described by the following equation:

$$\log k' = \text{constant} + M\,(\delta_m^2 - \delta_s^2)V_2/100 + S(\pi_s^* - \pi_m^*)\pi_2^*$$
$$+ A(\beta_s - \beta_m)\alpha_2 + B\,(\alpha_s - \alpha_m)\beta_2 \tag{399}$$

where the subscript 2 designates a solute property such as molar volume (V_2) polarizability-dipolarity (π_2^*), hydrogen bond acidity (α_2), and hydrogen bond basicity (β_2). Each solute property is multiplied by a term that represents the difference in complementary "solvent" properties of the mobile (see subscript m) and the stationary (see subscript s) phases. Thus α_m and α_s are the abilities of the phases (bulk or bonded) to donate a hydrogen bond. These properties complement the solute ability to accept a hydrogen bond (β_2). Similarly, σ_m^2 and σ_s^2, the squares of the Hildebrand solubility parameter or cohesive energies of the two phases, complement the solute molar volume.

Another recent theory that had an impact on QSRR is studies in the mean-field statistical theory of Dill,[353] applied to RPHPLC by Dorsey and his co-workers.[354,355] According to this theory, two main forces dominate the retention:

1. the free-energy change resulting from contact interactions of the solute and neighboring molecules of the stationary and mobile phases
2. ordering of the stationary phase hydrocarbon chains leading (at higher hydrocarbon bonding density) to an entropic exclusion of the solute from the stationary phase relative to that expected in an amorphous hydrocarbon–water partition system

However, there are two important consequences of this theory for retention prediction and other QSRR studies. One is that the retention in RPHPLC increases with the grafted stationary phase chain density up to a density value of about 3.0 µmol/m, where the retention reaches a plateau. Another conclusion derived from the theory concerns the nature of the slope and intercept of the rectilinear relationship between logarithm of capacity factor, log k', and composition of binary organic–water eluent. The slope was postulated[356,357] to be directly proportional to the size of the solutes, although measures of solute size such as van der Waals volume and molecular connectivity indices did not confirm the expectations. It has been argued that the RPHPLC distribution coefficient could be calculated from known values of the activity coefficients of the substance of interest in both chromatographic phases. A means for the assessment of activity coefficients is the UNIFAC group contribution method according to Fredeslund and co-workers.[358] The UNIFAC method transforms a solution of molecules into a solution of groups. The magnitude of a given group contribution to the activity coefficient depends on the van der Waals group volume and surface area. The number of distinct groups is limited, but is not so small as to neglect significant effects of molecular structure on physical properties. The parameters were tabulated by Fredeslund et al.[358]

There are theoretical approaches aimed at the prediction of RPHPLC retention using the substituent and or fragmental contribution to retention parameters. Some of the purely predictive applications of these methods (which form the basis of expert systems) are interesting attempts to identify and quantify mutual interactions between substituents or fragments. Smith and Burr[359] described the RPHPLC retention parameter I of disubstituted (X, Y) aromatic solutes employing the following equation:

$$I = I_P + I_{S,R} + \Sigma I_{S,Al-X} + \Sigma I_{S,Ar-X} + \Sigma I_{I,X-Y} \quad (400)$$

where I_P represents the retention parameter of a parent unsubstituted compound; $I_{S,R}$ is a contribution for saturated alkyl chains; $I_{S,Al-X}$ are contributions for substituents on saturated aliphatic carbons; $I_{S,Ar-X}$ are contributions for aromatic substituents; and $I_{I,X-Y}$ are terms accounting for any interaction between substituents caused by electronic, hydrogen bonding, and steric effects. The interaction terms are calculated by the following equation:

$$I_{I,X-Y} = (\sigma_X p_Y^* + \sigma_Y P_Y^*) + F_{HB}^* + F_0^* \quad (401)$$

where p^*, F_{HB}^*, and F_0^* are expressed in units of the retention parameter; p^* are the susceptibilities of X and Y to the modifying effects of Y and X on the Hammett constants of the substituents σ_X and σ_Y; F_{HB}^* is a term accounting for hydrogen bonding; and F_0^* is a term reflecting the ortho effect. It should be mentioned that the σ/p correction values, along with ortho effects, were demonstrated by Tsantili-Kakoulidou and his co-workers.[360]

6. Hydrophobicity Concept in Chromatography

Boyce and Millborrow[361] extrapolated retention parameters determined at various organic–water eluent compositions to a pure water eluent. It later became common practice to employ extrapolated data as measures of hydrophobicity.[361] The extrapolation based on the assumption of the linear Soczewinski–Wachtmeister relationship[362] between log k' and the volume fraction of the organic modifier in a binary aqueous eluent. It has been demonstrated that the rectilinear relationship in RPHPLC applies only over a limited solvent composition range that varies depending on the solute and the chromatographic system employed.[335,363] In effect, the values of the logarithm of the capacity factor extrapolated to a pure aqueous eluent — which is the intercepts in the Soczewinski–Wachtmeister equation denoted commonly by log k'_w — are usually different from those determined experimentally and depend on the organic modifier employed. Owing to this observation, some workers are inclined to believe that the extrapolation of capacity factors to 0% organic modifier is a manipulation and the values of log k'_w itself has no physical meaning.[364] Interpretation of log k'_w as the logarithm of the capacity factor corresponding to a pure water (buffer) eluent might be misleading, especially if the extrapolation is carried out over a considerable eluent composition range with a fitting function that is basically unreliable.[365] The parameter log k'_w is not devoid of merits, however, as it may be regarded as a means of normalizing retention.[366]

Isocratic capacity factors determined with various organic modifiers naturally depend on the properties of the modifier. One could expect the values extrapolated to pure water (log k'_w), however, to be independent of the organic modifier used. Different modifiers yield different chromatographic measures of solute hydrophobicity. There is no reason to assume that any one modifier provides a better measure of hydrophobicity than others. When the reference hydrophobicity scale is that of the 1-octanol–water partition system (log P), then individual organic modifiers appear to be advantageous. Braumann et al. strongly advocated the view that a general relationship between log P and log k'_w can only be expected for capacity factors determined in methanol–water eluents.[367] According to these authors, similar solute–solvent interactions operate in methanol–water and 1-octanol–water systems, whereas other organic modifiers (acetonitrile or tetrahydrofuran) introduce interactions that are not present in the 1-octanol–water system. Although Braumann et al. postulated the identity of log k'_w and log P with nonpolar solutes, such as chlorobipenyls and alkylbenzenes, separate regression of log P vs. log k'_w has to be developed for each class of compound.[368]

7. Correlations between RPHPLC Retention Parameters and 1-Octanol–Water Partition Coefficients

One can certainly expect close correlation between log P and RPHPLC retention parameters if the chromatographic system closely resembles the conventional 1-octanol–water partition system. Several authors have reported the procedures based on dynamically coating a stationary phase with 1-octanol and using a 1-octanol-saturated aqueous eluent.[369] The correlations were satisfactory but serious technical problems made the approach impractical. To achieve high correlations between retention parameters determined in a stable RPHPLC system and log P, various researchers have recommended specific treatment of the stationary phase before use and the presence of various additives in the mobile phase.[370]

The correlations of such chromatographic data with log P were good only as long as the solutes analyzed were more or less closely related (congeneric). In contrast to earlier trends in recent publications on relationships between log P and reversed-phase liquid chromatographic parameters (from both HPLC and TLC), only moderate correlations were reported.[371] For a series of congeneric solutes the correlation between log k'_w and log P reported by Clark et al.,[368] r = 0.953, appears reliable and realistic. However, poor correlations were reported even for pyrazine and oxazoline derivatives.[372]

To retain log P as the RPHPLC retention descriptor, some authors have introduced empirical correlations to log P or the hydrophobic substituent (fragmental) constant.[373,374] Although such correlations may be of some use for the prediction of retention, they did not offer much help with understanding the mechanism of separations. The same holds true if the correlation between log k' and log P is improved by the introduction of indicator variables or molecular refractivity into the regression equations.[375,376]

In a QSRR study Patel and co-workers[377] employed 1-octanol–water partition coefficients to account simultaneously for changes in solute structure and mobile phase composition. The ln k' was modeled by the following equation:

$$\ln k' = A + B(\log P/P_{sm}) + C(1/P_{sm^2}) \qquad (402)$$

or

$$\ln k' = A' + B'(\log P/P_{sm}) + C'(\log P/P_{sm^2}) \qquad (403)$$

where A, B, A', B', and C' are regression coefficients; P is the octanol–water partition coefficients of the solute; and P_{sm} stands for calculated 1-octanol–water partition coefficients of a solvent mixture containing water and methanol or acetonitrile or tetrahydrofuran. P_{sm} is calculated from the following equation:

$$\log P_{sm} = \Sigma \, (x_i \log P_{s,i}) \qquad (404)$$

where x_i is the mole fraction of the i-th solvent component; $P_{s,i}$ is the 1-octanol–water partition coefficient of the i-th solvent component.[377]

The limited performance of log P, determined in a liquid–liquid partition system, in modeling RPHPLC retention suggest differences in the two types of partition process.

There is evidence that with hydrocarbonaceous stationary phases the solute molecule can penetrate vertically into the bonded hydrocarbon layer.[378] In addition, retention in such phases is affected by the surface density of the bonded alkyl chains. In this situation the chromatographic process cannot be directly modeled by the bulk organic–water partitioning process, because the nonpolar stationary phase is an interphase (immobilized at one end) and not a bulk medium.

8. Correlation Analysis in Liquid Chromatography of Metal Chelates

Timerbaev et al.[379] discuss the choice of the most significant parameters to be used as variables for the multiple linear regression analysis. The fist selection principle utilized is to reject some less important parameters based on results obtained earlier in our one-dimensional correlations. Then, to facilitate the development of multivariate models, the set of the parameters were further reduced using one-dimensional correlation analysis for particular chelates and the separation system studied in this work. Finally, the remaining valuable parameters were termed solvophobic and hydrogen bonding groups. The molecular descriptors applied by the authors is listed in Table 87. The authors did not consider molar volume, which show a poor correlation with log k' values owing to inaccuracy in the evaluation of the metal increment and solubility, which requires special experiments to be performed. The retention values of metal dithiophosphates (as the chelates of the S,S type) in RPHPLC did not correlate with the effective charge or electronegativity of a metal atom, or with the ratio of electronegativity to the metal ion radius. The application of the electronic and steric constants of alkyl substituents, characterizing the hydrogen bonding capability of donor atoms, is limited by the comparatively small number of carbon atoms in a ligand (not more than four atoms). Both types of hydrophobic constants well describe the hydrophobic nature of a chelate molecule in RPHPLC. The majority of mobile phase parameters presented in Table 88 can be used to describe chelate retention in terms of one-dimensional linear models. The authors have stated that the solvent polarity parameter E_T was sensitive only to the proton-donating properties of the mobile phase and consequently was linearly correlated with log k' values only for water–alcohol eluents.

Table 89 shows the statistical criteria of the one-dimensional regression equation calculated from experimental retention data. Molar volumes show a slightly lower correlation with log k' values than do carbon numbers and hydrophobic constants, while the two latter parameters have identical predictive values owing to high intercorrelation between them. Therefore the authors limited the ligand descriptors to the number of carbon atoms in alkyl substituents. Of the parameters reflecting the ability of the phase to participate in intermolecular interactions of different types, π^* and proton-donating ability by Kosower will be considered below as those that are characterized by higher correlation factors. The list of eluent macroscopic parameters has been reduced to the volume concentration of the organic modifier and surface tension of the mobile phase based on the same principle.

Following the separation mechanism, the authors subdivided selected solute and eluent parameters into solvophobic and hydrogen bonding terms and then composed a number of multiparametric retention models given by

$$\log k' = a_0 + a_1 x_1 + a_2 x_2 + a_3 x_3 + a_4 x_4 \tag{405}$$

The parameters x_1 and x_2 characterize the hydrophobicity and hydrogen bond accepting ability of the chelates, respectively; x_3 and x_4 are the corresponding terms of the mobile phase (the eluent property complementary to solute hydrogen bond accepting ability is hydrogen bond donating ability).

TABLE 87
Structural Parameters of Metal Chelates Applicable to the Description Of Retention Behavior in RPHPLC

Molecular Parameters

Distribution constant (log K_D)
Stability constant (log β_n)
Molecular connectivity index
Molar volume (V_m)
Solubility in the mobile phase

Parameters of Metal Atom

Effective charge
Electronegativity
Ratio of electronegativity to ionic radius
Orbital electronegativity
Distribution coefficient
Metal increment in the distribution constant

Ligand Parameters

Carbon number of alkyl homologies (n_C)
Induction constant (σ^*)
Steric constant (E_S)
Hydrophobic constant, π and f
Molecular connectivity index of the ligand
Molar volume of the ligand

From Timerbaev, A. R., *J. Chromatogr.*, 648, 1993, 307–314.

TABLE 88
Parameters of Binary Water–Organic Mobile Phases used in RPHPLC of Metal Chelates

Macroscopic Parameters

Volume concentration of an organic modifier (c)
Molar concentration of an organic modifier (log C)
Hydrophobicity (log P_S)
Methylene selectivity (log α_{CH2})
Surface tension
Viscosity
Dielectric constant

Molecular Parameters

Solvent dipolarity/polarizability by Kamlet–Taft (π^*)
Acceptor strength by Dimroth–Reichardt (E_T)
Proton-donating ability by Kosower (Z)
Solvent strength by Brounstein (S)

From Timerbaev, A. R., *J. Chromatogr.*, 648, 1993, 307–314. With permission.

TABLE 89
Average Correlation Coefficient and Standard Deviation for One-Dimensional Correlations

Parameter	r	S.D.
n_C	0.9859	0.071
f	0.9859	0.071
V_m	0.9829	0.076
π^*	0.9890	0.080
Z	0.9820	0.119
S	0.9724	0.142
c	0.9890	0.080
log C	0.9874	0.096
log P_s	0.9801	0.212
Surface tension	0.9880	0.089
Dielectric constant	0.9871	0.089
α_{CH_2}	0.9868	0.099

Note: Statistical criteria calculated and averaged from six sets of retention data consisting of 28 and 21 data points for structural and eluent parameters, respectively.

From Timerbaev, A. R., *J. Chromatogr.*, 648, 1993, 307–314. With permission.

The authors obtained the following equations:

$$\log k' = -4.08 + 0.237 n_C + 0.008 E_n - 0.199 ST + 15.91 Z$$

$$n = 43; \quad r = 0.9736; \quad S.D. = 0.117 \tag{406}$$

$$\log k' = 7.80 + 3.31 M - 0.094 \log \beta_n - 0.752 ST + 30.60 Z$$

$$n = 14; \quad r = 0.9143; \quad S.D. = 0.193 \tag{407}$$

$$\log k' = 9.09 + 0.528 \log K_D - 0.665 ST + 21.94 \pi^*$$

$$n = 12; \quad r = 0.8878; \quad S.D. = 0.158 \tag{408}$$

$$\log k' = -7.10 + 3.22 \log K_D + 0.837 E_n - 0.075 ST + 0.21 Z$$

$$n = 12; \quad r = 0.9186; \quad S.D. = 0.127 \tag{409}$$

$$\log k' = -0.718 + 0.061 MC + 0.033 E_n - 0.339 ST + 17.45 \pi^*$$

$$n = 43; \quad r = 0.8716; \quad S.D. = 0.200 \tag{410}$$

$$\log k' = -12.98 + 0.038 MC + 0.020 \log \beta_n - 0.163 ST + 0.247 Z$$

$$n = 14; \quad r = 0.8533; \quad S.D. = 0.272 \tag{411}$$

where n_C is the carbon number of alkyl homologues; log K_D is the distribution constant; E_n is the orbital electronegativity; ST is surface tension; π^* is the solvent dipolarity/polarizability by Kamlet–Taft; Z is the proton-donating ability by Kosower; log β_n is the stability constant; M is the metal increment in the distribution constant; MC is the molecular connectivity index; n is the sample number; r is the correlation coefficient; S.D. is the standard deviation.

As demonstrated above, the multiparametric approach for analyzing the dependence of chromatographic retention of metal chelates on structure and on the composition of the mobile phase provides an adequate description and rational interpretation of RPHPLC data. Owing to sufficiently high statistical significance, correlation models can be widely used for forecasting the chromatographic behavior, including prediction of retention data, and searching for optimal separation systems.

9. Validation of Chromatographic Retention Models in Reversed-Phase High-Performance Liquid Chromatography

The development of chromatographic methology and the accumulation of numerous experimental data have been continuously followed with interest in the theory of chromatographic retention. This has resulted in various simplified models of mutual interactions between stationary phase and solutes to be chromatographed and components of the mobile phase. Such chromatographic systems are relatively complicated so any model can reflect only some aspects of the physicochemical processes significant to the retention behavior of a given solute in its environment. These models are most frequently characterized by the functional dependence between capacity factor (k') and the composition of the mobile phase. In RPHPLC the mobile phase is a mixture of water and a miscible organic solvent; the composition of the mobile phase is expressed by the volume fraction ρ or mole fraction X of the latter. Various models anticipate different relationships connecting k' with ρ or X and the relevant constants have different physical meaning.

Authors of individual approaches or researchers interpreting their experimental results in terms of a given model may state the goodness of fit of the experimental data to a functional dependence $k'(\rho)$ or $k'(X)$, which supports the theoretical premises on which this dependence is based.

The Snyder–Soczewinski displacement model[380–382] assumes monolayer adsorption of the solute or solvent on the stationary phase. The Scott–Kucera model also assumes monolayer adsorption and takes into account dispersion interactions between the solvent and solute in the mobile phase.[383,384] The Kemula–Buchowski equation the most universally adopted by RPLC exponential relationship for $k'(\rho)$ introduced first partition chromatography assumed to be empirical but this can be justified by consideration of the partition ratio of a substance between the stationary and mobile phases.[385,386] Dill developed a statistical thermodynamic theory underlying the affinity of the solute for the grafted chains by Ying et al.[354] The same quadratic dependence was deduced by Schoenmarkers and co-workers.[335,387] These authors introduced solubility parameters for the calculation of intermolecular forces, based on the Hildebrand theory of regular solutions.

10. Physicochemical Meaning of QSRR Equations

There are QSRR equations mostly aimed at retention prediction, which contain terms (descriptors) of obscure physicochemical meaning. One may argue that good retention prediction proves the validity of the descriptors present in the QSRR equation and an attempt should be made to try to discover the physical sense hidden in an effective structural descriptor.

Hasan and Jurs[389] described a representative QSRR equation developed from the multiparameter approach describing RPHPLC indices, I_R, of polyhalogenated biphenyls:

$$I_R = -66511\ (\pm 3646)\ \text{[fraction of positively charged surface area]}$$
$$- 2469\ (\pm 455)\ \text{[fraction of negatively charged surface area]}$$
$$- 72.9\ (\pm 18.8)\ \text{[number of ortho substituents]}$$
$$+ 3351\ (\pm 954)\ \text{[relative positive charge]}$$
$$- 15.8\ (\pm 7.0)\ \text{[path-3 kappa Kier index]}^3 + 840.2$$
$$n = 53;\ R = 0.968;\ s = 55;\ F_{6,48} = 285 \tag{412}$$

where n is the number of solutes used to derive the regression equation; R is the multiple correlation coefficient; s denotes the standard estimate error; F stands for the value of the statistical significance test (F-test) for the model; and the numbers in parentheses represent 95% confidence limits. This equation predicts relative retention on an ODS support, with pure methanol as the mobile phase, within the series of polyhalogenated biphenyls. However, it is difficult to assign physical meanings to the descriptors selected. The first two descriptors are defined as the surface areas of either positively or negatively charged positions of the molecule divided by the total surface area. It is not clear what might represent the relative positive charge descriptor, defined as the charge of the most positive atom in the molecule divided by the total charge of the molecule. Among the descriptors in this equation, the least significant is a molecular graph-derived index, path-3 kappa, proposed by Kier. Kappa indices are calculated by an algorithm that uses the number of atoms and the number of edge paths connecting the atoms in the graph. The path-3 kappa Kier index might encode the "general shape of the molecules," but it is difficult to decide what would be the "shape" raised to the third power. Regression equations containing indices derived from hydrogen-suppressed molecular graphs form a separate family of QSRR studies. However extended the family is, it is of little informative value regarding the retention mechanism, even though a high predictive potency in RPHPLC of several topological indices has often been claimed in the past and is still reported occasionally.[390,391]

In spite of the imagination of designers of the myriad of molecular graph-derived indices, it is difficult to assign a definite physical sense to the individual indices (not to mention their various transformations, such as squares, square roots, and reciprocals). Probably the only systematic structural information that can be extracted from molecular graph-derived indices is that concerning the bulkiness of a series of solutes, as was described by Osmialowski and Kaliszan.[392]

Solute bulkiness certainly strongly affects RPHPLC retention. The molecular bulkiness parameters used in QSSR studies may be considered reliable descriptors of dispersive interactions. This is verified by excellent correlation between these parameters and the retention data determined in a system in which dispersive interactions are decisive, when polar interactions are either meaningless or constant. Among the more recent examples of such QSRRs are those reported for a series of triazine derivatives.[393] The authors described the intercept of the linear function of log k' vs. volume fraction of acetonitrile in the moblie phase in terms of the water-accessible nonpolar surface area of solutes for energy minimized structures ($r = 0.942$). The slope of the function was poorly correlated ($r = 0.878$) with the difference between the nonpolar surface area minus water-accessible nonpolar surface area. The authors gave an equation with two parameters using both parameters independently, but the high intercorrelation between them ($r = 0.99$) invalidates the regression equation.

For solutes of equal molecular size and polarity, the differences in RPHPLC retention may arise from steric effects. To prove steric effects on retention by QSRRs one needs molecular form numerical measures. Obtaining one-dimensional shape descriptors is possible only for specific solutes. Such a group of solutes are planar polycyclic aromatic hydrocarbons (PAHs). The QSRR equation describing the RPHPLC retention of PAHs has been derived.[394,395] Sander and Wise[395] postulated a "slot" model of retention on hydrocarbon-bonded silica phases. According to this model, solute molecules would immerse in a hydrocarbon layer of stationary phase.

Various structural descriptors, more or less directly bound to molecular size, have been tested in QSRRs describing RPHPLC retention. However, none was able to account for the differences in structurally specific, polar properties of solutes. In this case multiparameter QSRRs were studied where bulkiness descriptors were accompanied by polarity parameters. The first parameter studied, the total dipole moment, performed poorly. As is known from the studies of Scott[302] for molecules such as 1,4-dioxane with an overall dipole moment of zero, a better explanation of relative retention was given by the assumption that the effective dipole moment was twice as high as that of diethyl ether. The two dipoles in 1,4-dioxane are in opposition and therefore cancel each other. In chromatography, however, single dipoles interact at close range with molecules forming the RPHPLC system.

In search of an effective polarity parameter, Kaliszan et al.[396,397] applied a submolecular polarity parameter Δ defined as the largest difference in electron excess charge on a pair of atoms in a given solute molecule. The test series of solutes were twelve mono- and disubstituted benzene derivatives with a range of functional groups. The compounds were investigated in three octadecylsilica stationary phases with different hydrocarbon coverages. For each stationary phase four or five compositions of methanol–water eluents were employed. To describe the retention of individual solutes in specific RPHPLC systems, a molecular size-related structural descriptor was used together with the polarity parameter Δ. As size descriptor, the total energy E_T from quantum chemical calculations was applied. Assuming a linear dependence of log k' on the fraction of methanol in the water–methanol eluent, X, and on an octadecyl coverage of the stationary phase, C, the following equation was derived describing the retention of solute i on phase j:

$$\log k_{i,j}' = [0.045(\pm 0.007)E_{i,j})E_{i,j} - 2.649(\pm 0.919)\Delta_i$$
$$- 0.105(\pm 0.067)C_j - 0.495(\pm 0.583)]X$$
$$+ [-0.038(\pm 0.04)E_{T,i}$$
$$+ 2.166(\pm 0.492)\Delta_i + 0.170(\pm 0.036)C_j$$
$$+ 1.296(\pm 0.312)] \tag{413}$$

The correlation between the calculated and experimentally determined values of 144 pairs of log k' was r = 0.986. Replacing Δ with the total dipole moment of solutes, μ, makes any retention prediction unreliable. This is not surprising because the correlation between μ and Δ is only r = 0.770. The advantage of the submolecular polarity descriptor, Δ, over the total dipole moment in predicting RPHPLC retention was also demonstrated in QSRR studies of RPHPLC data determined on a polybutadiene–encapsulated alumina stationary phase.[398] The series of solutes consisted of selected rigid and planar compounds. In this way, the possibility that the conformation of a solute interacting with the components of the chromatographic phases differs from the conformation for which structural descriptors are determined was eliminated. Multiple regression analysis in which 16 various size-related, molecular graph-derived, and quantum chemical descriptors were considered yielded the following equation:

$$\log k_w' = 0.089 MR - 2.505\Delta - 1.62$$

$$n = 21; \quad R = 0.909; \quad s = 0.50; \quad F = 43 \tag{414}$$

where MR was calculated as the sum of bond refractivities for all pairs of connected atoms according to Vogel.[399] When the parameter MR is interpreted as reflecting the ability of a solute to take part in nonspecific dispersive interactions with components of the chromatographic system, then Δ can be assumed to reflect the ability of a solute to participate in specific polar interactions. This equation indicates that the net effect of attractive–dispersive interactions of a solute with the stationary phase on the one hand, and with the mobile phase on the other, provides a positive input for retention parameters. The net effect on retention attractive polar interactions between a solute and molecules of the mobile phase and between a solute and the stationary phase is negative; the more polar the solute, the less it is retained.

Equation 414 rationalizes the mechanism of RPHPLC separation on polybutadiene-coated alumina phases[400] and on the nature of processes that determine RPHPLC retention on these stationary phase materials. But, this equation does not allow for the precise prediction of log k'_w for a given solute. The parameters MR and Δ are rough measures of the nonpolar and polar properties of solutes and cannot be claimed to be universal for all QSRR studies. The structural specificity of an individual set of solutes may give rise to the application of the better performing retention descriptors. The submolecular polarity parameter Δ performs better than the total dipole moment as a retention descriptor.

Lu and his co-workers[401] reported a successful description of RPHPLC data by a three-parameter regression equation containing the van der Waals volume, the square of the dipole moment, and the hydrogen bond energy. Unfortunately, no information on the significance of individual terms in the QSRR equation was provided. Similar statistical objections were described in the paper of Zou and co-workers[402] in which ln k'_w was determined in terms of substituent constants: Hansch, π; Hammett, σ; Taft, E_s; and Swain–Lupton, F. The RPHPLC retention of a series of benzodiazepine derivatives chromatographed on specially deactivated hydrocarbonaceous silica with methanol–buffer eluents has been investigated relative to various additive-constitutive molecular descriptors and to the descriptors derived by molecular modeling.[403]

$$\log k'_w = 1.823 + 0.444(\pm 0.046)f_{X+Y} - 1.187(\pm 0.312)C_3^*$$

$$- 0.0010(\pm 0.003)\mu^2 + 0.012(\pm 0.004)\text{MR}$$

$$n = 21; \quad R = 0.930; \quad F = 25.7; \quad p < 1.0 \cdot 10^{-6} \qquad (415)$$

where n is the number of samples; R is a multiple correlation coefficient; F is the value of the statistical significance test; and p is the significance level of the regression equation. The variables contained in the equation are the following: f_{X+Y} is the sum of hydrophobic fragmental constants of substituents at the aromatic part of the molecule, according to Taylor[404] (see Figure 22); MR is the molecular refractivity of the whole molecule calculated according to Vogel;[399] μ is the total dipole moment; and C_3^* is the electron excess charge on carbon atom C-3 of the 1,4-benzodiazepine system. Equation 415 shows that the hydrophobicity of the aromatic part of solutes as quantified by f_{X+Y} and solute bulkiness as described molecular refractivity provide net positive inputs to RPHPLC retention; the structural descriptors that may be related to solute polarity, C_3^* and μ^2, account for retention-decreasing effects.[405]

11. The S-Index in the Retention Equation in Reversed-Phase High-Performance Liquid Chromatography

A linear approximation of the retention equation to describe the effect of organic modifier concentration on the logarithm of the capacity factor in RPHPLC has been widely accepted in practical RPHPLC and no significant errors in retention prediction were found. It is expressed as

$$\log k' = \log k'_w - S\rho \qquad (416)$$

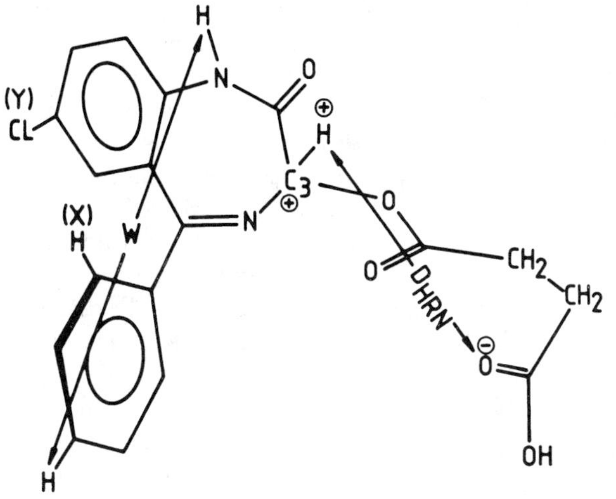

FIGURE 22
Determination of structural descriptors of benzodiazepine derivatives used in QSRR analysis. X and Y are substituents in aromatic rings; C^{+3} is the excess charge on carbon C-3; W is the width of the molecule along the phenyl substituents; D_{HRN} is the distance between hydrogen at C-3 and the most negatively charged atom of another substituent C-3. (From Kaliszan, R., *J. Chromatogr.*, 656, 1993, 417–435. With permission.)

where log k'_w and S are constant for a given column system. The S index is defined as the slope of log k' vs. volume fraction of organic modifier (ρ). A thermodynamic explanation of S can be attempted if the free energy of the interaction between the solute and solvent molecules is taken as a linear function of eluent composition. The combination of the thermodynamic method with an empirical relationship described the retention equation:

$$\log k'_w = \log \beta + (\Delta G^0_{A,C} - G^0_{A,L})/RT \tag{417}$$

where

$$S = (\Delta G^0_{A,C} - \Delta G^0_{A,B})/RT \tag{418}$$

R is the gas constant; T is the absolute temperature; β is the phase ratio; the subscripts A, B, C and L refer to solute, strong solvent, weak solvent and hydrocarbonaceous ligands, respectively; $\Delta G^0_{A,B}$ and $\Delta G^0_{A,C}$ are the nonelectrostatic free-energy changes for solute–strong solvent and solute–weak solvent, respectively; $\Delta G^0_{A,L}$ is the nonelectrostatic free-energy change for solute–hydrocarbonaceous ligands. As can be seen, S is determined mainly by the interaction in the mobile phase. The S index characterizes the properties of the mobile phase and approaches a constant for a certain solute with different C_{18} packing materials.

Chen et al.[406] studied the effect of the prolonged use of the columns, bonded phase density, column type, and temperature based on the empirical retention equation: log k' = log k'_w – S in RPHPLC. The authors indicated, based on the study of Smith and Burr[407] that the prolonged use of the columns does not affect the S values for a specific solute despite a considerable variation in the retention values for each of the compounds. The reproducibility of the S index for given solute within the same laboratory on a certain C_{18} column is generally on the order of 0.05. Chen et al.[406] established that the S index is independent of the bonded phase density for the solutes tested. The standard deviation of the S index for a specific solute area is about 0.05. The S index for a particular solute can make the standardization of different C_{18} packing materials possible. The difference between different C_{18} bonded phases can be compensated for using the S index, therefore the retention data with different compositions of the mobile phase on different C_{18} packings can be standardized. The S value that characterizes interactions in the mobile phase for a certain solute is independent of the different C_{18} packing materials with the same mobile phase. This allows the transfer of k' values from one packing material to another with different composition of the mobile phase with only one isocratic experiment, from which the difference in log k'_w for two kinds of C_{18} packing materials can be determined. The homogeneity of the bonded phases affects the linearity of the log k' vs. ρ relationship, which therefore leads to a change in S values. If solute retention is governed by a hydrophobic mechanism, the reproducibility of S values for a particular solute can be achieved in an eluent system by using energetically homogeneously distributed bonded phases. The reproducibility of the S index for a particular solute can serve as a useful parameter for comparison of the energetic homogeneity between different C_{18} bonded phases. When a compound is extremely polar or ionizable in an aqueous mobile phase, a parabolic shape of the log k' vs. ρ plot is observed, as described by El Tayar and co-workers,[408] which demonstrates different retention mechanisms over the whole concentration range of the mobile phase. This implies an inhomogeneous character of the stationary phase (silanophilic interaction) and of the molecular solute itself (ionic interaction), which means that S values are difficult to ascertain for these compounds. Therefore, during the practical separation of ionizable compounds, suppression agents or buffers should be added to the mobile phase to suppress the ionization of these compounds in order to enhance solute–hydrocarbonaceous ligand hydrophobic interactions, and therefore the pH should be controlled when determining S values for compounds. The presence of a buffer in the mobile phase does not affect the S values of nonelectrolytes.

In the temperature range studied the S values were observed to decrease with increasing column temperature, which can be explained by Equation 418. As the difference in the solute–weak solvent and solute–strong solvent free–energy change characterizes solute interactions in the mobile phase and approaches a constant value for a particular solute with different column temperatures, the value of ST approaches a constant value for non-ionic compounds as is demonstrated by the following equation:

$$\Delta G^0_{A,C} - \Delta G^0_{A,B} = S_1 T_1 = S_2 T_2 = S_3 T_3 \tag{419}$$

where S_1, S_2, and S_3 are S values at temperatures T_1, T_2, and T_3, respectively. Therefore, the S values decrease with increasing column temperature. The nonelectrostatic free-energy change can be separated into van der Waals interactions $\Delta G^0(van)$ and hydrogen bonding interactions $\Delta G^0(H)$:

$$\Delta G^0 = G^0(van) + G^0(H) \tag{420}$$

Dispersion, dipole–dipole (dipolarity), and dipole–induced dipole (polarization) interactions are included in the van der Waals interactions. These interactions can be characterized by using solvatochromic parameters, which were shown to be useful in evaluating and identifying the physicochemical properties governing aqueous solubilities.

The S value in Equation 416 plays an important role in the understanding of interactions in binary mobile phases and in computer simulations of RPHPLC. It has been suggested that S should be a constant, characterizing solvent strength;[409] however, it has been found that S is variable tending to increase with increasing solute retention, and there is a general trend of increasing S in RPHPLC as the molecular size of the solute increases. It has been observed that S values can be approximately related to the molecular weight (M) of the solute,[410] and therefore it may be questioned whether S is a characteristic constant of the solvent. In several other studies, a linear relationship between S and log k'_w was found.[344,411]

The study of Stadalius et al.[412] showed that S values increase in the following order for benzene derivatives: aniline < alkylbenzene < chlorobenzene < ether < aldehydes, ketones < nitriles < unsubstituted polyaromatics < nitro compounds < phtalates < phenylalkanols. The authors discovered that when the other factors remain the same, there is a tendency for more polar compounds to exhibit smaller values of S. All the above results show that S is a solute-related constant.

In linear solvation energy relationships (LSERs), solvent-dependent properties depend on three types of terms according to the following equation:

$$SP = SP_0 + \text{cavity term} + \text{dipolar term} + \text{hydrogen bonding terms} \tag{421}$$

The general LSER for solutes has taken the form:

$$SP = SP_0 + mv_w/100 + d\pi^* + b\beta + a\alpha \tag{422}$$

where v_w measures the cavity term and the van der Waals molecular volume; π^* is a measure of solute dipolarity/polarizability; α is the hydrogen bond donor (HBD) ability or HBD acidity; and β is the hydrogen bond acceptor (HBA) ability or HBA basicity. The solutes are characterized by the parameters v_w, π^*, β, and α, and the solvents are characterized by the coefficients m, d, and b. The solvent property complementary to solute HBA basicity is solvent HBD acidity. Solvent-dependent properties are given by the following equation:

$$SP = SP_0 + mv_w/100 + d\pi^* + b\beta + a\alpha \tag{423}$$

where

$$m = f(\delta^{2C} - \delta_B^2) \tag{424}$$

$$d = g(\pi_B^* - \pi_C^*) \tag{425}$$

$$b = h(\alpha_B - \alpha_C) \tag{426}$$

$$a = l(\beta_B - \beta_C) \tag{427}$$

where δ_C and δ_B are the solubility parameters for weak solvents and strong solvents, respectively; π_B^*, α_B, and β_B are solvatochromic parameters for strong solvents, π_C^*, α_C, and β_B are solvatochromic parameters for weak solvents; and f, g, h, and l are constants.

Chen et al.[413] investigated the effects of molecular structure on the S index in the retention equation in RPHPLC. The S index was quantitatively correlated with the solvatochromic parameters of the solutes. The coefficients in the correlation of S with the solvatochromic parameters were discussed based on the solvatochromic comparison method. The authors described the following equation:

$$S = (1.09 \pm 0.14) + (4.55 \pm 0.15)v_w/100 - (0.25 \pm 0.13)\pi^*$$

$$- (2.50 \pm 0.17)\beta + (0.0948 \pm 0.15)\alpha$$

$$n = 49; \quad r = 0.987; \quad S.D. = 0.156 \tag{428}$$

where n is the number of data points in the regression; r denotes the coefficient of regression; and S.D. stands for the standard deviation.

The sign of the coefficients is determined by whether the term represents an exoergic or endoergic factor in the retention process. The coefficients of the m, d, b, and a terms have the expected signs. The value of solubility parameter for water is higher than that for methanol, whereas the dipolarity of water ($\pi_C^* = 1.09$) is higher than that of methanol ($\pi_B^* = 0.60$), which leads to m having a positive sign and d a negative sign.

Equation 428 shows that increasing hydrogen bonding interaction results in a decrease in S when other conditions do not change. This is consistent with the practical observations that when other factors remain the same, more polar compounds have lower S values, whereas increasing the size of the solute leads to an increase in S. Therefore, there is a general trend that as the solute becomes increasingly hydrophobic, S will become increasingly positive. In contrast, as the solute becomes more hydrophilic and more polar, S will decrease when other conditions remain unchanged.

For homologous series v_w can be written as

$$v_w = n\Delta v_w(CH_2) + n_e\Delta v_w(e) \tag{429}$$

where n and n_e are the number of methylene groups and end groups, respectively; $v_w(CH_2)$ and $v_w(e)$ are the van der Waals volume contributed by methylene groups and end groups, respectively. The average contribution of a methylene group to S for alkylbenzenes is defined by

$$S(CH_2) = [S(n\text{-alkylbenzene}) - S(\text{benzene})]/n \tag{430}$$

where S(n-alkylbenzene) is the experimental S value for n-alkylbenzenes and S(benzene) is the experimental S value for benzene. The calculated S values for these alkylbenzenes is based on the linear relationship between S and carbon number.

Chen et al.[414] studied the effects of molecular structure on the log k'_w index (hydrophobic index) and linear S–log k'_w correlation in RPHPLC. The coefficients in the correlation of log k'_w with the solvatochromic parameters of the solutes are determined mainly by the properties of water and quasi-liquid chemical bonded phases. Accordingly, the distribution or partition between a pair of solvents takes the form:

$$SP = SP_0 + mV_2/100 + d\pi_2 + b\beta_2 + a\alpha_2 \tag{431}$$

where SP denotes solvent-dependent properties; V_2, π_2, β_2, and α_2 characterize the solutes; and m, d, b, and a characterize the solvent water and the quasi-liquid stationary phase. When two or three of the solute parameters were constant and V_2 varied, a linear relationship between S and log k'_w was obtained. The parameter SP_0 in the correlation equation referring to the various reversed-phase columns with different carbon loadings was linearly correlated with the logarithm of surface coverage.

QSRRs based on linear solvation energy relationships (LSERs) and the solvatochromic comparison method have been the subject of several publications. Carr et al.[351] investigated log k' data from RPHPLC determined for a set of benzene derivatives in five acetonitrile–water systems at four different temperatures. The informative equation obtained was

$$\log k'_i = c + mV_2/100 + s\pi_2^* + a\alpha_2 + b\beta_2 \tag{432}$$

where i denotes individual eluent composition and temperature; V_2 is the molar volume of the solute; π_2^* is its polarizability/dipolarity; α_2 represents hydrogen bond acidity; and β_2 is the hydrogen bond basicity of the solutes. With the data set considered, not every variable of this equation appeared statistically significant. The meaningful equation obtained contained the $V_2/100$, β_2, and either π_2^* or α_2 terms. Carr and his co-workers opted for π_2^* as the third regression parameter. The predictive efficiency of the three-variable regression obtained was very high. But the authors excluded (for reasons not explained) several outlier solutes when deriving their individual regression equations. The highest number of solutes considered in the regression was 21 out of a total of 26. For nine of the reported equations, the number of solutes taken into consideration was only 15 to 17 from a total of 20.

Park et al.[415] applied the same approach to determine log k' for eight small aromatic solutes on ODS support with methanol–water (6:4 v/v) as the eluent, and they obtained QSRRs relating retention to the V_2 and β_2 parameters. The small size of the solute series limited the possibility of demonstrating the significance of other variables in this equation for HPLC.

Other researchers employed solvatochromic parameters in QSRR equations describing the slope of the relationship of log k' versus volume fraction of organic solvent in a binary aqueous mobile phase.[413,416] Analysis of the significance of the coefficients of individual regression variables, V_2, π_2^*, β_2, and α_2, shows that the term representing hydrogen bond acidity, α_2, is statistically significant. This is in spite of using a sufficiently large group of solutes (49 solutes) for the QSRR study. Thus the authors obtained basically the same form of relationship as that reported by Carr and his co-workers.[351] In view of the reported QSRRs, it is difficult to judge whether solvatochromic parameters are more suitable and convenient for the description of the RPHPLC retention of chemical compounds than the parameters obtained theoretically by molecular modeling and quantum chemical calculations. It is known that the solvatochromic parameters are empirical. Respective data are available for an increasing number of compounds.[417] Theoretical descriptors can be obtained readily for any structure owing to the common access to software for chemical calculations. The descriptors used in QSRR studies may be of empirical, additive-constitutive semiempirical, and theoretical nature. A typical empirical descriptor of solutes is the logarithm of the partition coefficient in 1-octanol–water liquid–liquid partition system, log P. In general, there is some correlation between RPHPLC retention data and log P for a given series of solutes. This relationship proves certain common features of the liquid–liquid and chromatographic partition systems. It supports the assumption that in RPHPLC the partition, rather than adsorption processes, are decisive for retention. However, each RPHPLC system provides a distinct individual hydrophobicity measure of solutes. Specific aspects of information of solute hydrophobicity obtained from different RPHPLC systems may be unsuitable for the prediction of log P but may be of use for the evaluation of other hydrophobicity-related properties, for example, bioactivity.

QSRR models of RPHPLC based on theoretically calculated structural parameters, which are assumed to reflect the abilities of solutes to participate in fundamental intermolecular interactions with components of chromatographic systems, allow the interpretation of retention mechanisms in simple rational terms. The models are of comparable predictive potency to those based on solvatochromic parameters and linear solvation energy relationships. Their advantage is that a variety of structural descriptors can easily be generated by standard molecular mechanics and quantum chemical calculations.

12. Functional Group Contributions to the Retention of Analytes in Reversed-Phase High-Performance Liquid Chromatography

The functional group contribution (τ_X) to the retention of a compound can be determined from the difference of two solutes, which differ by the presence and absence of X, the functional group of interest. It can be defined the following equation:

$$\tau_X = \log k'_{R-X} - \log k'_{R-H} = \log k' (k'_{R-X}/k'_{R-H}) \tag{433}$$

Early studies in HPLC were reported by Riley et al.,[418] who showed that the τ values were directly related to Hansch partition constants (π values) and Rekker f constants, and the solvophobic theory provided a general framework for rationalization of many observations. For standardized functional group contribution, which is independent of the mobile phase composition, the τ values in mixed organic–water eluents are often extrapolated to give τ_w or τ_{water} values in 100% water.

Hafkenscheid and Tomlinson[419] examined the retention of a number 1,4-disubstituted benzenes and measured the extrapolated τ_w values for combinations of methyl, chloro, nitro, amino, and carboxylic acid groups using methanol–water eluents on an Hypersil ODS column. These investigations showed a close correlation with the corresponding π values. The τ_w values for the monosubstituted groups (methyl = 0.54, chloro = 0.66, nitro = –0.14, hydroxyl = –0.70, amino = –1.19 and carboxylic acid = –0.13) compared well with those obtained in other studies. 4-Nitrophenol and 4-nitroaniline showed great differences and nitro group contributions were 0.40 and 0.45, respectively, with similar significant changes in the values of the hydroxyl and amino groups, pointing to the presence of strong interaction.

In order to obtain a comparison between retentions and octanol–water partitions coefficients, Unger et al.[420] determined a number of log k'_w values using an ODS column and a mobile phase of pH 7.0 buffer saturated with octanol. The corresponding τ_w values derived from monofunctional solutes can be calculated and show good comparison with the values obtained earlier.

Riley et al.[418] found good correlations between τ values and π values for a range of substituents on phenylazapurines, phenyltriazines, and benzoic acids, although the correlation was poorer in the case of 2-substituents, which could undergo hydrogen bonding. The τ_w values obtained by Altomare et al.[421] for substituents on benzenesulfonamides were correlated with π values and significant correlation was found (r = 0.953). No significant differences were found, however, between 3- or 4-substituents, which suggested that no significant crossing interactions were taking place. Gago et al.[422] used measured log k' values for 3- and 4-substituted pyridines to calculate log P values and found that these correlated well with literature log $P_{o/w}$ values, suggesting that the τ values and group contributions to log P were closely related. Lurie and Allen[423] examined τ values for methyl, alkyl, and fluoro groups on fentanyl homologues and analogues and correlated changes with calculated connectivity values. Good correlation (r = 0.9710) was found between the values of the aqueous group contributions (τ_w) from three modifiers determined by Smith[424] and the Hansch contribution π (see Figure 23). The values for group contributions for the individual eluents, which would give similar capacity factors, were also compared with the Hansch contributions (see Figure 24). Correlations for the methanol–buffer (1:1) were good (r = 0.9850), but there were greater deviations for acetonitrile–buffer (4:6) (r = 0.9645) and tetrahydrofurane–buffer (3:7) (r = 0.9537) and most of the changes seemed to be occurring between the more hydriphilic groups. In both acetonitrile and tetrahydrofurane, the noticeable groups, which had a relatively more negative group contribution than other groups with similar π values, were the carboxamide ($CONH_2$), N-acetyl

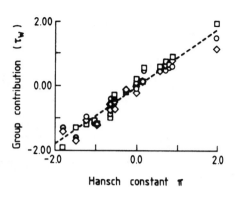

FIGURE 23
Overall relationship between functional group contribution in water derived from three different organic modifiers and Hansch π constants: (□) methanol; (○) acetonitrile; (◇) THF. (From Smith, R. M., *J. Chromatogr.*, 656, 1993, 381–415. With permission.)

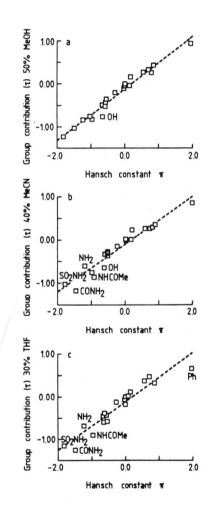

FIGURE 24
Relationship between functional group contribution derived from three different organic modifiers and Hansch π constants: (a) methanol; (b) acetonitrile; (c) THF. (From Smith, R. M., *J. Chromatogr.*, 656, 1993, 381–415. With permission.)

($NHCOCH_3$), hydroxyl (OH), and methylenehydroxyl (CH_2OH) groups. Amino and sulfonamide groups were apparently unaffected. Smith[424,425] compared the group contributions for aliphatic substituents on ethylbenzene with corresponding Rekker (fragmental) constants f. In this case the most noticeable outlier was the methoxyl group (OME; see Figure 25).

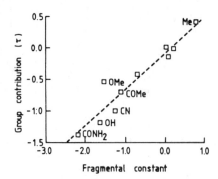

FIGURE 25
Correlation of functional group contributions for aliphatic substituents and Rekker f constants. Eluent: methanol–pH 7.0 buffer; correlation = 0.9282. (From Smith, R. M., *J. Chromatogr.*, 656, 1993, 381–415. With permission.)

Some researchers have calculated contributions as ln α values and these have been converted to τ values where appropriate.

$$\alpha = k'_{R-X}/k'_{R-H} \qquad (434)$$

such that log $\alpha = \tau_X$. The τ values are closely related to substituent contributions ($I_{S,X}$) to the retention indices of solutes in HPLC proposed by Smith and Burr.[425] These values were determined from the difference in the retention indices of solutes on the addition of a substituent, such that

$$I_{S,X} = 100(\tau_X/\tau_{CH2}) \qquad (435)$$

The authors indicated that the τ values can be used to predict the retention of simple substituted compounds based on a parent structure. Some problems may arise with polyfunctional compounds when interaction between substituents can markedly alter group contribution because of electronic, steric, or hydrogen bonding effects.

13. Application of Various Physicochemical Parameters for the Evaluation of Chromatographic Retention Data in RPHPLC

Valkó and Slégel[426] suggested the application of a new chromatographic index (φ) based on the slope and intercept of the log k' vs. organic phase concentration plot. The parameter φ is defined as the organic phase concentration (methanol or acetonitrile) in the mobile phase that is required for log k' = 0 (this means that the retention time of compounds is double the dead time), that is, the molar fraction of compounds is identical in the mobile and the stationary phase. The φ values, therefore, range from 0 to 100%, and the higher the value the more hydrophobic is the compound. The authors indicated that the value of φ is characteristic of the compound and depends only on the type of organic modifier, pH, and temperature. It is independent of the RP column type and length, flow rate, and the mobile phase composition, where actual retention measurements are carried out.

The advantages of φ are that it can be precisely measured, as it has a concrete physical meaning, namely the organic phase concentration of the mobile phase at which the retention time is exactly double the dead time, and it is independent of the linear or quadratic function of the log k' vs. φ relationships. The φ values not only reflect the hydrophobic character of compounds, but also provide a valuable means for method development in RPHPLC as they reveal a mobile phase composition with known retention time values. The authors described that the φ values obtained with methanol and acetonitrile showed excellent correlation with each other and they found significant correlations between the φ values and the logarithm of 1-octanol–water partition coefficient (log P).

The authors[426] defined the φ as

$$\varphi = -\log k'_w / S \qquad (436)$$

where $\log k'_w$ means the log k' value extrapolated to pure water as mobile phase; S (slope) can also be written as the log k' change caused by changing the organic phase concentration by 1%. With help of Eq. 436 the φ values can be calculated from the experimental data. When the measured log k' values are close to zero, the application of the linear fit to the log k' vs. plot for the calculation of φ does not lead to great errors. The correct φ value is that obtained at lower organic phase concentrations, when only hydrophobic interactions govern the retention. The φ value belonging to higher organic phase concentration is caused by a dual retention mechanism (hydrophobic and silanophilic), thus it cannot be regarded as the chromatographic hydrophobicity index.

A graphical illustration of the calculation of φ values for values for various compounds is shown in Figure 26. The figure shows situations when the log k' vs. φ plots are straight lines (compound 1), quadratic (compound 2), and cross each other (compounds 2 and 3) and a dual retention mechanism (compound 4). When the measured log k' values are close to zero due to mobile phase composition, the error of the linear extrapolation for the calculation of φ is negligible. The φ values for over 500 compounds were calculated and presented in Reference 426.

The correlation of φ_{ACN} values with the calculated log P for the data for 140 compounds can be described by the equation

$$\varphi_{ACN} = 9.31 \log P + 37.94$$

$$n = 140; \quad r = 0.88; \quad s = 12.8 \qquad (437)$$

where n is number of investigated compounds; r is the correlation coefficient of the equation; s is the standard error of estimate. This equation shows significant correlation between the two parameters, although owing to the relatively low correlation coefficient and the high standard error of the estimate it cannot be used for measurements of log P values.

Similarly, a statistically significant correlation could be found between the φ_{MEOH} and the log P values for the data for 448 compounds:

$$\varphi_{MEOH} = 7.08 \log P + 42$$

$$n = 448; \quad r = 0.787; \quad s = 13.48 \qquad (438)$$

The authors[426] ascribed the high standard error of the estimate to the error in the calculation of the partition coefficient, to errors of φ values, and also to the error caused by differences in the applied chromatographic conditions. The authors found a significant correlation between the two types of chromatographic hydrophobicity indices for 72 compounds as described by the equation:

$$\varphi_{MEOH} = 0.82 \, \varphi_{ACN} + 20.46$$

$$n = 72; \quad r = 0.96; \quad s = 5.0 \qquad (439)$$

This equation suggests calculation of hydrophobicity indices with methanol from those obtained with acetonitrile. The standard errors of the estimate shows that in spite of the fact that the data for the 72 compounds were obtained with different RP columns

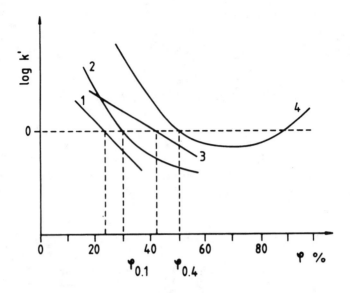

FIGURE 26
Graphical illustration of the determination of the chromatographic hydrophobicity index (ϕ). Numbers refer to hypothetical compounds for which the log k' vs. ϕ plots are straight lines (1), cross each other (2 and 3), or show a dual retention mechanism (4). (From Valkó, K. and Slégel, P., *J. Chromatogr.*, 631, 1993, 49–61. With permission.)

and buffers or pure water, the hydrophobicity index values obtained with acetonitrile as organic modifier can be used to calculate those derived with methanol, with only ± 5% error. Buszewski et al.[427] studied the sorption properties of alkylamide phases under reversed-phase conditions, using different compositions of methanol–water as eluents. Their study has revealed that the specific structural properties of alkylamide phases (AA) have a greater influence on solute retention and selectivity than was observed for conventional alkyl bonded phases. Also, it has been demonstrated that the composition of solvents in the stationary phase can be changed significantly by the presence of a specific interaction site in the bonded ligands.

A significant excess of water in the AA phases with short alkyl chains strongly affects their chromatographic selectivity. Figure 27 shows separations of derivatives of aniline under totally aqueous conditions on the AA phases. The best resolution between individual peaks was obtained for the AA phases with longer alkyl chains, that is, for the AA_3 and AA_2 packings. However, an extension of the alkyl chain resulted in an increase in the analysis time. Thus, the alkylamide phases with medium length alkyl chains seem to be particularly suitable for the separation of polar compounds such as amines. It has been established that the advantage of the AA phases in comparison to the conventional alkyl packings is that good resolution can be achieved under fully aqueous conditions. The authors[427] summarized the K_{W0} values of the investigated systems on the basis of following equation:[428,429]

$$(s_w - s)^{-1} = (s_w - s)^{-1} + K_{W0} (s_w - s_0)^{-1} \rho_w^1 / \rho_0^1 \tag{440}$$

where K_{W0} is the equilibrium constant that describes the displacement process between molecules of water and organic solvent for ideal mobile and surface phases; s_w represents a hypothetical methylene increment that characterizes retention for the pure aqueous mobile phase; ρ_w and ρ_0 are volume fractions of water and methanol in the mobile phase. According to the equation described by Gilpin and co-workers,[429] based on a

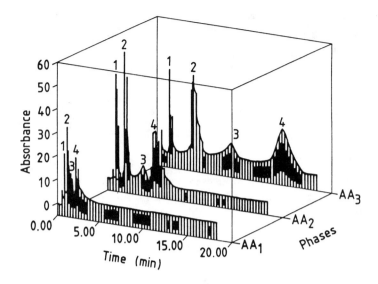

FIGURE 27
Separations of alkyanilines under fully aqueous conditions on the AA phases at 298 K. Peaks: 1, aniline; 2, methylaniline; 3, dimethylaniline; 4, diethylaniline. (From Buszewski, B. et al., *J. Chromatogr.*, 668, 1994, 293–299. With permission)

partition-displacement model for the solute's retention, the dependence of $(s_w - s)^{-1}$ on the ρ_w^1/ρ_0^1 ratio should be linear.

Buszewski et al.[427] stated that in case of the octadecyl phases the K_{W0} values decrease by increasing carbon loading, for example, $K_{W0} = 1.08$ for the C_{18} phase by 11.1% C and 0.81 for the phase by 23.2% C. Based on the model of competitive adsorption for ideal solutions on energetically homogeneous solids, when $K_{W0} > 1$ water is preferentially intercalated into the bonded phase, whereas for $K_{W0} < 1$ the organic solvent is preferentially sorbed. Thus, since for the high coverage density C_{18} packing the K_{W0} value is lower than unity, the data indicate that methanol is preferentially intercalated into the chemically bonded phase. However, the preferential sorption of water can occur for octadecyl phases with lower coverage densities. Comparison of the K_{W0} values for the alkylamide and octadecyl phases with similar carbon loadings show that although methanol is preferentially intercalated into both these phases, its sorption is slightly smaller for the AA_3 packing. The latter effect is probably due to the presence of the specific interaction site in alkylamide phases, which has a stronger interaction with water molecules.

Kowalska et al.[430,431] developed a new retention model primarily established for RPHPLC systems employing the methanol–water mobile phase that has been adapted for the description of solute retention in analogous systems using methanol–phosphate buffer mixtures. The basic physical relationship of this approach was very carefully tested upon experimental material, and excellent agreement was found between theory and practice of the phenomenon considered. The authors assumed self- and mixed association of the mobile phase components, basically due to intermolecular interactions through H bonds. Within the framework of our new model quantification of these complex phenomena was made with the use of the Ostwald dilution law.

The authors proposed the following relationship for the R_F factor of a solute:

$$R_{F3} = X_1^{1/2} \cdot A + X_2^{1/2} \cdot B + C \tag{441}$$

where X_1 and X_2 are, the volume fractions of methanol and water (or the volume fractions of methanol and the buffer, respectively). A, B, and C are the equation constants with following physical meanings:

$$A = (K_1/c_1)^{1/2} \cdot R_1^{*0} \quad (442)$$

$$B = (K_2/c_2)^{1/2} \cdot R_2^{*0} \quad (443)$$

$$C = R_{1...2}^{*} \quad (444)$$

where R_1^*, R_2^*, and $R_{1...2}^*$ are the hypothetical retardation coefficients of the solute chromatographed separately in each moiety of the applied mobile phase. R_1^* and R_2^* can also be defined in the following way:

$$R_1^* = R_1^{*0} \cdot \beta_{s1} \quad (445)$$

$$R_2^* = R_2^{*0} \cdot \beta_{s2} \quad (446)$$

where β_{s1} and β_{s2} are the degrees of dissociation of self-associated methanol and water, respectively, and R_1^{*0} and R_2^{*0} are the equation constants. K_1 and K_2 are the dissociation constants of methanol and water multimers, respectively; c_1 and c_2 are the hypothetical molar concentrations of pure methanol and pure water, respectively, appearing exclusively as associative multimers.

In RPHPLC the chromatographic retention is governed by hydrophobic forces, therefore various RPHPLC retention data have been suggested for calculating the log P values of investigated compounds. There are three main approaches. The first is the use of RPHPLC log k' values obtained on a given column with given mobile phase composition. The approach is to use log k' values extrapolated to 0% organic modifier concentration (log k'$_w$). The log k'$_w$ values can be directly obtained only for a relatively small number of compounds, and therefore some means of predicting this value must be utilized. Some authors[432] used the linear extrapolation from log k' vs. organic modifier concentration plot to predict log k'$_w$ values. However, several results[433] showed that the linearity of the plot is not valid for a wide organic modifier concentration range, and the log k'$_w$ values are not the same when they were derived from data obtained by using acetonitrile or methanol as the organic modifier. Schoenmarkers et al.[434] described quadratic relationships between log k' and x = log P values. Wells and Clark[435] suggested the application of the solvophobic theory proposed by Horváth et al.[312] for the prediction of log k'$_w$. For compounds without hydrogen-bonding functional groups, their log P values can often be calculated by taking into account the additive property of substituent hydrophobicity constant π. However, if the compounds include polar groups, such calculations tend to yield erroneous log P values, and the use of experimental values is required.[436]

Yamagami et al.[437] investigated log P values in some heteroaromatic series (pyridines and diazines) and found that the partition behavior was greatly affected by hydrogen-bond abilities of the ring heteroatoms and the substituent on the heteroring.[436] In this work the authors prepared several systematic series of ester and amide derivatives of typical heteroaromatic compounds and measured their capacity factors. The relationships between log P and log k' were compared with those for the parents and their alkylated compounds to survey the hydrogen-bond effects of ester and amide groups in both chromatographic and octanol–water partitioning systems. To separate the hydrogen-bond effects of investigated compounds they were divided into subgroups depending on the hydrogen-bond types as follows: system N, parent compounds furan, benzofuran, 1-methyl-pyrrole, and 1-methyl-indole and their alkyl derivatives and substituents (subgroups) CO_2R, CONHR, and $CONH_2$; and system H, parent compounds pyrrole and indole and their alkyl derivatives and substituents (subgroups) CO_2R CONHR, and $CONH_2$. In system H the aromatic rings have H donors, and in system N the aromatic rings are non-hydrogen-bonders or very weak H acceptors. Alkyl derivatives were included in parent compounds because they exhibit no hydrogen-bond effect. Classification of the furan ring into this group is rationalized also by a theoretical approach: preliminary semiempirical molecular orbital calculations by the PM3 method suggested that the interaction energy between furan and water molecules was relatively insignificant. Analyses were made by using the hydrogen-bond indicator variables as follows: $HB_H = 1$ for compounds of system H and $HB_H = 0$ for those of system N; $HB_E = 1$ for esters and $HB_E = 0$ for the others; $HB_{AM} = 1$ for amides and $HB_{AM} = 0$ for the others.

First, preanalysis in terms of the following equation were made step-by-step on the assumption that the hydrogen-bond effects attributable to each functional group are additive:

$$\log k_w = a \log P + \Sigma b_i HB_i + c \tag{447}$$

where $\log k_w$ is the log k' extrapolated to 0% organic modifier; HB_i represents one of the HB parameters given above. The coefficients and constant values can be obtained by regression analysis. By this treatment the b_i value should reflect the hydrogen-bond effect contributed from each hydrogen-bonding group. In order to study the hydrogen-donor effect attributable to the ring NH group the authors analyzed both N and H systems of the parent compounds, obtaining an excellent correlation with an improvement (Eq. 449) compared with the single correlation (Eq. 448):

$$\log k_w = 1.023 \log P - 0.190$$

$$n = 15; \quad r = 0.990; \quad s = 0.101; \quad F = 627 \tag{448}$$

$$\log k_w = 1.012 \log P - 0.152 \, HB_H - 0.0907$$

$$n = 15; \quad r = 0.997; \quad s = 0.062; \quad F = 843 \tag{449}$$

where n is the number of compounds used for calculations; r denotes the correlation coefficient; s stands for the standard deviation; and F is the value of the F-ratio between and residual variances. To estimate the contribution of the H-acceptor effect attributable to the ester group, analysis for the parent compounds and the esters of system N were carried out and Eqs. 450 and 451 were obtained.

$$\log k_w = 0.938 \log P - 0.223$$

$$n = 16; \quad r = 0.948; \quad s = 0.191; \quad F = 124 \tag{450}$$

$$\log k_w = 1.019 \log P - 0.152 \, HB_E - 0.097$$

$$n = 16; \quad r = 0.995; \quad s = 0.061; \quad F = 674 \tag{451}$$

Addition of the parent and ester compounds of system H to the above data set gave the following correlations:

$$\log k_w = 0.998 \log P - 0.042$$

$$n = 31; \quad r = 0.956; \quad s = 0.210; \quad F = 309 \tag{452}$$

$$\log k_w = 0.990 \log P + 0.334 \, HB_E - 0.175 \, HB_H - 0.041$$

$$n = 31; \quad r = 0.997; \quad s = 0.059; \quad F = 1437 \tag{453}$$

Similarly, the contribution of amide groups was described by Eq. 455 derived from the parent compounds and the amide derivatives of system N:

$$\log k_w = 0.918 \log P + 0.186$$

$$n = 28; \quad r = 0.990; \quad s = 0.111; \quad F = 1306 \tag{454}$$

$$\log k_w = 0.976 \log P + 0.181 \, HB_{AM} - 0.020$$

$$n = 28; \quad r = 0.994; \quad s = 0.089; \quad F = 1026 \tag{455}$$

Addition of the compounds of system H produced Eqs. 456 and 457.

$$\log k_w = 0.895 \log P + 0.166$$

$$n = 45; \quad r = 0.987; \quad s = 0.129; \quad F = 1555 \tag{456}$$

$$\log k_w = 0.957 \log P + 0.173 \, HB_{AM} - 0.140 \, HB_H + 0.010$$

$$n = 45; \quad r = 0.995; \quad s = 0.081; \quad F = 1327 \tag{457}$$

As preanalysis gave very stable correlations, the authors attempted analysis with the combined data set including all the compounds studied and obtained a very excellent correlation (Eq. 459), which was much improved in comparison to Eq. 458.

$$\log k_w = 0.950 \log P + 0.145$$

$$n = 61; \quad r = 0.981; \quad s = 0.165; \quad F = 1478 \tag{458}$$

$$\log k_w = 0.960 \log P + 0.340 \, HB_E + 0.178 \, HB_{AM} - 0.147 \, HB_H + 0.008$$

$$n = 61; \quad r = 0.996; \quad s = 0.077; \quad F = 1759 \tag{459}$$

The authors stated that the log k_w values calculated by Eq. 458 agreed well with the observed values.

As can be seen in Eq. 458, the log k_w value involves a positive contribution from the H acceptor ester group and a negative contribution from the ring NH group (H donor), leading to overestimated log P values for esters and underestimated values for compounds with the ring NH group, provided that the log P values were estimated from their log k_w values. The overall hydrogen-bond effect of amides could be treated as the sum of effects of –C(=O)NH (H acceptor) and –C(=O)NH– (H donor) moieties. Interestingly, the sum of the coefficients of HB_E and HB_H terms is approximately equivalent to the coefficient of the HB_{AM} term in all eluents.

Further investigations will be needed to elucidate the hydrogen-bond effects of amphiprotic solutes on the correlation between chromatographic hydrophobicity and octanol–water partition coefficients.

Korakas et al.[438] determined the retention parameters of diastereomeric mixtures of 19 lipidic acid conjugates on Spherisorb ODS and Supercosil LC-ABZ stationary phases by changing the acetonitrile concentration in acetonitrile–0.1% trifluoroacetic acid–water mobile phases. The chemical structure of the compounds was characterized by calculated log P values (logarithmic values of the octanol–water partition coefficients) and the dissociation constants (pK_a). Stepwise regression analyses were carried out on the data, and the calculated hydrophobicity values showed good correlations between slope values (s) and pK_b values, as described by the following equations for the data obtained on the Sperisorb ODS and Supercosil LC-ABZ columns, respectively:

$$\log P = 1.936(\pm 0.186)S_{sph} + 0.23(\pm 0.08)pK_b + 0.12$$

$$n = 30; \quad r = 0.910; \quad s = 0.517; \quad F = 64.7 \tag{460}$$

$$\log P = 2.086(\pm 0.279)S_{sup} + 0.26(\pm 0.10)pK_b - 1.86$$

$$n = 30; \quad r = 0.848; \quad s = 0.660; \quad F = 34.5 \tag{461}$$

where n denotes the number of compounds; r is the multiple regression coefficient; s is the standard error of the estimate; and F is the Fisher-test value (significant at 0.01% level). The $\log k'_0$ values are generally expected to show good correlation with the partition coefficients of compounds. Similar, significant correlations can be found in our case as well, but the mathematical statistical parameters are slightly worse than these equations:

$$\log P = 2.124(\pm 0.22) \log k'_{0sph} + 0.368(\pm 0.082)pK_b - 0.738$$

$$n = 30; \quad r = 0.896; \quad s = 0.552 \tag{462}$$

$$\log P = 1.940(\pm 0.249)\log k'_{0sup} + 0.346(\pm 0.095)pK_b - 0.386$$

$$n = 30; \quad r = 0.857; \quad s = 0.642 \tag{463}$$

The authors indicated that slightly better correlations were obtained on the Spherisorb ODS column (again, data where the silanol effect could be observed were not included in the calculations), which proves that not only hydrophobic interactions took place on the Supercosil LZ-ABZ. The retention of the compounds was governed by hydrophobicity on the Spherisorb ODS column and the effect of the free silanol groups could be observed for basic derivatives.

Cupid et al.[439] studied the HPLC retention times for 47 substituted aromatic molecules in three solvent mixtures. Steric and electronic properties of these compounds were derived using semiempirical molecular orbital and theoretical methods.

A subset of the experimental data was used to derive property–retention time relationships and the remaining data were then used to test the predictive capability of the methods. The authors showed that good retention time prediction was possible using derived regression equations for the individual solvents, and after including parameters it was possible to predict retention for all solvents by use of a single equation. This method showed that the most useful properties were calculated log P and the calculated dipole moment of the solutes, and the calculated solvent polarizability. In addition, 90% of the data were applied to set up an artificial neutral network and the remaining 10% of the data used to test the network. Excellent prediction was obtained; the neutral network approach proved to be as successful as regression analysis.

The authors established the effect of 41 theoretical molecular properties on the retention time of different aromatic compounds in three solvent mixtures.

Methanol–water

$$\log k' = 0.393 \text{ CLOGP} + 0.104 \text{ DIP} + 0.114 \text{ PEAX} - 0.328 \text{ QON} - 1.851$$

$$r = 0.932 \tag{464}$$

Acetonitrile–water

$$\log k' = 0.436 \text{ CLOGP} + 0.119 \text{ DIP} - 0.328 \text{ QON} + 0.182 \text{ APPDIAM} - 1.765$$
$$r = 0.950 \tag{465}$$

Tetrahydrofuran–water

$$\log k' = 0.408 \text{ CLOGP} + 0.006 \text{ DIP} + 0.167 \text{ HOMOE} - 4.055 \text{ ESDL1} - 1.896$$
$$r = 0.952 \tag{466}$$

where DIP is the dipole moment, CLOGP denotes the calculated octanol–water log partition coefficient, APPDIAM is approach diameter, HOMOE is the energy of the highest occupied molecular orbit, ESDL1 is the electrophilic superdelocalizabilities for the six aromatic carbons, and PEAX stands for the principal ellipsoid axes about x.

A mixture of electronic factors (DIP and QON), shape (PEAX and APPDIAM), and CLOGP are all important. CLOGP is a complex property reflecting shape, polarity, and size. The interparameter correlations make it difficult to interpret the relevance of individual parameters as the coefficient magnitude reflects both contribution and the individual scale of each parameter. The similarity between the methanol and acetonitrile regression equations indicates a similarity of the physicochemical properties of the solutes important for chromatographic separation. The results suggest that methanol–water and acetonitrile–water solvent systems may have the same mechanism of retention.

The tetrahydrofuran–water solvent system regression differs somewhat from the other two. The largest parameter in this equation is ESDL1, which can be described as the ease of electrophilic attack at the aromatic carbon nearest the carboxylate group. The fourth term, HOMOE, relates to the energy of the electrons most available for donation. There are no significant interparameter correlations between ESDL1 and HOMOE and between DIP, QON, PEAX, APPDIAM, and CLOGP. Hence the chromatographic separation with tetrahydrofuran–water solvent system appears to depend on the electronic physicochemical properties of the molecule more than the methanol– and acetonitrile–water solvent systems.

The RPLC retention behavior of some small molecules on β-CD-bonded silica was compared with that on an octadecylsilica based on the linear solvation energy relationship by Park and co-workers.[440] In this work the values of the Kamlet–Taft hydrogen bonding (HB) donor acidity (α) of cyclodextrins were conveniently calculated from the correlation of proton NMR chemical shifts of some aliphatic primary and secondary alcohols with their α values. The investigated molecules on β-cyclodextrin-bonded silica phase and ODS column were the following: mesitylene, toluene, benzene, anisole, methyl-benzoate, *o*-toluidine, *m*-toluidine, aniline, and *n*-methylaniline. The mobile phase was methanol–water (6:4 v/v). The authors examined data based on the LSERs to see how different the two stationary phases were in terms of the type and strengths of interactions with the solutes. As cavity formation processes are involved in any transfer process and β-CD is dipolar and HB donating, the authors started with regressing the capacity factors of the β-CD-bonded silica using three parameters in the equation including the $V_1/100$, π^*, and β parameters:

$$\log k' = -0.92(\pm 0.43) - 0.03(\pm 0.39)V_1/100 + 2.77(\pm 0.56)\pi^* - 2.33(\pm 0.31)\beta$$
$$n = 9; \quad r = 0.964; \quad \text{S.D.} = 0.088 \tag{467}$$

The authors stated that the coefficient for the cavity formation term ($V_1/100$) is statistically zero, indicating that the cavity formation process is not affecting the retention of the solutes under study on the β-CD-bonded silica. It is not likely that the cavity formation processes do not really affect the retention process. This model may be incomplete because of omission of the solute α parameter. When the authors included the α parameter they obtained

$$\log k' = 0.53(\pm 1.31) - 0.27(\pm 1.03)V_1/100 + 2.26(\pm 1.68)\pi^*$$
$$- 1.78(\pm 1.72)\beta - 0.87(\pm 2.68)\alpha$$
$$n = 9; \quad r = 0.965; \quad S.D. = 0.097 \tag{468}$$

There is no improvement in the goodness of fit and the coefficients for both the $V_1/100$ and α parameter turned out to be statistically zero. This fact indicates that the two-parameter equation (π^* and β) is appropriate for representation of the retention behavior on the β-CD-bonded phase:

$$\log k' = -0.90(\pm 0.25) + 2.76(\pm 0.48)\pi^* - 2.31(\pm 0.27)\beta$$
$$n = 9; \quad r = 0.964; \quad S.D. = 0.079 \tag{469}$$

Retention data on ODS were examined in a fashion similar to those on the β-CD-bonded silica. The resulting LSER equation was

$$\log k' = -0.46(\pm 0.27) + 2.52(\pm 0.50)V_1/100 - 2.35(\pm 0.22)\beta$$
$$n = 8; \quad r = 0.980; \quad S.D. = 0.08 \tag{470}$$

As was previously observed by various authors,[351,441] retention on ODS is mainly determined by cavity formation and the solute HB-acceptor basicity.

It can be seen that the types of intermolecular interactions affecting retention are different on the two stationary phases and hence there are disparate chromatographic selectivities towards a given set of solutes. On ODS the cavity formation term is the major factor affecting retention whereas on β-CD-bonded silica this term has no effect. On ODS the dipolar interaction term has no effect on retention, but on β-CD-bonded silica this term increases the retention. Similar trends in the sign and size of the coefficients in the LSER equations were also observed for retention data in mobile phases with different methanol compositions. As mentioned above, for comparison of the chromatographic properties of the two stationary phases to be validated, a greater number of solutes of widely varying sizes and chemical properties than used in this study must be employed. The retention of 45 barbituric acid derivatives was determined[442] on a polyethylene coated silica column in unbuffered methanol–water eluent mixtures at various organic phase concentrations. Linear correlations were calculated between the log k' values and the methanol concentration in the eluent and the parameters of the relationship were correlated with the various hydrophobic and electronic parameters of barbituric acid derivatives using stepwise regression analysis. Significant linear correlations were found between the log k' values and the concentration of the organic mobile phase in the eluent. The significant linear relationship between the lipophilicity and specific hydrophobic surface area of barbituric acid derivatives indicates that these solutes can be regarded as a homologous series of compounds from the chromatographic point of view. Stepwise regression analysis indicated that the retention of barbituric acid derivatives is mainly governed by the molecular lipophilicity and to a lesser extent by steric effects of the various substituents.

Stepwise regression analysis found significant linear correlation between the combined retention parameter (intercept/slope)[426] and physicochemical parameters of barbituric acid derivatives:

$$\log k'_0/b = -12.87 - (8.12 \pm 1.87)\pi + (0.95 \pm 0.42)Es$$

$$n = 38;\ r^2 = 0.4363;\ F_{calc.} = 13.54;\ F_{99.9\%} = 8.75 \tag{471}$$

Only the lipophylicity (π; path coefficient 65.62%) and steric effect (Es) of substituents (path coefficient 34.38%) of barbituric acid derivatives have a significant impact on the retention behavior of barbituric acid derivatives on PEE columns. The fact that the selected independent variables account for the relatively low ratio (43.63%; see r^2 value) of change of retention behavior indicates that other physicochemical parameters not included in the calculation may have considerable influence on the retention. The results further indicate that the retention behavior of the PEE column slightly differs from that of the traditional RP column because not only lipophylicity (π), but also steric parameter (Es) of barbituric acid derivatives have significant effects on the retention.

The retention of ethoxylated nonylhylphenyl surfactants was determined in RPHPLC using various supports (C1, C2, C6, C8, C18, polyethylene-coated silica, and polyethylene-coated alumina). The legends were as follows: C1 (Column A), C2 (Column B), C6 (Column C), C8 (Column D), C18 (Column E), RP supports as well as polyethylene-coated silica (Column F), and alumina (Column G).[443] A spectral mapping technique was applied for the calculation of the retention strength of surfactants taking into consideration simultaneously their retention on each RP column and for the calculation of the retention capacity of columns taking into consideration simultaneously the retention of each surfactant. To elucidate the influence of the length of the apolar alkyl chain on the retention capacity linear correlation was calculated between these two variables. To assess the effect of the number of ethyleneoxide groups of surfactants on their retention strength correlations between these variables were also calculated. To find the similarities and dissimilarities between the retention behavior and hydrophobic molecular parameters of surfactants and RP chromatographic systems PCA was applied. The surfactants were the variables and the hydrophobicity (R_{M0}), specific hydrophobic surface area (b_1), number of ethyleneoxide groups per molecule (n_e), and the capacity factors of the surfactants on the seven columns were the observations.

The differences between the retention capacity of hydrocarbon bonded silicas are very high, indicating the marked impact of the alkyl chain length on the retention of nonionic surfactants (see Table 90). The significant linear correlation between the retention capacity of alkyl-bonded silica columns and the length of alkyl chain indicates that the separation mechanism of these columns may be similar (Figure 28).

The retention strength of surfactants depended nonlinearly on the number of ethyleneoxide groups per molecule (Figure 29). This finding indicates that the polar ethyleneoxide chains are in a folded state in the eluent and their contact surface with the support do not increase linearly with the increasing length of the ethyleneoxide chain.

In principal component analysis the first three principal components explained the overwhelming majority of variance (92.49%), indicating that the retention behavior of each surfactant is similar in each RP chromatographic system, and it can be described by three background variables (Table 91). As the chromatographic systems with alkyl bonded silica and the physicochemical parameters of surfactants have high loadings in the first principal component the authors assume that the first background variable corresponds to the molecular hydrophobicity that governs the retention of surfactants in these RP systems. The two-dimensional nonlinear map of PC loadings and the cluster dendograms of columns and physicochemical parameters are shown in Figures 30 and 31, respectively.

TABLE 90
Retention Capacity of Reversed-Phase HPLC Columns and Retention Strength of Nonionic Surfactants (Arbitrary Units). Results of Spectral Mapping Technique

Column Capacity	Retention	Surfactant Strength	Retention
A	0.06	A5	0.58
B	0.66	A6	0.56
C	1.12	A8	0.57
D	1.47	A9	0.55
E	2.69	A10	0.52
F	−0.74	A11	0.52
G	−0.95	A15	0.53
		A23	0.53
		A30	0.52

Note: For symbols see text.
From Cserháti, T., *Anal. Lett.*, 27, 1994, 2615–2637. With permission.

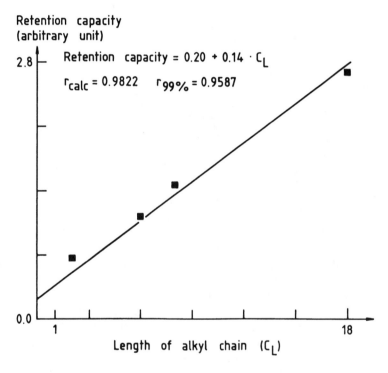

FIGURE 28
Relationship between the retention capacity of alkyl-bonded silica supports and the length of alkyl chain. (From Cserháti, T., *Anal. Lett.*, 27, 1994, 2615–2637. With permission)

14. Investigation and Characterization of RPHPLC Phases

Homologous series are in principle useful for the study of retention mechanisms and the characterization of reversed phases for liquid chromatography. They are potentially attractive for the characterization of such phases, for they allow differentiation between the nonspecific contribution of reversed phases to retention, caused by a regular increase in the length of the aliphatic chain (CH_2) of the solutes being used, and the specific

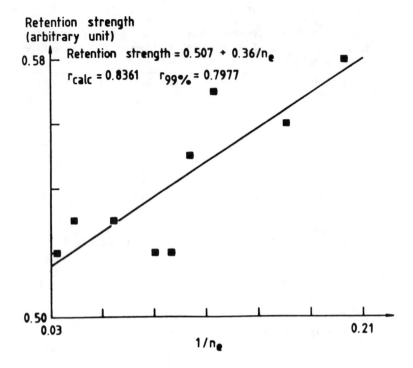

FIGURE 29
Relationship between the retention strength of surfactants on reversed-phase high-performance liquid chromatographic columns and the number of ethyleneoxide groups per molecule (n_e). (From Cserháti, T., *Anal. Lett.*, 27, 1994, 2615–2637. With permission.)

contribution of reversed phases to retention, caused by the molecular residue of the homologous series applied.

Jandera[444] used the interaction indices of appropriate test compounds with solvent molecules for the calibration of the RP system. In this model the nonpolar, RP stationary phase is considered to be a passive solute receptor. However, for widely used silica-based RP stationary phases, many researchers have shown that these phases consist of networks of alkyl ligands and residual silanol groups. The manner of attaching the ligands to the silica surface and the nature of the residual silanol groups determine the final properties of these phases. Reversed phases may mainly show significantly strong interactions with polar solutes, and therefore it is very doubtful whether reversed phases can behave as passive acceptors. It seems more realistic to assume that the structure and composition of RP stationary phases play an active role in the separation process and specific stationary phase interactions will contribute significantly to the retention.

Breuer et al.[445] characterized the RP stationary phases based on retention of homologous series. The authors determined the retention parameters of homologous series of alkylbenzenes as a function of the nature and the composition of the eluent on reversed phases. The resulting retention data were used in first- and second-order equations of the model, resulting in specific characterization values, which should describe the status of a stationary phase. Finally, the authors compared the results of both these approaches.

The retention model in RPLC based on interaction indices can readily be applied to the separation of the members of a homologous series. The following quadratic equation can be used to calculate the capacity factor of the members of homologous series in binary aqueous mobile phases, containing an organic solvent, with volume fraction x:

$$\log k' = \log k_0 + a_1 n_c - n_0 x - m_1 n_c + d_0 x^2 + d_1 n_c x^2 \tag{472}$$

TABLE 91
Relationship between the Retention of Nonionic Surfactants on Reversed-Phase HPLC Columns and Their Physicochemical Parameters. Results of Principal Component Analysis

No. of Principal Component	Eigenvalue	Variance Explained %	Total Variance Explained %
1	6.16	61.58	61.58
2	2.28	22.80	84.37
3	0.81	8.12	92.49
4	0.45	4.46	96.95

Principal Component Loadings

Parameter	No. of Principal Component		
	1	2	3
R_{M0}	0.91	−0.20	0.00
b_1	0.93	−0.19	0.00
n_e	−0.97	0.21	−0.04
Column A	−0.96	0.00	−0.03
Column B	0.57	0.28	0.73
Column C	0.85	0.39	−0.28
Column D	0.87	0.25	0.08
Column E	−0.82	−0.29	0.42
Column F	0.33	−0.90	0.06
Column G	−0.09	0.98	0.07

From Cserháti, T., *Anal. Lett.*, 27, 1994, 2615–2637. With permission.

where n_c is the number of carbon atoms in the aliphatic saturated chain of a homologous series and a_0, a_1, m_0, m_1, d_0, and d_1 are the equation parameters. This equation can also be written as

$$\log k' = a - mx + dx_2 \tag{473}$$

$$a = a_0 + a_1 n_c \tag{474}$$

$$m = m_0 + m_1 n_c \tag{475}$$

$$d = d_0 + d_1 n_c \tag{476}$$

or

$$\log k' = (a_0 + a_1 n_c)(1 - px) - qx + d_0 x^2 + d_1 n_c x^2 \tag{477}$$

with

$$p = m_1 / a_1 \tag{478}$$

$$q = m_0 - m_1 (a_0/a_1) \tag{479}$$

$$m = q + pa \tag{480}$$

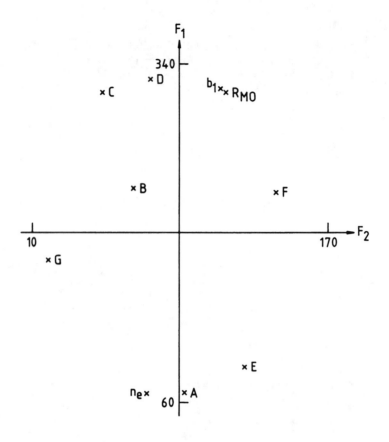

FIGURE 30
Similarities and dissimilarities between the retention characteristics of RPHPLC columns and the physicochemical parameters of surfactants. Two-dimensional nonlinear map of principal component loadings. No. of iterations, 94; maximal error, 1.7×10^{-2}. For symbols see text. (From Cserháti, T., *Anal. Lett.*, 27, 1994, 2615–2637. With permission.)

These equations all resemble a second-order polynomial model.

Jandera stated that the parameters d_0 and d_1 can be neglected over a limited eluent composition range often applied in practical separation, for $x \geq 0.5$. If these parameters are omitted the following equations are obtained successively, which resemble a first-order polynomial model.

$$\log k' = a_0 + a_1 n_c - m_0 x - m_1 n_c x \tag{481}$$

$$\log k' = a + mx \tag{482}$$

$$\log k' = (a_0 + a_1 n_c)(1 - px) - qx \tag{483}$$

The equation parameters can be calculated after calibration of the retention scale with members of a suitable homologous series (*n*-alkylbenzenes). These calibrations were carried out using multivariate linear regression, with x and n_c as independent variables. In this approach a matrix least-squares method is used to fit the mathematical models (Eqs. 472 and 483) to the experimental data obtained from homologous series, resulting in estimates of equation parameters.

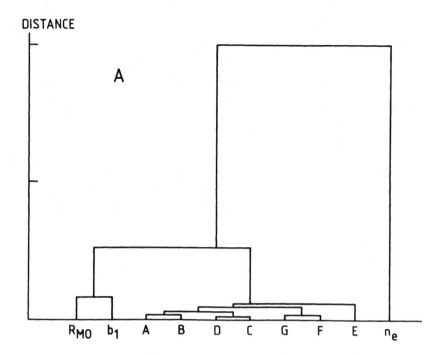

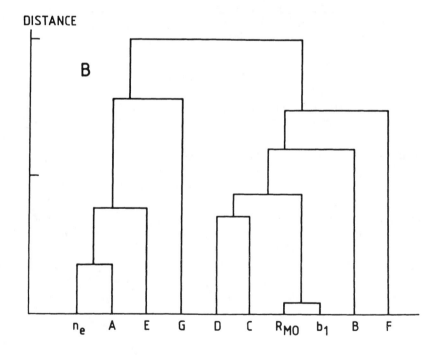

FIGURE 31
Cluster dendograms of RPHPLC columns and physicochemical parameters of surfactants calculated from the original data matrix (A) and from the principal component loadings (B). For symbols see text. (From Cserháti, T., *Anal. Lett.*, 27, 1994, 2615–2637. With permission.)

D. Graphitized Carbon Packing Materials

1. Hydrophobic Properties of Carbon Packing

A number of stationary phases, including donor–acceptor-bonded silica or carbon packings, capable of more positive interactions with analytes than silica C_{18} and polymer packings, involving charge-transfer, dipole–π, dipole–dipole, and steric interactions, are available for the separation of compounds with structural similarity. Graphitized carbon is one of the extremes as stationary phase for RPLC, possessing rigid, planar surfaces and functions capable of dispersion and charge-transfer interactions.

Table 92 indicates that the retention increase caused by one methylene group, a measure of the hydrophobic property of the stationary phase, is always greater on the carbon phase than on alkyl- or aryl-bonded silica phases. The free-energy change associated with the transfer of one methylene group from water to the stationary phase was about –900 cal/mol with the carbon phase compared with –810 cal/mol on silica C_{18}.

TABLE 92
Hydrophobic Properties of Packing Materials

Stationary Phase	Methanol %				
	0	10	30	50	80
C_{18}	3.84	3.62	2.94	2.25	1.54
PYE[a]	3.27	3.15	2.81	2.13	1.45
Carb.[b]	4.50	4.09	3.67	2.83	2.10
Carb.[c]	4.53	4.49	3.74	2.87	2.11

Note: The values given are $\alpha(CH_2)$, calculated from the retention times of alkanols $C_nH_{2n+1}OH$ (n = 2 to 5), except in 80% methanol, where alkylbenzenes (ethylbenzene to amylbenzene) were used.

[a] 2-(1-Pyrenyl)ethyldimethylsilylated silica.
[b] Hypercarb (Shandon Sci. Ltd.).
[c] Carbonex (Tonen).

From Tanaka, N. et al., *J. Chromatogr.*, 656, 1993, 265–287. With permission.

The difference in methylene-binding energy between C_{18} and carbon phases was even greater at a higher methanol concentration. The free-energy change associated with the transfer of one methylene group from water to organic liquid phase was –820 to 850 cal/mol for polar solutes and –884 cal/mol for alkane–water partitioning.[310] These results are striking, if the rigid planar graphite surfaces are assumed to be the binding sites for alkyl groups. Some solventlike behavior of C_{18} stationary phase is expected to reduce the contact between the hydrophobic C–H surface of a solute and water as in aqueous–organic liquid–liquid partition systems, which lowers the free energy of the system, although smaller in magnitude. The rigid planar carbon surface would not be able to completely surround the alkyl chain of the solute. The results suggest the presence of positive interactions between the stationary phase and the solute with dispersion forces, even for C–H groups.[446] The retention process can be described as hydrophobic partitioning with C_{18} where the analyte–stationary phase interaction is not very positive. In the case of a mechanism including dispersion forces, any molecular mass increase in solutes, either in hydrophilic or in dipolar moieties, will tend to cause a retention increase in comparison with other chromatographic systems. Then the close proximity of the molecular surface of a solute and the stationary phase made possible by mutual steric compatibility should be a critical factor in attaining a more favorable retention, leading to pronounced steric selectivity.[320]

2. Steric Selectivity of Carbon Packing

Alkanes and cycloalkanes are suitable for the investigation of the steric selectivity of various types of stationary phases, as only dispersion forces are expected to play a role in their interactions with the stationary phase and solvents. The retention on a carbon phase seems to be determined by how much contact is possible between a solute and carbon surface.

The separation factors $k'_{n\text{-hexane}} / k'_{cyclohexane}$ and $k'_{decalin} / k'_{adamantane}$ on carbon phase were very large compared with silica C_{18}.[325] This is exactly what is expected from the contribution of dispersion forces to retention, since a greater dispersion interaction is expected for more planar solutes with rigid planar graphite surfaces than with flexible C_{18}. The carbon atoms in n-hexane can assume a completely planar arrangement whereas those in cyclohexane cannot adopt a stable conformation. Also, decalin can have more points of contact with a flat surface than adamantane at a similar molecular mass. A slightly greater separation factor between decalin and adamantane on 2-(1-pyrenyl)ethylsilated silica (PYE in Table 92) than C_{18} is an indication of the selective retention of the more planar hydrocarbon in each pair, as with the carbon phase, taking into account the lower hydrophobic selectivity of PYE than C_{18}. As mentioned earlier, polymer gels showed a selective retention of more rigid, compact compounds, i.e., cycloalkanes over linear alkanes. The results on a carbon phase are different from those on the polymer-based phase and on C_{18}. Other researchers discussed the retention tendency of xylenes based on the points of contact with the carbon surface.[447,448] The results with cycloalkanes, which are electronically most inert, support the conclusion on the importance of matching atoms in surface and solute molecules.

Graphitized carbon packing showed retention characteristics based on the major contribution of dispersion forces, and hence steric factors of solutes. Solute–stationary phase interactions on carbon phase are much greater than on any silica-based packing materials, resulting in the preferential retention of planar compounds. The results clearly show the utility of carbon and PYE phases with rigid, planar interacting surfaces to provide steric selectivity for the separation of compounds with similar hydrophobicities, which are difficult to separate with C_{18} phase, although in some instances the mechanism of separation is not necessarily similar. Silica-based PYE phase showed properties between those of carbon and alkyl type silica-based stationary phases, and also provided a better column efficiency than carbon packing. The steric selectivity and excellent chemical stability of the carbon phase under extreme pH values are obvious advantages. Under favorable conditions much better separation can be obtained in a much shorter time.[449] Stronger eluents with dipolar properties such as tetrahydrofurane and acetonitrile or a proton donor should be used with the carbon phase to avoid peak tailing and/or long retention times for late-eluting substances.

3. Retention Strength of Porous Graphitized Carbon Column

Porous graphitized carbon (PGC) support shows unique retention characteristics. Separations on PGC columns use typical reversed-phase eluents (water and organic modifiers miscible with water); however, the retention order of solutes generally does not follow the order of hydrophobicity. Molecular hydrophobicity influences but does not determine the elution order of any set of solutes. It has been proved many times that electrostatic interactions may occur between the planar ring substructures of solutes and the hexagonal graphite molecules on the PGC surface. These interactions have a considerable impact on the retention capacity and selectivity of PGC. Porous graphitized carbon support allows the separation of many solutes (both polar and apolar type). It seems to be especially suitable for the separation of positional isomers containing polar substructures. Due to its highly inert surface the PGC column can be used throughout the complete pH range without deterioration of column efficiency. Therefore it is well suited for the separation of both basic and acidic solutes. The effect of eluent buffering on the retention is not clearly understood. Sometimes good separation of solutes with highly dissociable polar groups has been achieved; in other instances the pH of the eluent exerted a considerable impact on the retention. Similarly to other reversed-phase supports TFA has proved to be an excellent general-purpose mobile phase additive for the separation of anions and other electron-rich species and can be used as an ion-pairing agent for cations. Enantiomer separation can also be carried out on the PGC column with slight modification of the separation methods developed for traditional supports.

Active carbon has long been used as an adsorbent in classical liquid chromatography.[450] Due to its well-known high pH stability and neutrality much effort has been devoted to the development of graphitized carbon black (GCB) support and for the assessment of its application both in gas chromatography and HPLC. Graphite is composed of layers of hexagonally arranged carbon atoms. The carbon atoms are covalently bonded, while the layers themselves are held together by weak van der Waals forces. To all intents and purposes, the graphitic sheets are giant aromatic molecules containing functional groups only at their peripheries. It has been established that the adsorption of molecules on the nonspecific and nonporous surface of GCB is mainly determined by the intermolecular dispersion forces between the solute molecules and the graphite surface. Unfortunately, the otherwise attractive GCB support was unsuitable for routine chromatographic work owing to its extremely poor mechanical stability.[451]

PGC support was produced by impregnating a 7-μm spherical porous silica gel template with the melt of phenol and hexamine. The impregnated materials were heated gradually to 150°C to form phenol-formaldehyde resin within the pores of the silica gel. The resin was carbonized by heating slowly to 900°C. After dissolution of the silica template with aqueous KOH, the carbon was heated to 2500°C in oxygen–argon atmosphere.[447,452–454]

Characteristics parameters of the PGC support are listed below.

1. Sufficient hardness to withstand high pressures (7000 psi)
2. A well-defined, reproducible and stable surface that shows no change during chromatographic work or storage
3. A specific surface area in the range of 150 to 200 m^2/g to give adequate retention of solutes and maintain reasonable linear sample capacity in a fairly great concentration range
4. A mean pore diameter of ≥10 nm and the absence of micropores to ensure rapid mass transfer of solutes into out of the particles
5. Uniform surface energy to give linear adsorption isotherm[455]

4. Eluotropic Strength of Organic Solvents on PGC Column

With selected solute samples, an attempt has been made to establish the eluotropic strengths ($\epsilon°$) of some organic solvents on PGC. The results showed that the actual value of $\epsilon°$ for a particular solvent considerably depends on the chemical character of the solute or solute group used to determine it, and the differences in the eluotropic strength are usually small. In general, methanol and acetonitrile showed low, while dioxane and tetrahydrofuran showed higher eluotropic strengths for the majority of the solutes. It was further established that ethyl acetate, dichloromethane, butylchloride, and hexane have eluotropic strengths that varies the most strongly with solute. It has been concluded that an eluotropic series comparable to that of silica and alumina is difficult if not impossible to establish for PGC support.[456]

5. Retention Behavior of the PGC Column

Earlier studies considered PGC as a "pure" reversed-phase support for it did not contain free silanol groups not covered by the hydrophobic ligand or any other polar adsorptive centers on the surface. A study on the retention behavior of 24 substituted aromatic compounds in nonpolar eluents has led to the conclusion that direct electronic interactions may arise either from the admixture of HOMO or LUMO orbitals of the solute with the surface electrons of PGC or from charge transfer between the solute and PGC. It has been demonstrated by a statistical approach that PGC primarily behaves as an electron pair acceptor for substituted aromatic solutes capable of n-donation under nonpolar conditions. The retention order of solutes closely followed the basicity of the lone electron pair of the solutes. These data indicate that physicochemical parameters other than molecular hydrophobicity may influence solute retention on PGC.[457,458]

Other studies using different sets of solutes led to similar conclusions. Using quantative structure–retention relationship (QSAR) calculations as a tool for the elucidation of the retention mechanisms of PGC, the retention of 29 ring-substituted phenol,[459] 23 ring-substituted aniline,[460] and 45 barbituric acid derivatives[461] were determined in unbuffered acetonitrile–water (except for barbiturates) and methanol–water eluent systems and the relationship between log k' and the concentration of organic modifier (C vol.%) was calculated separately for each solute and eluent system:

$$\log k' = \log k'_0 - bC \qquad (484)$$

The significant relationships between the retention parameters (log k'_0 and b values) and the electron-withdrawing power (σ), proton donor capacity (H-Do), and steric effects (E_s) of substituents proved that the retention behavior of these compounds are mainly governed by these polarity parameters. The results suggest that the graphite surface is sensitive to changes in the solute electron density caused by the electron-donating and -withdrawing ability of solute substituents and the number and position of electron dense bonds in the solute. It can be further concluded that PGC is highly sensitive to steric changes that disturb the electron density of the solute molecule and the resulting interaction of the solute with the graphite surface. The position of substituents on the solute ring determines how the solute can approach and interact with the PGC surface. The lipophilicity of the compounds mentioned above did not affect significantly the retention parameters log k'_0 and b. This finding and the fact that the retention order of solutes on PGC and ODS columns was not correlated emphasizes again the marked differences between the retention characteristics of PGC and ODS columns, although the eluents used on PGC column are typical reversed-phase eluents generally used also on ODS column.[462]

Retention behaviors of PGC and ODS columns were compared with the aid of canonical correlation analysis by Forgács et al.[463] Physicochemical parameters included in the calculation were π, Hansch–Fujita's substituent constant characterizing hydrophobicity; H-Ac and H-Do, indicator variables for proton acceptor and proton donor properties, respectively; M-RE, molar refractivity; F and R, Swain–Lupton's electronic parameters characterizing the inductive and resonance effect, respectively; σ, Hammett's constant, characterizing the electron-withdrawing power of the substituent; Es, Taft's constant, characterizing steric effects of the substituent; B_1 and B_4, Sterimol width parameters determined by the distance of the substituents at their maximum point perpendicular to attachment. Calculations proved that marked differences can be detected between the retention characteristics of PGC and ODS columns, and the electronic parameters of phenol derivatives have the highest impact on their retention.

Canonical correlation analysis (I) found only two significant correlations. The specific adsorption surfaces (slope values) did not depend significantly on the physicochemical parameters of the phenol derivatives. This result indicates that the compounds turn towards the PGC and ODS supports with different molecular substructures (for example, with the hydrophobic ring for ODS and with the polar substituents for PGC). The parameters of significant correlations are compiled in Tables 93 and 94. When the slope and intercept values were taken into consideration (Table 93), the relative weight of the intercept values was higher than that of the slopes. This finding indicates again that the similarity between PGC and ODS column is in the adsorption capacity (intercept values) and not in the molecular surface connecting the support (slope values). The electronic parameters (H-Do, F, and R values) have the highest impact on the retention of the compounds, followed by steric (B_1 and B_2) and lipophilicity (π values). When only the intercept values were included in the calculation the result was similar (see Table 94, canonical correlation II). The adsorption capacity (intercept value) of the phenol derivatives is also influenced by the electronic parameters (F, R, H-Ac, and σ values) followed by the steric parameter (B_1). The individual contribution of each compound is similar in both canonical correlations ($r_{calculated}$ = 0.9777). In both cases the highly bulky di-*tert*-butyl derivatives exert the greatest influence on the retention. Alkyl and amino substituents also have a relatively large impact on the retention of both PGC and ODS columns.

6. Application of PGC Column

PGC is more hydrophobic than octadecylsilica, therefore a higher ratio of organic modifier is needed for the elution of the same solutes. However, the electronic interactions

TABLE 93

Relationship between the Physicochemical Parameters and Retention Characteristics of Phenol Derivatives. Result of Canonical Correlation Analysis I

$r^2 = 0.9837$; $\chi^2 = 121.23$; $\chi^2_{95\%} = 79.10$

Variable	Canonical Coefficient	
	Standard	Weighted (%)
Chromatographic Parameters		
(PGC) log k'_0 (ACN)	−0.4023	34.23
(PGC) b (ACN)	0.0119	0.00
(PGC) log k'_0 (MET)	0.5388	13.54
(PGC) b (MET)	0.6519	0.15
(ODS) log k'_0	0.1486	52.08
(ODS) b	0.3337	0.00
Physicochemical Parameters		
π	0.9647	15.02
H-Ac	0.0003	0.01
H-Do	0.3202	22.37
M-Re	−0.4057	0.82
F	0.2299	14.35
R	0.2716	17.43
σ	−0.0532	2.63
Es	−0.1569	1.92
B_1	−0.8835	14.48
B_2	0.7822	10.97

From Forgács, E. et al., *Chromatographia*, 36, 1993, 19–26. With permission.

between the surface of the PGC support and the solute may markedly modify the retention of the latter, therefore these interactions must also be taken into consideration.

Due to its neutral surface PGC is especially suitable for the separation of basic solutes, as was demonstrated by the separation of remoxipride and FLA-981, two potential neuroleptic agents. Two methods were compared: the separation of remoxipride and FLA-981 by ion suppression at pH 10 (50% acetonitrile in 0.1% M ammonium hydroxide) and by ion pairing with TFA (50% acetonitrile in 0.1% TFA). The superiority of the TFA mobile phase system was established, remoxipride and FLA-981 being eluted with convenient retention times (remoxipride ≈ 3 min, FLA-981 ≈ 12 min). With ion suppression, excessive retention was observed for both compounds (remoxipride ≈ 18 min, FLA-981 < 60 min).[449.] The separation of a thiconazole derivative with marked antihypoxia activity from its impurities was not adequate on ODS column. Using PGC and eluent tetrahydrofuran–water (7:3 v/v) with 1% ammonia the thioconazole derivative (retention time 3.5 min) was successfully separated from its impurities (6 and 9 min).[464] Various substituted propargylamine derivatives are promising therapeutic compounds as monoamine oxidase inhibitory drugs. The retention time of 17 monoamine oxidase inhibitory drugs were determined on PGC column using unbuffered methanol–water and acetonitrile–water mixtures at various concentrations and the log k'_0 and b values were calculated. The difference between the retention times of any pair of solutes at any concentration of organic modifier can be calculated:

$$t_1 - t_2 = t_0 (10^{a_1+b_1 C} - 10^{a_2+b_2 C}) \tag{485}$$

where a and b are intercept (log k'_0) and slope values for solutes 1 and 2 at C organic phase concentration. The eluent composition corresponding to the maximum retention

TABLE 94

Relationship between the Physicochemical Parameters and Retention Characteristics of Phenol Derivatives. Result of Canonical Correlation Analysis II

$r^2 = 0.9573$; $\chi^2 = 67.39$; $\chi^2_{95\%} = 59.72$

Variables	Canonical Coefficients	
	Standard	Weighted (%)
Chromatographic Parameters		
(PGC) log k'_0 (ACN)	−0.5156	14.26
(PGC) b (MET)	0.4661	3.88
(ODS) log k'_0	0.7189	81.86
Physicochemical Parameters		
π	0.0156	0.14
H-Ac	0.5051	11.83
H-Do	0.1061	22.37
M-Re	0.0893	0.10
F	0.4645	16.22
R	0.6893	25.36
σ	−0.9334	26.82
Es	−0.2403	1.68
B_1	1.2274	11.53
B_4	0.2062	1.66

From Forgács, E. et al., *Chromatographia*, 36, 1993, 19–26. With permission.

time difference can also be calculated. The first derivative of this Equation must be zero, with the organic phase concentration expressed accordingly:[465]

$$C = (a_1 - a_2 + \log b_1/b_2)/(b_2 - b_1) \tag{486}$$

The separation capacity of PGC columns for important agrochemicals, the chlorophenoxy-acetic acid congeners, has also been explored using dioxane–water mobile phase without additives and with added sodium acetate, acetic acid, or lithium chloride. The results indicated that acetic acid had the greatest effect on the retention, emphasizing the considerable role of the degree of dissociation on the retention on a PGC column. Retention parameters (log k_0' and b) are given for each chlorophenoxy-acetic acid congener and eluent system.[466]

Retention behavior of 30 commercial pesticides was determined on a PGC column using dioxane and water mixtures as eluents. Linear correlations were calculated between log k' values and dioxane concentration in the eluent. The intercept log k_0' and slope (b) values were regarded as the best estimations of the retention capacity and the contact surface area of the pesticides with the stationary phase. Both the intercept and slope values showed high variations, suggesting that PGC can be successfully used for the separation and quantitative determination of pesticides. The retention capacity of pesticides significantly depended both on their lipophilicity (R_{M0}) and specific hydrophobic surface area (HSA) determined by reversed-phase thin-layer chromatography, indicating that the lipophilicity parameters of solutes have a significant impact on their retention capacity on PGC. The relatively low ratio of variance, which was determined by principal component analysis, suggested that factors other than hydrophobicity parameters may have a considerable influence on the retention.[467]

The extremely hydrophobic surface of PGC has led to difficulty in eluting many hydrophobic solutes, particularly those with planar structures that fit better to the flat hexagonal structures on the surface of PGC. However, the surface of PGC can be modified by adsorption of surfactants, hydrophobic molecules, or high molecular weight polymers. These modifications result in lower retention capacity and changed retention characteristics of PGC, allowing the separation of compounds that would be otherwise totally retained. For example, it has been proved that a monolayer of Tween 80 (polyoxyethylene sorbitol stearate) adsorbed onto the surface of PGC reduces by 15 to 20% the log k' values of hydrophobic molecules. PGC coated with a hydrophilic polymer (polyvinal alcohol, PVA) was used for size exclusion chromatography. Proteins including thyroglobulin, transferrin, ovoalbumin, myoglobin, and cytochrome-C were eluted in roughly reverse order of molecular weight on PGC column coated with PVA multilayers (10 µmol m^{-2} of OH) using 20 mM phosphate buffer as an eluent.[456]

E. Other Metal Oxide-Based Supports

1. Chemistry of Zirconia

Zirconium dioxide exists in many crystallographic and amorphous forms. Zirconia is available in four forms: amorphous, tetragonal, cubic, and monoclinic. The optical, thermal, and electrical and chromatographic properties of zirconia depend on the structure. Crystallographic forms may change as follows:

$$\text{monoclinic} \xrightarrow{1170°C} \text{tetragonal} \xrightarrow{2680°C} \text{cubic}$$

Amorphous precipitates usually transform to the metastable tetragonal phase upon thermal treatment and then convert to the monoclinic phase.

$$\text{amorphous} \rightarrow \text{tetragonal} \rightarrow \text{monoclinic}$$
$$\text{metastable}$$

The transformation of tetragonal to monoclinic zirconia can also be brought about by pressure. Prolonged mechanical treatment (grinding) gave a pure monoclinic phase from polymorphs containing various amounts of tetragonal and monoclinic materials.

The chemical environment of zirconium and oxygen will influence their surface chemical properties and thus it is important to understand the crystal structure. Monoclinic zirconia is a crystalline substance in which all zirconium atoms are heptacoordinated to oxygen atoms. As can be seen in Figure 32, two different oxygen atoms are present in monoclinic zirconia: tricoordinated oxygen is marked O(1) and tetracoordinated oxygen is denoted O(2).[468] The structure of monoclinic zirconia is illustrated in terms of ZrO_2. Closer inspection of the structure reveals that the oxygen atom layers inside the crystal contain twice as many oxygen atoms as does the surface layer. The ratio of oxygen atoms in the surface layer to the zirconium atoms in the immediate underlayer is 1:1. Now we consider a layer of surface oxygen and a layer of zirconium atoms, bearing in mind that a second layer of oxygen is below the zirconium atoms. Taking the above points into account we identify two extreme cases:

1. The top layer is composed of O1 type oxygen atoms.
2. The top layer is composed of O2 type oxygens.

We assume that the Zr and O atoms on the surface preserve as far as possible their coordination configurations found in the bulk crystal. In case 1 the zirconium atoms in surface 2 (see Figure 33) will also be trigonally coordinated, bearing in mind that the other four coordination bonds are satisfied by type 2 oxygens (O2) in the plane beneath.

In our assumption, in the layer of type 1 oxygen (O1) every second oxygen atom is coordinated by two bonds to a surface zirconium atom and every second type 1 oxygen (O1) is coordinated to only a single zirconium atom. Some coordination valencies of both oxygen and zirconium atoms are unsatisfied. Oxygen atoms bear a negative charge while a positive charge will accumulate on the zirconium atoms. This is the fundamental origin of the Lewis basicity and acidity of zirconium oxide.

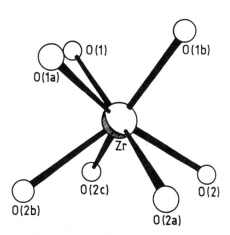

FIGURE 32
Coordination sphere of zirconium atom in monocyclic zirconia. Three type O1 (O1, O1a, and O1b) and four type O2 (O2, O2a, O2b, and O2c) oxygen atoms are coordinated to each zirconium atom. (From Nawrocki, J. et al., *J. Chromatogr.*, 657, 1993, 229–282. With permission.)

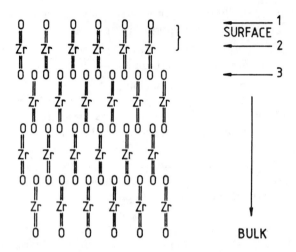

FIGURE 33
Schematic view of atom layers perdendicular to the crystallographic x axis (100 plane) in terms of zirconia molecules. 1, surface layer of oxygen atoms; 2, surface layer of zirconium atoms; 3, layer of bulk oxygens. (From Nawrocki, J. et al., *J. Chromatogr.*, 657, 1993, 229–282. With permission.)

In case 2 a similar analysis can be performed. Here the top layer is comprised exclusively of O2 type oxygens. In this case four coordination valences of the zirconium atoms are projected toward the top layer of oxygen. As is known, three coordination bonds of Zr are satisfied by the subsequent layer of type 1 (O1) oxygen. In our assumption only two of the zirconium valences can be satisfied by type 2 (O2) oxygens. The formation of Lewis acid sites can be observed on the surface, since the unsatisfied valences are equivalent to an accumulation of a positive charge on the zirconium atoms. In the next layer the type 2 (O2) can satisfy only two (of four) coordination valences, and, thus, a negative charge will accumulate on the oxygen atoms. This is the origin of Lewis base sites on the zirconia surface. In both cases all surface zirconium atoms are Lewis acids and all oxygen atoms Lewis bases.

The specific surface area of a support is generally accepted as one of its most important chromatographic parameters. The specific surface area and density of zirconia depend strongly on the thermal history of sample. Surface area decreases sharply between 300 and 500°C. Generally, particles heated above 500°C will have a specific surface area below 100 m²/g. Two different processes are responsible for the decrease in surface area:

1. microcrystallite growth
2. intercrystallite sintering[469]

The apparent density of zirconia samples increases as the sample is subjected to increasingly higher temperature treatment. The apparent density of zirconia is about three to four times higher than that of silica. Taking this into account, we can compare the surface area of a zirconia column to a silica column. Due to the higher packing density a 30-m²/g zirconia column has a surface area equivalent to a 90- to 120-m²/g silica column.

The pore volume of zirconia also depends strongly on its sample history. Thermal treatment at temperatures higher than 200°C gradually decreases the pore volume from 0.25 to 0.01 g/cm³. The pore volume of various zirconias is generally lower than that of silicas. The pore volume of sintered zirconia particles depends on the size of the microcrystallites.

Particles in the 3- to 10-μm range have been routinely prepared, although size classification (due to the high density of zirconia) represents a special challenge. There are no comparable data in the literature on the mechanical strength of chromatographic-grade silica, alumina, and zirconia particles. It is clear that porous microparticulate zirconias can and have been prepared with the desirable physical properties associated with lower high-performance inorganic supports such as silica and alumina. Amorphous zirconia beads have been packed at pressures up to 50 MPa without cracking;[470] commercially available, nonporous zirconia particles were safely packed at 45 MPa; while monoclinic zirconia particles have been packed at pressures up to 9000 psi.

2. Purity of Zirconia

As is known the chromatographic properties of silica depends on the presence of metallic impurities. Mercera et al. indicated that hafnium is the main impurity in zirconia.[469] However, traces of Cu, Fe, Ti, K, and Si were also detected by X-ray fluorescence. Trüdinder and his co-workers[470] analyzed spectrally pure $ZrOCl_2 \cdot 8H_2O$ and found 4 ppm Si, 3 ppm Al, 1 ppm Ca, 1 ppm Fe, and Mg and Mn both less than 1 ppm. Holmes and his co-workers[471] found the following impurities in the zirconia: 100 ppm Al, 400 ppm Ca, 40 ppm Cr, 10 ppm Cu, 3000 ppm Fe, 20 ppm Mg, 50 ppm Mn, 400 ppm Si, and 10 ppm Ti.

Porous zirconia and titania supports with many desirable properties have been prepared by sol-gel type processes.[472] For obtaining relatively uniform, small (2 to 25 μm), spherical, rigid particles, which can be used for HPLC, several different methods of wet-chemical preparation of zirconia particles can be employed.[473] Two processes are most promising:

1. Sol-gel microsphere synthesis
2. PICA process (also termed coacervation or microencapsulation)[474]

The sol-gel processes can be carried out in two ways:

1. by hydrolyzing zirconium compounds
2. by using commercially available zirconia colloids

The PICA process consists of:

1. Generation of a stable sol of uniform sized colloidal particles
2. Generation of polymerizable and water-soluble material
3. Initiation of polymerization of the organic material in the mixture to cause polymerization-induced colloid aggregation into substantially spherical particles
4. Burning off the organic material

The PICA methods produce extremely uniform zirconia particles.

3. Chemical Stability of Zirconia

An appropriate starting point for zirconia as an HPLC support is its chemical stability. The chemical stability of zirconia particles was measured under extreme pH concentrations and compared to that of silica[475] and alumina.[476] The excellent chemical stability of zirconia was confirmed in both studies.

The stability of zirconia under alkaline conditions has been further tested by exposing a column of zirconia to a mobile phase of 1 M sodium hydroxide at 100°C for 3.25 h at a flow rate of 1 ml/min. No zirconium was detected in the column effluent after this treatment. In contrast, about 15% of the alumina present in a comparable column was dissolved and eluted under these extreme conditions. Zirconia offers a distinct advantage relative to either silica or alumina in terms of its chemical stability.[476]

4. Application of Zirconia in Liquid Chromatography

Zirconia is an amphoteric material with anion exchange properties in neutral and acid solutions, and cation exchange properties in alkaline solutions. The difference in the amphoteric acid–base chemistry of zirconia and titania surfaces relative to that of silica is clearly reflected in the chromatography of basic analytes. Strongly basic solutes are typical of high retention retained on silica and yield severely tailed peaks. In contrast, basic compounds with pK_b values as high as 11 have been separated on titania in normal phase mode. As the zirconia surface contains many adsorption sites and is capable of ion and ligand exchange, it should, in most cases, be modified. Generally, there are three methods of surface modification of zirconia:

1. Dynamic chemical modification — in this case a mobile phase containing a strongly interacting Lewis base is used: a great variety of such systems can be imagined.
2. Permanent chemical modification — for example, sylation of the zirconia surface
3. Physical screening — the zirconia surface is coated with a polymer or carbon layer.

a. Dynamic Chemical Modification

Blackwell and Carr[477] examined the retention of a variety of benzoic acid derivatives with pK_a values ranging from 2.8 to 5.4 in the pH range of 4.8 to 9.3. (The acids were almost completely ionized over the entire range of pH values tested. Thus, no pH dependence based on changes in the degree of ionization of the solute can be expected.) Capacity factors were correlated with both the pK_a of the solute and the pH value of the mobile phase, which led to the following equation:

$$\log k' = 0.67 \, pK_a - pH + C \qquad (487)$$

where C is a constant. Similar behavior of benzoic acid derivatives has been observed [478] on alumina:

$$\log k' = 0.33 \, pK_a - pH + C \qquad (488)$$

The weaker dependence on pK_a means that Al is not as strong a Lewis base as Zr. The higher selectivity of zirconia over alumina toward benzoic acid derivatives has been noted. Chromatographic selectivity toward benzoic acids was much higher on zirconia surfaces than on the anion exchanger, as their ionic properties do not differ significantly while their Lewis basicities differ greatly. Isomeric selectivity was also much higher for zirconia than for a conventional anion exchange column, which did not reveal any isomer selectivity. Lewis acidity of zirconia can, however, be confirmed by chromatography where ligand exchange at Lewis sites was observed.[477]

The most interesting application of zirconia in chromatography is the separation of proteins. Zirconia as a base-stable stationary phase not only allows separation of those molecules at high pH values, but also provides an easily "cleaned" surface. This stability offers obvious chromatographic advantages. It is also significant in terms of the use of zirconia support for the downstream processing of proteins because the use of hot alkaline media is a routine sanitization procedure in these applications. Blackwell and Carr[479] showed that a variety of proteins can be separated on zirconium oxide supports. To elute the proteins, a mobile phase must contain an appropriate concentration of a hard Lewis base. The initial concentration of the base appears to be a critical parameter in achieving acceptable retention and peak shape. During the separation the eluent should contain phosphate, fluoride, polyvalent carboxylates, or organophosphate ligands. The chemical flexibility of the zirconia surface was demonstrated by the application of a borate-modified zirconia surface to the separation of proteins. The chromatographic behavior of this type of stationary phase correlated well with the properties of a borate-modified surface as derived from data obtained by static liquid–solid interface studies.

Blackwell and Carr also demonstrated[480] an application of the ion exchange properties of zirconia for protein separations. The zirconia surface can be loaded with Cu^{2+} ions (3.9 μmol/m^2).

b. Permanent Chemical Modification

Trüdinger and his co-workers[470] reported a successful silanization of the zirconia surface with a trifunctional silane. The phase was extremely stable, and it was able to withstand 500 h at pH 12. Yu and his co-worker[481] silanized nonporous zirconia particles with mono- and trichlorooctadecylsilanes. The authors established that the monomeric octadecyl phase was very unstable at pH 2 as well as pH 12. An essentially polymeric octadecyl phase was also unstable, although its properties seemed to stabilize after 1000 to 2000 column void volumes of mobile phase. Due to extremely low surface area, no coverage values are provided. Several applications of the polymeric octadecyl zirconia are provided by different authors: separation of PAHs, *p*-nitrophenyl-malto-oligosaccharides, dansylated amino acids, peptides, and proteins.

c. Physical Screening

i. Polybutadiene-Coated Zirconia

Chemically stable modification of the chromatographic performance of zirconia was accomplished by deposition and cross-linking of polybutadiene reported by Schomburg and co-workers for the modification of silica and alumina.[482,483] This modification results in a remarkably stable reversed-phase support.

ii. Polystyrene-Coated Zirconia

Particles of zirconia coated with polystyrene were used to separate several mixtures of basic compounds.

iii. Polyethylene-Coated Zirconia

The retention time of methylbenzoate, toluene, and naphtalene was determined on RPHPLC columns (polyethylene-coated silica, alumina, and zirconia and octadecyl-coated silica and alumina). The zirconia supports contained various quantities of carbon loadings (4.6 v/v%, 9.0 v/v%, 12.95 v/v%). Retention capacity and retention selectivity of the columns were separated by the spectral mapping technique followed by two-dimensional nonlinear mapping. The data of traditional elemental analysis clearly showed that under identical conditions the carbon loading of alumina support was significantly lower than that of the silica support both in the case of polyethylene or octadecyl hydrophobic ligand. This finding can be explained by the fact that the number of reactive adsorption centers of the surface of alumina particles is considerably lower than on the surface zirconia and silica. The spectral mapping technique indicated that both the retention capacity and selectivity of the columns considerably depends on the type of the original uncoated support and on the quality and quantity of the hydrocarbon ligand.[484]

The retention of ten nonylphenyl ethyleneoxide oligomer surfactants on polyethylene-coated zirconia support was determined in eluents containing methanol–water mixtures in various volume ratios.[485] The relationship between the log k′ and organic phase concentration was linear. The log k' values (related to the retention capacity) were markedly lower than those determined on alkyl-bonded silica supports. This phenomena might be explained by the assumption that the linear polyethylene chains lie parallel with the surface of the zirconia support, the end groups being in close contact with the adsorption centers of the support. Only the surface of polyethylene coating pointing to the eluent is available to the surfactants. It is a hydrophobic layer similar to that of C1 coating in thickness and it differs considerably from brushlike coatings where alkyl chains more or less penetrate the mobile phase and more than one $-CH_2-$ group can interact with the hydrophobic substructure of surfactants. Both retention parameters (retention capacity and specific hydrophobic surface area) depend significantly on the length of the ethylene oxide chain. These results can be explained as follows: not only the apolar nonylphenol moiety, but also the ethylene oxide chain interact with the hydrophobic surface of the polyethylene-coated zirconia and the retention observed is the sum of the interactive forces.

iv. Octadecyl-Coated Zirconia

Microspherical zirconia particles were synthesized and surface modified with octadecylsilane compounds for RPHPLC. Monomeric and "polymeric" octadecyl-zirconia-bonded stationary phases were obtained by reacting the support with octadecyl-dimethylchlorosilane or octadecyltricholorosilane, respectively. The surface coverage of the zirconia-based stationary phases with octadecyl functions was approximately the same as that of octadecyl-silica sorbents. These phases were evaluated in terms of reversed-phase chromatographic properties with nonpolar, slightly polar, and ionic species over a wide range of mobile phase composition and pH. Monomeric octadecyl-zirconia with end-capping exhibited some metallic interactions with both basic and acidic solutes, but these interactions were greatly reduced in the presence of competing agents (for example, tartarate ions) in the mobile phase. "Polymeric" octadecyl-zirconia sorbents exhibited higher retention than monomeric ones among the various solutes investigated, and their residual adsorptivities toward acidic solutes were much lower. The retention of nonpolar and slightly polar aromatic compounds was quasi-homoenergetic on both types of octadecyl-zirconia stationary phases. Stability studies conducted under extreme pH conditions (pH 2.0 and pH 12.0) showed "polymeric" octadecyl-zirconia sorbents are more stable than their monomeric counterparts. These stationary phases were quite useful in the separation of polycyclic aromatic hydrocarbons, alkylbenzene and phenylalcohol homologous series, oligosaccharides, dansyl-amino acids, peptides, and proteins.[481]

v. Carbon-Coated Zirconia

Over the past few years carbon-based HPLC supports have been the subject of intensive study. These supports also have much greater chemical stability over a wide range of pH and temperature values than generally accepted bonded phases. These supports, however, have the following drawbacks: poor mechanical stability, low surface area and heterogeneous surface and therefore low loading capacity, and nonuniform pore structure. Zirconia particles coated with carbon layer were developed by Funkenbush et al.[486]

The process of carbon coating is carried out by passing organic vapors over the zirconia particles at an elevated temperature and reduced pressure. The resulting chemical vapor deposited (CVD) carbon-coated material is useful as a reversed-phase support for liquid chromatography. This support demonstrates much greater selectivity for the separation of both nonpolar and polar isomers than do conventional reversed-phase supports.[487]

Carbon-clad zirconia supports are stable at elevated temperature and pH values.

vi. Polymer-Coated Carbon-Clad Zirconia

When the carbon-clad zirconia is covered by a polymer, some of the unique properties of the carbon-clad material are lost. The resulting phase can be considered as a composite material with high chemical and mechanical stability. The solute-adsorbent interactions are significantly weakened.[488]

The polybutadiene coating improves the efficiency of the packing and the loading capacity is increased by mass transfer characteristics; some selectivity is, however, lost.

5. *Aluminum Oxide-Based Liquid Chromatography Support*

Alumina offers still another alternative to silica because of its inherent higher pH stability. However, in contrast to its extensive use as a medium in column chromatography for purification purposes or for separations in the normal-phase mode, there are still relatively few reports involving alumina-based materials in the reversed-phase mode.[489]

Cserháti[489] determined the retention of 29 aniline derivatives by high-performance liquid adsorption chromatography using an alumina column and dichloromethane–n-hexane mobile phases in various volume ratios. The objectives of this work were to study the retention behavior of some aniline derivatives on an alumina column and to find correlation between the physicochemical parameters of aniline derivatives and their retention. It was assumed that due to their strongly polar character the aniline derivatives are suitable to improve the separation efficiency of alumina. To establish the physicochemical parameters of the aniline derivatives that influence their retention on alumina, the author determined their lipophilicity (R_{M0}) and specific hydrophobic surface area values (B) by reversed-phase thin-layer chromatography, using methanol, acetone, and acetonitrile as the organic mobile phase, and their adsorption capacities (ADS) on silica were correlated with the log k' values of aniline derivatives at 10 vol% n-hexane concentration and with the slope (b) and intercept values (log k'_0) of linear correlation between the log k' values and eluent composition. The choice of the log k' values determined at a mobile phase concentration of 9:1 (v/v) dichloromethane–n-hexane was motivated by the fact that most aniline derivatives showed reasonable retention at this eluent composition. As the character of the correlation between physicochemical and retention parameters of the aniline derivatives had not been previously established, stepwise regression analysis was applied to elucidate the problem outlined above. Calculation was carried out three times, using log k' values at 10 vol% n-hexane concentration, and slope and intercept values of the linear correlation between the log k' values of aniline derivatives and dichloromethane concentration as dependent variables. In each case the independent variables were R_{M0} and B values determined by using methanol, acetone, and acetonitrile as the organic mobile phase and the adsorption capacities (ADS) of the aniline derivatives (altogether seven variables).

The results of stepwise regression analysis are compiled in Table 95. It can be concluded from these data that in our case alumina behaves as a normal adsorption phase. The path coefficients indicate that the lipophilicity value (R_{M0}) has a higher impact on the log k' value than the specific hydrophobic surface area (B).

The good positive correlation between the retentions of aniline derivatives on silica and alumina may be due to the following facts:

1. The surface pH value of alumina is similar to that of the silica surface. Otherwise the retention of the strongly polar aniline derivatives would be highly different on sorbents having different surface pH values.
2. The polarity of the aniline derivatives in the strongly apolar mobile phase is suppressed to such an extent that only the adsorption capacity of the supports governs the retention and the effect of the surface pH value of the material is negligible.

The application of alumina support in HPLC was motivated by its high isoelectric point and high stability in the alkaline pH values.[490]

The theoretical plate number, capacity factor, peak asymmetry and separation factor of an alumina HPLC column were determined using nonylphenyl ethylene oxide oligomers as solutes and eluents containing n-hexane, dichloromethane, dioxane, tetrahydrofuran, chloroform, and ethyl acetate.[491] The relationship between the chromatographic characteristics and physicochemical parameters of the stronger component in the eluent was calculated by multivariate mathematical-statistical methods. It has been established that each solvent influences differently the chromatographic characteristics. Principal component analysis (PCA) followed by two-dimensional nonlinear mapping and cluster analysis were used for the elucidation of the similarities and dissimilarities between the elution characteristics of the various eluent systems. The theoretical plate number (variable I), capacity factor (II), the asymmetry (III) and separation factors (IV), and the physicochemical parameters of the stronger component in the eluent were the variables. The physicochemical parameters included in the calculation were π, Hansch–Fujita's substituent constant characterizing hydrophobicity (V); H-Ac (VI) and H-Do (VII), indicator variables for proton acceptor and proton donor properties, respectively; M-RE, molar refractivity (VIII); F (IX) and R (X), Swain–Lupton's electronic parameters characterizing the inductive and resonance effect, respectively; σ (XI), Hammett's constant, characterizing the electron-withdrawing power of the substituent; Es (XII), Taft's constant, characterizing steric effects of the substituent; B_1 (XIII) and B_4 (XIV), Sterimol width parameters determined by distance of substituents at their maximum point perpendicular to attachment; dielectric constant (XV); dipole moment (XVI). The five eluent systems listed above were the variables.

TABLE 95

Correlation between the Retention Characteristics of Aniline Derivatives on the Alumina Column and Their Physicochemical Parameters. Results of Stepwise Regression Analysis

Parameter	Eq.(1)	Eq.(2)	Eq.(3)
n	26	12	12
a	-9.4×10^{-2}	-1.17	-1.00
b_1	-1.7×10^{-2}	5.76	5.93
$S_{b(1)}$	4.8×10^{-3}	1.42	1.52
Path coeff.%	59.93		
b_2	-1.10		
$S_{b(2)}$	0.46		
Path coeff.%	40.07		
r	0.7330	0.8093	0.7932
F	12.77		

Note: The three equations are

$$\log k' = a + b_1 + R_{M0} + b_2 B$$

$$\log k'_0 = a + b_1 ADS$$

$$b = a + b_1 ADS$$

where R_{M0} is the lipophilicity values of aniline derivatives determined in methanol mobile phase; B is the specific hydrophobic surface area of aniline derivatives determined in methanol mobile phase; ADS is the adsorption capacity of aniline derivatives on silica.

From Cserháti, T., *Chromatographia*, 29, 1990, 593–596. With permission.

Each eluent system separated well the ethoxylated nonylphenol derivatives; that is, each of them can be successfully used for the separation of similar surfactants on alumina support (Figure 34). The capacity factors were the lowest for eluents containing dioxane and chloroform followed by tetrahydrofuran, dichloromethane and ethyl acetate, although the theoretical solvent strength was the same for each eluent. This finding indicates that the solvent strength of these organic modifiers is different using ethoxylated nonylphenol derivatives as solutes (see Table 96). The effective order of solvents depended on the chromatographic parameter. Thus, for theoretical plate number and asymmetry factor the order was ethyl acetate > chloroform > tetrahydrofuran > dichloromethane > dioxane and chloroform > tetrahydrofuran > dichloromethane > dioxane > ethyl acetate, respectively. The separation factor was the highest with n-hexane–dichloromethane mixture and the lowest with n-hexane–dioxane, indicating that the best separations can be achieved by using n-hexane–dichloromethane eluent.

The results of PCA are summarized in Table 97. Four principal components explain the total variance indicating that the 16 original variables can be substituted by four background (abstract) variables without any loss of information. The chromatographic parameters do not have the highest loadings in the same PC, indicating that their information content is slightly different. The distribution of variables on the two-dimensional nonlinear map (Figure 35A) and cluster dendogram (Figure 35B) of PC loadings show the similarities and dissimilarities between the chromatographic and physicochemical parameters. Chromatographic parameters form a loose cluster (cluster A), indicating that they may depend differently on the various physicochemical parameters of solvents. Separation

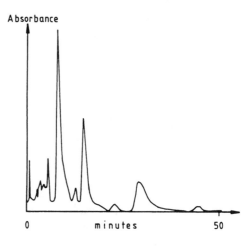
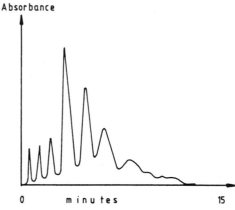

FIGURE 34
Separation of ethoxylated nonylphenyl derivatives on alumina support. Flow rate; 1.0 ml/min; detection wavelength, 280 nm. (A) n-Hexane–dichloromethane, 17:83 v/v; (B) n-hexane–dioxane, 38:62 v/v. (From Forgács, E. and Cserháti, T., *Anal. Lett.*, 29(2), 1996, 321–340. With permission.)

TABLE 96

Theoretical Plate Number (N) and Logarithm of the Capacity Factors (log k') of Ethoxylated Nonylphenol Derivatives on Alumina Column Using Different Eluent Systems.
I, n-Hexane–Dichloromethane 17:83 v/v;
II, n-Hexane–Dioxane 38:62 v/v;
III, n-Hexane–Tetrahydrofuran 22:78 v/v;
IV, n-Hexane–Chloroform 12:88 v/v;
V, n-Hexane–Ethyl Acetate 40:60 v/v

	N	log k'
First fraction		
I	1955	0.339
II	1716	0.365
III	948	–0.102
IV	2253	–0.653
V	2100	0.339
Second fraction		
I	1521	0.562
II	1500	0.076
III	2022	0.110
IV	5786	0.157
V	4892	0.639
Third fraction		
I	1923	0.822
II	1306	0.248
III	2042	0.223
IV	3512	0.246
V	3932	0.878
Fourth fraction		
I	1377	1.116
II	976	0.366
III	2086	0.363
IV	2821	0.359
V	2412	1.098

From Forgács, E. and Cserháti, T., *Anal. Lett.*, 29(2), 1996, 321–340. With permission.

factor (point IV) forms a distinct cluster with the steric characteristic (XII), dipole moment (XV), and dielectric constant (XVI) of the stronger component in the eluent. This finding suggests that the separation of ethoxylated nonylphenol oligomers can be easily influenced by changing the bulkiness, dielectric constant, and dipole moment of the stronger component in the eluent. Solvents are widely separated both on the two-dimensional nonlinear map and cluster dendogram of PC variables (Figure 36A and B), indicating that each solvent has different impact on the chromatographic parameters. Only the bulky solvents dioxane and tetrahydrofuran are relatively near to each other, supporting again the previous conclusions concerning the considerable role of molecular dimensions in the determination of eluent characteristics.

The retention behavior of tributylphenol ethyleoxide oligomer surfactants was studied on an alumina HPLC column using ethyl acetate–n-hexane mixtures as eluents. The surfactants contained various numbers of ethyleneoxide groups per molecule and tributylphenol isomers as hydrophobic moieties.[492] The column separated the surfactants according to the length of the ethylene oxide chain and the position of butyl substituents in one run, demonstrating the good separation power of alumina. The tributylphenyl ethyleneoxide sample was separated into 15 fractions on the alumina column (Figure 37.)

TABLE 97

Similarities and Dissimilarities between the Chromatographic and Physicochemical Parameters of Solvents on Alumina Column. Results of Principal Component Analysis

No. of Component	Eigenvalue	Variance Explained %	Sum of Variance Explained %
1	6.95	43.46	43.46
2	4.86	30.35	73.80
3	2.67	16.71	90.51
4	1.52	9.49	100.00

Principal Component Loadings

	No. of principal component			
Parameter	1	2	3	4
I	−0.33	0.37	−0.80	0.33
II	−0.88	0.39	0.26	0.12
III	0.84	0.27	−0.35	0.31
IV	−0.05	0.99	0.00	0.10
V	0.71	0.68	−0.01	0.18
VI	−0.21	−0.93	0.00	0.31
VII	−0.85	−0.05	−0.47	0.23
VIII	0.68	−0.71	0.19	0.00
IX	0.59	0.46	−0.51	−0.43
X	0.95	−0.02	0.29	0.05
XI	0.75	0.27	−0.45	−0.41
XII	−0.40	0.35	0.73	−0.39
XIII	0.98	0.10	0.14	0.13
XIV	0.82	−0.32	0.36	0.31
XV	−0.17	0.88	0.44	−0.09
XVI	0.07	0.63	0.36	0.69

Note: Roman numbers indicate chromatographic and physicochemical parameters in the text.

From Forgács, E. and Cserháti, T., *Anal. Lett.*, 29(2), 1996, 321–340. With permission.

The peaks were symmetric and they were clustered in groups of three peaks. The authors assume that each "triad" represents tributylphenyl derivatives with identical ethyleneoxide number, and the three members of the triad represent the possible structural isomers of tributylphenol moiety. These results indicate that the number of ethyleneoxide groups and the isomery of the hydrophobic moiety govern the retention of the surfactants on alumina. Steric considerations make it probable that the sterically less hindered 2,4,6-tributylphenol isomers are present in the greatest quantity in the sample. Due to the electron-withdrawing character of substituents the authors postulated the existence of 2,4,5- and 2,3,5-tributylphenyl isomers. To elucidate the impact of various molecular substructures on the retention parameters (slope and intercept values of tributylphenol ethylene oxide oligomer) the relationship between the retention parameters and the number of ethyleneoxide groups per molecule (n) and the position (PI) of butyl substituents was determined. Calculations were carried out by stepwise regression analysis.

Both retention parameters of the tributylphenol ethyleneoxide oligomer surfactants depended on the number of ethylene oxide groups per molecule and the position of the butyl substituents (see Table 98). The equations fit well to the experimental data; the significance level was over 99.9% (see F values). The number of ethylene oxide groups per molecule (n) and the position of butyl substituents (PI) of solutes account for the 91.50% and 97.31% of the change of the dependent variable (see r^2 values). The path

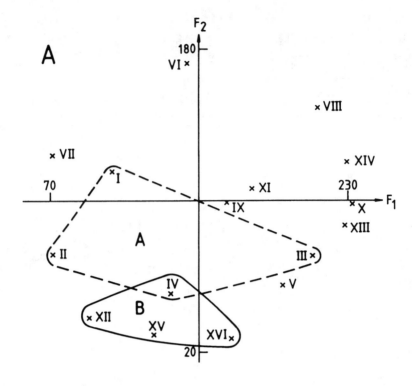
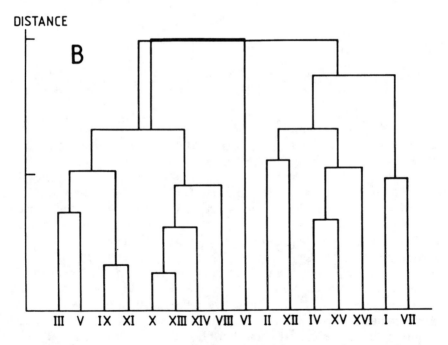

FIGURE 35
Similarities and dissimilarities between the chromatographic and physicochemical parameters of the stronger component in the eluents. (A) Two-dimensional nonlinear map of principal component loadings; (B) cluster dendogram of principal component loadings. For symbols see text. (From Forgács, E. and Cserháti, T., *Anal. Lett.*, 29(2), 1996, 321–340. With permission.)

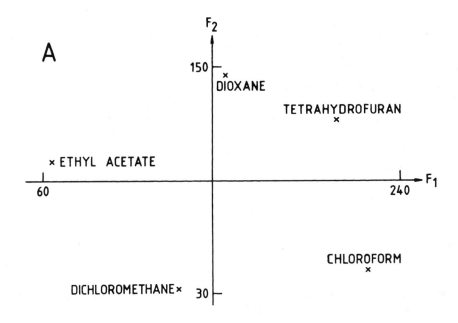
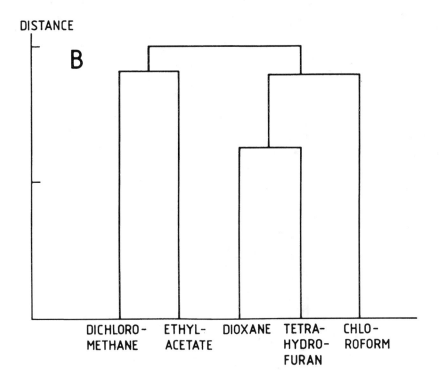

FIGURE 36
Distribution of the stronger components in the eluents on a two-dimensional nonlinear map (A) and on a cluster dendogram of principal component loadings (B). (From Forgács, E. and Cserháti, T., *Anal. Lett.*, 29(2), 1996, 321–340. With permission.)

coefficients indicate that the number of ethyleneoxide groups per molecule has a higher impact on the retention than the position of the butyl substituents of the tributylphenol moiety (see b_1% and b_2% values). These results support the conclusions that the hydrophilic ethyleneoxide chains of surfactants are in direct (probably electrostatic) interaction with the polar alumina surface, and the solvation state of the hydrophobic tributylphenyl moiety only modifies the strength of the interaction.

Chromatographic properties of an octadecyl-bonded alumina (ODA) column as high-performance liquid chromatographic stationary phase were described and compared with those of commonly used octadecylsilica (ODS) stationary phases.[493] Selectivity of the ODA column, was similar to that of the ODS column resulting in similar elution orders for most compound mixtures. The higher stability of the ODA material to alkaline solvents and the absence of acidic silanol sites on its surface allow more efficient separation of organic bases. The correlation of the capacity factors of 31 compounds on ODA and ODS stationary phases was compared. This comparison indicated a higher degree of hydrogen bonding solute–stationary phase interaction on ODA than on ODS. Aromatic compounds with multiple rings are more strongly adsorbed on the ODA support than on ODS. The differences in solute–stationary phase interactions on the two phases result from the chemical properties of their silica and alumina backbones.

Buchberger and Windsauer[494] demonstrated some favorable aspects of alumina as stationary phase in ion chromatography. It is unlikely that alumina will replace existing ion-exchange stationary phases, but it turned out to be a valuable complement, which, in some instances, can facilitate the determination of ions in difficult matrices.

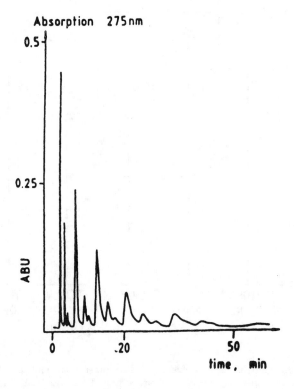

FIGURE 37
Separation of tributylphenol ethylene oxide oligomer surfactants on an alumina column. Eluent, ethylacetate-*n*-hexane (1:1 v/v); flow rate, 1 ml/min, room temperature; detection wavelength: 275 nm; ABU, absorbance unit. (From Forgács, E. and Cserháti, T., *J. Chromatogr.*, 661, 1994, 239–243. With permission.)

The investigated columns packed with alumina were combined with common anion-exchange columns and applied in ion chromatographic determination of the sulfate in brines and biological fluids and in trace determination of iodide in mineral waters and fruit juice samples. Alumina was found to be a highly selective stationary phase for the preconcentration of sulfate from complex matrices. Furthermore, owing to its selectivity, which is different from that of R_4N^+ type anion-exchange columns, it is well suited for on-line column coupling techniques. By application of this method, sample cleanup can be minimized and the sensitivity of the chromatographic system allows the determinations of iodide down to low ppb ranges.

Haky and Vemulapalli[493] compared the log P–log k' correlations obtained with octa-decyl-bonded alumina (ODA), octadecyl-bonded silica (ODS), polybutadiene-coated alumina (PBD), and octadecyl-modified polystyrene divinylbenzene (ACT-1) and determined the partition coefficients of some basic pharmaceutical compounds. The mobile phases consisted of methanol and aqueous buffers. The correlation between the octanol water partition coefficients (log P) of compounds of various chemical classes and the logarithms of their chromatographic capacity factors (log k') was found to be superior to that obtained with columns packed with octadecylsilica, polybutadiene-coated alumina, or octadecyl-derivatized polystyrene-divinylbenzene copolymers. In contrast to the results obtained with other columns, it was not necessary to treat phenols and other hydrogen-bonding compounds as a separate data set on the ODA column to obtain good correlation between log k' values and log P values. The resistance of ODA to degradation by alkaline solvents allowed the use of a basic mobile phase (pH > 10) for suppressing ionization and determining the lipophilicities of organic bases that could not be evaluated within the stable pH range of ODS (pH 2 to 8).

The relative Lewis acidities of alumina and zirconia supports were compared by Blackwell.[495] Studies with benzoic acid derivatives show that zirconia is significantly more selective towards benzoic acid derivatives than alumina under the same conditions. As solute Brönsted acidity increases, the capacity factor increases more significantly on the zirconia phase than on the alumina phase. Efficiencies, as determined by the dominant

TABLE 98

Relationship between the Retention Parameters of Tributylphenol Ethyleneoxide Oligomers and Number of Ethylene Oxide Groups per Molecule (n), and Position of the Butyl Substituents (PI). Results of Stepwise Regression Analysis (Sample Number = 15)

I. $b = a + b_1 \cdot n + b_2 \cdot PI$

II. $\log k_0' = a + b_1 \cdot n + b_2 \cdot PI$

Parameters	Number of equation	
	I	II
a	3.56	-3.41×10^{-1}
b_1	7.80×10^{-1}	3.56×10^{-1}
S_{b1}	7.30×10^{-1}	1.17×10^{-2}
b_2	5.21×10^{-1}	1.28×10^{-1}
S_{b2}	1.27×10^{-1}	3.02×10^{-2}
r^2	0.9150	0.9731
$b_{1\%}$	78.17	82.79
$b_{2\%}$	27.83	17.21
F	64.65	217.65

From Forgács, E. and Cserháti, T., *J. Chromatogr.*, 661, 1994, 239–243. With permission.

desorption rate constants, are generally comparable between the two phases, with alumina showing slightly higher efficiencies. Secondary interactions between the solutes and the stationary phase are nearly absent on zirconia, but are very apparent on the alumina phase.

F. New RPHPLC Stationary Phase Materials

1. Donor–Acceptor Type Bonded Stationary Phase

A number of charge-transfer type packing materials have been utilized in liquid-chromatography. These include donor type, naphthalene or pyrene bonded, as well as acceptor type, nitroaromatic groups bonded. The charge-transfer type stationary phases have already been reviewed.[496]

The nitroaromatic stationary phase undergoes very effective dipole–dipole interactions to discriminate very closely related aromatic compounds, which was not possible with any other stationary phases in RPLC.[497] The underlying mechanisms with all the packings in RPLC include hydrophobic and solute–solvent interactions, resulting in increased retention of compounds with greater hydropbobic surface area, and lower retention for compounds containing polar (hydrophilic) functionality. In addition to these universal interactions observed with the use of any packing material, the following have been observed:

- Polymer packing shows the size-exclusion effect even for low-molecular-mass compounds, which results in the preferential retention of rigid compact aromatic solutes; polymer packings containing ester or ether functionalities show preferential retention of dipolar and/or aromatic compounds based on dipole–π and dipole–dipole interactions.
- Electron donor–acceptor type bonded silica phases show a selectivity that can be explained in terms of charge-transfer, dipole–π, and dipole–dipole interactions between stationary phases and solutes.
- Carbon packing shows a dominant contribution of dispersion forces, leading to the preferential retention of planar molecules which may be termed as hydrophobic adsorption.

The silica C_{18} phase normally undergoes none of these positive interactions except for the weak steric effect based on the ordered structure of long alkyl chains. The simple understanding of the retention process, partitioning of solutes between the mobile phase and bonded alkyl groups containing organic solvents extracted from the mobile phase, may explain the effects of mobile phase composition and stationary phase structure. Alkyl chain length and surface density primarily determine the phase ratio between the mobile and stationary phases in a column, orderliness of alkyl chains, as well as the solvent content (or polarity) of the stationary phase, resulting in different selectivity on stationary phases with different surface coverage or alkyl chain length.

2. Determination of Hydrophobicity with New RPHPLC Phases

For many years octadecyl-bonded silica stationary phases were commonly employed in hydrophobicity studies. However, the retention data obtained with nominally the same type of reversed-phase columns under identical mobile phase conditions are hardly comparable.[296,498] The chromatographic determination of the hydrophobicity of nonionized forms of organic bases cannot be performed directly on silica-based materials.

In attempts to provide a universal, continuous chromatographic hydrophobic scale (not necessarily mimicking log P), several RPHPLC materials have been recently tested. These materials are claimed to be devoid of the major problems of alkyl-bonded silicas (they have no accessible free silanols and are chemically stable over a wide pH range).

Poly(styrene-dininylbenzene) (PS-DVB) copolymers were found to be stable over the pH range of 1 to 14 and to promote moderate correlations with log P, which hold usually only within subgroups of congeneric solutes.[499,500] However, columns packed with PS-DVB are characterized by low efficiency and the material suffers from excessive shrinkage and swelling.[501]

Recently, several polymeric phases having a chemically bonded octadecyl moiety have been tested in hydrophobicity determinations. Phases such as octadecylpolyvinyl copolymer or rigid macroporous polyacrylamide with bonded octadecyls did not undergo swelling or shrinkage and offered the possibility of having reasonable flow rates without undesirable pressure increases at the column inlet.[502] Depending on the specific phase used, the reported correlations with log P of test solutes were found to be low or at best as good as those obtained with ODS phase. There is evidence, however, that individual polymeric phases provide a specific input to retention. The octadecylpolyvinyl copolymer was reported to be less hydrophobic than alkylsilicas and strongly retained some specific compounds.[503]

Recently, great progress has been achieved in the technology of the silica-based reversed-phase materials. Owing to high C_{18} bonding densities, significant protection of octadecylsilica phases against hydrolysis was attained.[504]

Hydrocarbonaceous silica phases exhibiting a high level of silanol deactivation became commercially available and these phases proved valuable for hydrophobicity determinations of drug solutes.[505,506]

When alumina-based reversed-phase materials appeared in chromatographic practice, interest in them focused mainly on hydrophobicity parametrization.[507] Alumina support is stable over a wide pH range and possesses no interfering silanol groups. Polybutadiene chemically encapsulated alumina (PBA) reversed-phase material was introduced by Bien-Vogelsang et al.[508] Owing to the chemical stability of PBA, nonionized forms of acids, bases and neutral species can be analyzed in the same HPLC system operated at an appropriately adjusted pH. Hence a continuous hydrophobicity scale may be obtained in an easier, faster, and more reproducible manner than is the case with the octanol–water system.

Carbon support for RPHPLC may also be interesting for comparative hydrophobicity studies.[509] It has good chemical stability over a wide pH range, the carbon supports do not exhibit peak tailing for amines, as do silica-based supports, and do not adsorb phosphates or carboxylates, as do alumina and zirconia polymer-coated phases. The selectivity of carbon support could provide information on specific features of the hydrophobicity of the solutes. This information could be of value for structure–activity studies and thus the reported lack of correlation of log k' as determined on graphitic carbon with log P should not be discouraging.[510]

The hydrophobic effect is assumed to be one of the "driving forces" for passive diffusion of xenobiotics through biological membranes and drug-receptor binding. If the hydrophobicity measuring system is to model a given biological phenomenon, then close similarity of the component entities is a prerequisite. Hence the partition system expected to model the transport through biological membranes should be composed of an aqueous phase and an organized phospholipid layer (bilayer). Miyake et al.[511] derived HPLC hydrophobicity parameters employing a column of silica gel coated physically with dipalmitoyl-phosphatidylcholine (DPPC). Disregarding the inconveniences associated with their preparation and stability, the systems with DPPC adsorbed on silica probably do not emulate the lipid dynamics of biological membranes because the adsorbed lipids are not organized in a similar manner to natural (or artificial) membranes.

Recently introduced immobilized artificial membranes (IAM) as chromatographic packing materials appear to be more reliable and convenient models of natural membranes.[512] The IAM surfaces are synthesized by covalent binding of the membrane-forming phospholipids to solid surfaces. They are colfluent monolayers of immobilized membrane lipids, wherein each lipid molecule is covalently bound to the surface. Membrane lipids possess polar head groups and two nonpolar alkyl chains. One of the alkyl chains is linked to the solid surface. The immobilized head groups protrude from the stationary phase surface and the first contact site between solutes and IAM.

Correlations between log k' data determined on IAM type columns and experimental log P values are low. The log k' values with non-end-capped columns determined with an eluent composed of sodium phosphate buffer (pH = 7.0)–acetonitrile (8:2, v/v) correlated with log P (correlation coefficient = 0.81) and for a column end-capped with methylglycolate (IAM.PC.MG) operated with an eluent composed of buffer–acetonitrile (75:25, v/v) the corresponding value of the correlation coefficient was 0.76. There was very low correlation between log k' values from an IAM column and log k'_w values determined on a deactivated hydrocarbonaceous silica column. This means that retention data determined on IAM columns contain information properties of solutes that are distinct from those provided by hydrocarbonaceous silica reversed-phase columns and by the 1-octanol–water slow equilibrium systems.

The IAM columns are easy to operate although there is a problem with column stability. The recommendation of the manufacturer that alcohols be avoided as components of the mobile phase also causes some inconveniences. It can be assumed that the IAM columns facilitate hydrophobicity characteristics and they are most suitable for modeling the pharmacokinetics of drugs.

3. β-Cyclodextrin-Coated Silica Column

Cyclodextrin and various cyclodextrin derivatives have found growing acceptance and application in many fields of chromatography. They have been used in reversed-phase thin-layer chromatography to study their interaction with various bioactive compounds. Cyclodextrins modify the effective mobilities of various inorganic ions in isotachophoresis, improve separation of peptides in capillary electrophoresis, and enhance the efficiency of enantiomeric separation in gas chromatography. Cyclodextrins are used in two different manners in HPLC either by adding cyclodextrins to the eluent or by covalently bonding cyclodextrins to the silica surface.

Separations on silica columns with covalently bonded CDs are generally carried out in aqueous eluents similarly to the separations on traditional reversed-phase columns; however, CD derivatives for application in the adsorption separation mode of enantiomers have been synthetized too.[513]

The authors[514] determined the retention of 45 barbituric acid derivatives on a β-cyclodextrin polymer-coated silica column (β-CD column) in various eluents systems (methanol–water, acetonitrile–water, ethanol–water, dioxane–water and tetrahydrofuran–water at 50 vol% organic modifier concentrations). To find the physicochemical parameters of solutes that significantly influence their retention behavior stepwise regression analysis was applied. In these calculations the dependent variables were always the logarithm of capacity factors and independent variables were the different physicochemical parameters of barbituric acid derivatives.

Stepwise regression analysis was carried out five times, the log k' values determined in the five eluent systems being separately the dependent variables. The physicochemical parameters included in the calculation were π, Hansch–Fujita's substituent constant characterizing hydrophobicity; H-Ac and H-Do, indicator variables for proton acceptor and proton donor properties, respectively; M-RE, molar refractivity; F and R, Swain–Lupton's electronic parameters characterizing the inductive and resonance effect, respectively; σ, Hammett's constant, characterizing the electron-withdrawing power of the substituent; Es, Taft's constant, characterizing steric effects of the substituent; B_1 and B_4, Sterimol width parameters.

To separate solvent strength and solvent selectivity the "spectral map" technique was applied. The data matrix consisted of log k' values of barbituric acid derivatives determined in five eluent systems. Calculations were carried out twice, the eluent systems and barbituric acid derivatives being the variables and observations, respectively. The potency values of eluent systems and solutes were considered as the solvent strength and retention capacity, respectively. To visualize the spectral characteristics of eluents and solutes two-dimensional spectral maps of barbituric acid derivatives and eluent systems were separately calculated. The relationship between the physicochemical parameters of the organic modifiers on their solvent strength was calculated by stepwise regression analysis. Stepwise regression analysis were carried out three times (termed calculations A, B, and C below).

For calculation A the independent variables were the various physicochemical parameters of organic modifiers (π, H-Ac, H-Do, M-RE, F, R, σ, Es, B_1 and B_4, and ratio of B_4/M-RE). The inclusion of the combined variable B_4/M-RE was motivated by the assumption that the surface/volume ratio may influence the complex formation of solutes with the CD cavities on the support surface resulting in significant impact on the retention. The dependent variable was the solvent strength of organic modifiers (potency values determined with the spectral map technique).

The independent variables for calculations B and C were as in calculation A, and the dependent variables were the first (B) and second (C) coordinates of the two-dimensional spectral map of organic modifiers. The other parameters of calculations were the same as in point A. The parameters of the correlations describing the relationships between retention parameters and physicochemical characteristics of barbituric acid derivatives are compiled in Table 99. The lipophilicity values and steric parameters of substituents have the highest impact on the retention characteristics of barbiturates on the β-CD polymer-coated silica column (see b% values in Table 99). The preponderant role of solute lipophilicity and steric parameters can be explained by the assumption that the retention of barbiturates on β-CD support is influenced by the following interactions:

1. Interactions of solutes with the cyclodextrin cavity. It is generally accepted that these interactions are determined both by the lipophilicity and the size of the guest molecules. The steric parameters define the capacity of the guest molecule to enter the cyclodextin cavity, and the lipophilicity of the guest molecule determines the strength of interaction with the hydrophobic inner surface of the CD cavity. The results in Table 99 prove the importance of these interactions.

2. Polar interactions between solutes and the polar groups on the β-CD polimer surface or between the polar substructures of solutes and free silanol groups not covered by β-CD polymer. These interactions are probably influenced by the electron-withdrawing and electron-donor properties of the polar substructures of barbituric acid derivatives.

The parameters of equations describing the influence of organic modifiers on the retention behavior of β-CD columns are shown in Table 100. The solvent strength of organic modifiers is determined by the surface/volume ratio (see Eq. 1 in Table 100). As the inclusion complex formation depends markedly on the capacity of the guest molecule to enter the CD cavity (that is, on the steric parameters), this result suggests the involvement of inclusion complex formation in the retention mechanism of β-CD columns. The selectivity of organic modifiers depends on their electronic parameters (σ) and molar refraction (bulkiness; see Eqs. 2 and 3 in Table 100). These results emphasize again the impact of polar interactions and that of inclusion complex formation between the molecules of organic modifiers and the surface of β-CD polymer-coated silica.

The chromatographic properties of 17 monoamine oxidase inhibitory drugs (pro-parlgylamine derivatives) were investigated by the application of HPLC and TLC methods.[515] The relative strength of interaction between the drugs and a water soluble β-cyclodextrin polymer was determined by charge-transfer chromatography carried out on reversed-phase TLC layers. The relationship between capacity factors, physicochemical parameters, and inclusion complex-forming capacity of the monoamine oxidase inhibitory

TABLE 99

Effect of Various Physicochemical Parameters of Barbituric Acid Derivatives on Their Retention Behavior on β-CD Column. Results of Stepwise Regression Analysis (Number of Samples = 45)

$$\log k' = a + b_1 \cdot x_1 + b_2 \cdot x_2$$

	I	II	III	IV	V
a	−0.47	0.98	−0.71	−0.62	−0.69
b_1	0.02	0.19	0.17	0.05	0.06
x_1	M-RE	Σπ	Σπ	Σπ	Σπ
S_{b2}	0.01	0.04	0.03	0.03	0.01
$b_{1\%}$	71.27	—	52.16	34.73	—
b_2	−0.29	—	0.31	0.01	—
x_2	σ	—	H-Ac	M-RE	—
S_{b2}	0.12	—	0.05	0.01	—
$b_{2\%}$	28.72	—	47.84	65.24	—
r^2	0.5321	0.3060	0.6127	0.4569	0.1421
F	21.61	18.08	31.64	16.83	6.79

Note: Eluent (1:1 v/v): I, methanol–water; II, acetonitrile–water; III, ethanol–water; IV, dioxane–water; V, tetrahydrofuran–water. a, intercept; b_1 and b_2, regression coefficients; s_{b1} and s_{b2}, standard deviations of regression coefficients b_1 and b_2; $b_{1\%}$ and $b_{2\%}$, path coefficients (dimensionless numbers indicating the relative impact of the individual independent variables on the dependent variable); r^2, coefficient of determination (indicates the ratio of variance explained by the independent variables); F, calculated value of the Fisher significance test.

From Forgács, E. and Cserháti, T., *J. Chromatogr.*, 668, 1994, 395–402. With permission.

TABLE 100

Relationships between Solvent Strength and Selectivity of Organic Modifiers and Their Physicochemical Parameters. Results of Stepwise Regression Analysis

$$y = a + b \cdot x$$

	No. of Equation		
Parameter	1	2	3
a	7.05	102.10	−39.91
b	19.63	273.07	8.01
x	B_4/M-RE	σ	M-RE
S_b	5.00	67.93	1.49
r	0.9147	0.9183	0.9516

Note: 1. y = solvent strength of eluent systems (roman numbers refer to eluent systems in Table 99): I, 2.39; II, 1.57; III, 0.19; IV, 1.17; V, 2.83. 2. y = 1. coordinate of selectivity map of eluent systems. 3. y = 2. coordinate of selectivity map of eluent systems. For symbols see Table 99.

From Forgács, E. and Cserháti, T., *J. Chromatogr.*, 668, 1994, 395–402. With permission.

drugs were evaluated by stepwise regression analysis and by principal component analysis (PCA) followed by two-dimensional nonlinear mapping and varimax rotation. Stepwise regression analysis indicated that the retention of monoamine oxidase inhibitory drugs significantly depend on their calculated lipophilicity:

$$\log k' = -0.60 + (0.13 \pm 0.02)\pi$$

$$r_{calc.} = 0.8639; \quad r_{99.9\%} = 0.7246$$

Standard error of the estimate = 0.09 \hfill (489)

Equation 489 clearly shows that the calculated hydrophobicity of solutes has the highest impact on their retention, that is, the retention mechanism of β-CDP support may be similar to that of the traditional alkyl-bonded silica phases. However, the relatively low ratio of variance explained (74%) indicates that factors other than molecular hydrophobicity may also have a significant influence on the retention. The results of stepwise regression analysis slightly contradict the results in Figure 38. This discrepancy can be tentatively explained by the supposition of a slight difference between the measured and calculated hydrophobicity values.

The results of PCA are summarized in Table 101. The log k' values — together with the measured hydrophobicity, complex-forming capacity, and steric and electronic parameters of drugs — have high loading in the second PC, indicating the marked influence of these parameters on the mode of retention of the β-CDP support. The distribution of variables on the two-dimensional nonlinear map of PC loadings supports our previous conclusions (Figure 39); the log k' values form a loose cluster with the measured hydrophobicity, complex-forming capacity, and steric and electronic parameters of drugs. This finding indicates again the mixed retention mechanism of the β-CDP support.

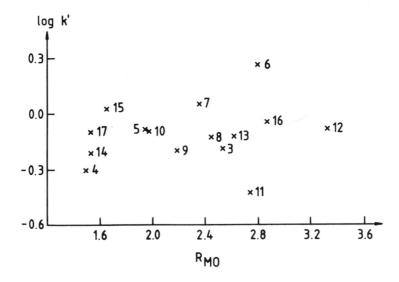

FIGURE 38
Relationship between the retention of monoamine oxidase inhibitory drugs on β-cyclodextrin polymer-coated silica column and the measured hydrophobicity value (R_{M0}). (From Forgács, E., *J. Pharm. Biomed. Anal.*, 13, 1993, 525–532. With permission.)

TABLE 101
Similarities and Dissimilarities between the Physicochemical Parameters of Monoamine Oxidase Inhibitory Drugs and Their Retention on β-Cyclodextrin Polymer-Coated Silica Column. Results of Principal Component Analysis

No. of Component	Eigenvalue	Total Variance Explained %
1	5.45	36.38
2	3.11	57.15
3	2.64	74.53
4	1.47	84.56
5	0.93	90.79

Principal Component Loadings

	1	2	3	4	5
π	0.12	0.14	−0.55	0.18	0.70
H-Ac	0.89	−0.38	0.17	0.05	0.01
H-Do	−0.82	0.29	0.29	−0.25	−0.02
M-Re	−0.61	0.34	0.17	0.50	0.20
F	0.81	−0.15	0.43	0.27	−0.01
R	0.40	0.68	−0.18	−0.52	0.02
σ	0.53	−0.46	−0.60	0.15	−0.21
Es	0.45	0.62	−0.30	−0.47	0.16
B_1	−0.82	−0.07	0.08	0.17	0.29
B_4	0.42	0.36	0.77	0.19	0.17
b_1	0.22	−0.06	0.90	−0.20	0.06
b_2	0.19	−0.59	−0.28	0.61	−0.24
R_M	0.79	−0.51	0.04	0.17	−0.04
log k'	0.11	0.90	0.03	0.17	−0.15

From Forgács, E., *J. Pharm. Biomed. Anal.*, 13, 1993, 525–532. With permission.

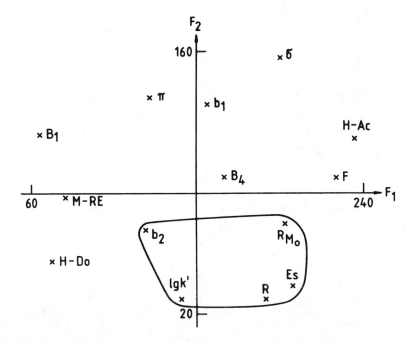

FIGURE 39
Two-dimensional nonlinear map of PC loadings. Number of iterations 86; maximum error, 6.29×10^{-3}. For symbols see text. (From Forgács, E., *J. Pharm. Biomed. Anal.*, 13, 1993, 525–532. With permission.)

Appendix

CHEMOMETRICS IN CHROMATOGRAPHY

The development of new calculation methods to interpret large retention data matrices has been one of the major advances in chromatography during the last decade. The most decisive features of this rapidly evolving field are the various automated chromatographic instruments, high-speed computers, and a variety of mathematical-statistical methodologies. The evaluation of data sets containing a considerable number of information (i.e., retention times of a great number of compounds determined on several chromatographic columns using a wide variety of eluents) is practically impossible by the traditional linear regression model. The modern multivariate mathematical-statistical methods make possible the simultaneous assessment of a practically unlimited number of variables (generally chromatographic parameters) that highly facilitates the solution of both theoretical and practical problems. Multivariate methods have been mainly used in chromatography to identify basic factors that have a significant impact on solute–solvent interactions and to cluster solutes, supports, and solvents into groups exhibiting similar retention characteristics. However, the application of multivariate mathematical-statistical methods in various fields of chromatography is hampered by the fact that these methods have been developed for the evaluation of other than chromatographic data matrices, therefore their successful application considerably depends on the expertise of the chromatographer in mathematical statistics. As each multivariate mathematical-statistical method highlights only one aspect of the chromatographic problem under investigation, the use of more than one method is rather a rule than an exception. The application of various mathematical-statistical methods for quantitative structure–retention relationship studies has been many times reviewed.[516-518]

STEPWISE REGRESSION ANALYSIS

The application of various multilinear (or nonlinear) regression methods requires one dependent variable (i.e., retention time or any other retention parameter) that may or may not depend significantly on the independent variables (various physicochemical parameters of solvents such as solvent strength, dielectric constant, dipole moment, etc.; solute substructures, solute hydrophobicity, and many other parameters that may have a significant impact on the change of the dependent variable). However, in the traditional multivariate regression analysis, the presence of independent variables (chromatographic parameters) that exert no significant influence on the dependent variable (retention behavior) lessens the significance level of the independent variables that significantly influence the dependent variable. To overcome this difficulty, stepwise regression analysis automatically eliminates from the selected equation the insignificant independent variables, increasing in this manner the information power of the calculation.[519] When we are interested in the relative impact of the individual independent variables on the dependent variables (i.e., the concentration of the organic modifier or the column temperature have a higher effect on the retention) we have to calculate the normalized slope values (path coefficients). The path coefficients are dimensionless numbers indicating the relative importance of a given independent variable (chromatographic parameter). The calculation programs generally contain the F values. When our calculated F value is higher than the corresponding tabulated one found in each statistical handbook the relationship between the variables is significant. The r^2 value indicates the ratio of the change of dependent variable explained by the change of the independent variables. In other words, $r^2 = 0.9200$ indicates that 92% of the change of the dependent variables can be explained by the change of the independent variables.

CANONICAL CORRELATION ANALYSIS

Sometimes the scientists are interested in the simultaneous dependence of more than one retention parameter or more than one chromatographic characteristic on a high number of chemical (eluent composition, eluent pH, etc.), physical (temperature, flow rate, etc.), and physicochemical (connectivity, sterical indices, etc.) variables. In these instances multilinear regression methods cannot be applied because we have more than one dependent variable. Canonical correlation analysis (CCA) has been developed to find a relationship between two different data matrices.[520] First, CCA extracts theoretical factors that explain the maximal variance in the matrix containing the lower number of variables (matrix II, i.e., more than one retention parameter of a set of solutes). The number of factors explaining 100% of the variance in matrix II is equal the number of original variables in matrix II. Then, CCA extracts from the other matrix (matrix I, i.e., various physicochemical parameters of the same solutes) variables that have the highest correlation with the factors extracted from matrix II. CCA calculates the standard canonical coefficients, weighted canonical coefficients, loadings of the variables of matrix I in the canonical variates from matrix I, loadings of the variables of matrix I in the canonical variates from matrix II, loadings of the variables of matrix II in the canonical variates from matrix II, loadings of the variables of matrix II in the canonical variates from matrix I, variances in matrix I explained by the variates extracted from matrix I, variances in matrix II explained by the variates extracted from matrix I, variances in matrix II explained by the variates extracted from matrix II, variances in matrix I explained by the variates extracted from matrix II, and the χ values of the equations. Standard canonical coefficients are similar to the regression coefficients of traditional multivariate regression analysis: they can be used to calculate and predict dependent variables previously not determined. The weighted canonical coefficients are similar in character to the normalized slope values (path coefficients): they contain information about the relative impact of the individual variables (both dependent and independent ones) in the equations selected by CCA. The various loadings and variances explained refer to the importance of the variables in the equation and to the capacity of the equation to explain the variance of the original data matrix, respectively. The calculated χ values can be compared by the similar tabulated values. When the calculated χ value is higher than the tabulated one corresponding to a given significant level, the relationship is significant.

PRINCIPAL COMPONENT AND FACTOR ANALYSIS

In many cases the scientist is not interested in the dependence of one (see above stepwise regression analysis) or more (see canonical correlation analysis) retention parameters on the other chromatographic or physicochemical characteristics, but rather wishes to find the relationship between all parameters without one or more being the dependent variable. Both factor analysis (FA)[521–523] and principal component analysis (PCA)[524–527] complies with these requirements. The main advantages of these methods in chromatography are

1. clustering of variables according to their relationship (clustering chromatographic systems or solutes according to their retention behavior);
2. the possibility of the extraction of one or more background variables having concrete physicochemical meaning for the theory and practice of chromatography;
3. decrease in the number of variables (a decrease in the number of chromatographic systems or solutes to the minimum necessary for the solution of a problem);
4. combination with visualization methods such as two-dimensional nonlinear mapping[528] or cluster analysis[529] considerably facilitates the evaluation.

PCA can be carried out both on the correlation and covariance matrices of the original data set. The matrix for PCA and FA is generally composed of a set of measured (retention time and capacity, peak symmetry, theoretical plate number, etc.) and calculated chromatographic and/or physicochemical parameters (hydrophobicity, specific hydrophobic surface area, electron-withdrawing capacity, etc.) of homologous or nonhomologous series of solutes. The parameters are taken as variables and the solutes as observations. It is preferable that the number of observations is three times the number of variables.

The basis of these methods is the extraction of a component or factor that explains the highest ratio of variance in the original data matrix (first principal component or first factor). Then the calculation (extraction of principal components or factors) continues as long as the ratio of variance determined previously by the scientist is explained. The aim of the calculation is the continuous extraction of information as much as the possible extracted components or factors are orthogonal. These methods calculate the eigenvalues of principal components (or factors), the variance explained by them, the loadings and variables. Eigenvalues are numbers related to the variance explained, but they are not identical to them. It is generally accepted that only components with eigenvalues over 1 are valid for further consideration. Loadings are relative numbers referring to the impact of individual variables on the given component or factor. High values indicate that the variable has a marked influence on the component. Unfortunately, these methods have no significance test, and therefore the evaluation and interpretation of the results considerably depends on the consideration of the scientist.

Because the human brain is limited to visualizing no more than three dimensions, the evaluation of the multidimensional matrices of loadings and variables is fairly difficult. Many methods have been developed to overcome this difficulty. The nonlinear mapping technique projects the points from a multidimensional space in a two-dimensional plane in such a manner that the relative distances of the points on the two-dimensional plane is similar to the distances in the multidimensional space. The points representing chromatographic parameters or solutes near to each other on the two-dimensional map can be considered to be related to each other. Cluster analysis is a similar method to decrease the dimensionality of data matrices; however, in contrast to the nonlinear mapping technique the points are equidistantly distributed in one dimension and only the other dimension contains information about their similarity or dissimilarity. As was previously mentioned, components or factors form a multidimensional and orthogonal structure. This orthogonality can be modified by the varimax rotation where the axes are constructed in such a manner that they explain the highest ratio of variance.

REFERENCES

1. Giddings, J. C. and Eyring, H., A molecular dynamic theory of chromatography. *J. Phys. Chem.*, 59, 1953, 416–421.
2. Giddings, J. C., Stochastic considerations on chromatography dispersion. *J. Chem. Phys.*, 26, 1957, 169–173.
3. Giddings, J. C., The random downstream migration of molecules in chromatography. *J. Chem. Educ.*, 35, 1958, 588–591.
4. MacQuarrie, D. A., On the stochastic theory of chromatography. *J. Chem. Phys.*, 38, 1963, 437–445.
5. Scott, D. M. and Fritz, J. S., Model for chromatographic separation based on renewal theory. *Anal. Chem.*, 56, 1984, 1561–1566.
6. Dondi, F. and Remelli, M., The characteristic function method in the stochastic theory of chromatography. *J. Phys. Chem.*, 90, 1986, 1885–1891.
7. Woodbury, C. P., Jr., A stochastic model of chromatography. *J. Chromatogr. Sci.*, 32, 1994, 339–348.
8. Jönsson, J. A. (Ed.), *Chromatographic Theory and Basic Principles*. Marcel Dekker, Inc., New York and Basel, 1987.
9. Heftmann, E. (Ed.), *Chromatography, Part A: Fundamentals and Techniques*. Elsevier, Amsterdam, 1983.
10. Berezkin, V. G., Adsorption in gas-liquid chromatography. *J. Chromatogr.*, 159, 1978, 359–396.
11. Conder, J. R., Existence of gas-liquid interfacial adsorption in solutions of alcohols and ketones in saturated hydrocarbons. *Anal. Chem.*, 48, 1976, 917–918.
12. Eon, C. and Guiochon, G., Surface activity coefficients as studied by gas chromatography. *J. Colloid Interface Sci.*, 45, 1973, 521–528.
13. Castells, R. C., Arancibia, E. L., and Nardillo, A. M., Surface and bulk activity coefficients of non-electrolytic mixtures studied by gas chromatography. *J. Colloid Interface Sci.*, 90, 1982, 532–535.
14. Riedo, F. and Kováts, E. S., Effects of adsorption on solute retention in gas-liquid chromatography. *J. Chromatogr.*, 186, 1979, 47–62.
15. Jönsson, J. A. and Mathiasson, L., Mixed retention mechanism in gas-liquid chromatography. III. Determination of the gas-liquid adsorption coefficient for diiso-propyl ether on n-octadecane. *J. Chromatogr.*, 206, 1981, 1–5.
16. Karger, B. L., Castells, R. C., Sewell, P. A., and Hartkopf, A., Study of the adsorption of insoluble and sparingly soluble vapors at gas-liquid interface of water by gas chromatography. *J. Phys. Chem.*, 75, 1971, 3870–3879.
17. Tóth, J., Gas-(Dampf-)Adsorption auf festen Oberflachen inhomogener Aktivität. III. *Acta Chim. Acad. Sci. Hung.*, 32, 1962, 39–57.
18. Tóth, J., Rudzinski, W., Waksmundski, A., Jaroniec, M., and Sokolowski, S., Adsorption of gases on heterogeneous solid surfaces: The energy distribution function corresponding to a new equation for monolayer adsorption. *Acta Chim. Acad. Sci. Hung.*, 82, 1974, 11–20.
19. Snyder, L. R., *Principles of Adsorption Chromatography*. Marcel Dekker, New York, 1968.
20. Eon, C. H., Study of liquid-powder interfaces by means of solvent strength parameter measurements. *Anal. Chem.*, 47, 1975, 1871–1873.
21. Scott, R. P. and Kucera, P., Solute-solvent interactions on the surface of silica gel. *J. Chromatogr.*, 149, 1972, 93–110.
22. Slaats, E. H., Kraak, J. C., Brugman, W. J. T., and Poppe, H., Study of the influence of competition and solvent interaction on retention in liquid-solid chromatography by measurement of activity coefficients in the mobile phase. *J. Chromatogr.*, 149, 1978, 255–270.

23. Slaats, E. H., Markowski, W., Fekete, J., and Poppe, H., Distribution eqilibria of solvent components in reversed-phase liquid chromatographic columns and relationship with the mobile phase volume. *J. Chromatogr.*, 207, 1981, 299–323.
24. Busev, S. A., Zverev, S. I., Larionov, O. G., and Jakubov, S., Study of adsorption from solutions by column chromatography. *J. Chromatogr.*, 241, 1982, 287–294.
25. Selim, M. I., Parcher, J. F., and Lin, P. J., Adsorption of polar solutes on liquid-modified supports. *J. Chromatogr.*, 239, 1982, 411–421.
26. Wahlund, K. G. and Beijertsen, I., Adsorption of 1–pentanol on alkyl-modified silica and its effect on the retention mechanism in reversed-phase liquid chromatogrpahy. *Anal. Chem.*, 54, 1982, 128–132.
27. Hill, T. L., Statistical mechanics of adsorption. VI. Localized unimolecular adsorption on a heterogeneous surface. *J. Chem. Phys.*, 17, 1949, 762–771.
28. Rudzinski, W., Jagiello, J., and Grillet, Y., Physical adsorption of gases on heterogeneous solid surfaces: Evaluation of the adsorption energy distribution from adsorption isotherms and heats of adsorption. *J. Colloid Interface Sci.*, 87, 1982, 478–491.
29. Rudzinski, W., Oscik, J., and Dabrowski, A., Adsorption from solutions on patchwise heterogeneous solids. *Chem. Phys. Lett.*, 20, 1973, 444–447.
30. Rudzinski, W. and Partyka, S., Adsorption from solutions on solid surfaces: Effects of surface heterogeneity on adsorption isotherms and heats of immersion. *J. Chem. Soc. Faraday Trans.*, 1, 77, 1981, 2577–2584.
31. Rudzinski, W., Narkiewicz-Michalek, J., and Partyka, S., Adsorption from solutions on solid surfaces: Effects of topography of heterogeneous surfaces on adsoprtion isotherms and heats of immersion. *J. Chem. Soc. Faraday Trans.*, 1, 78, 1982, 2361–2368.
32. Snyder, L. R. and Glajch, J. L., Solvent strength of multicomponent mobile phases in liquid-solid chromatography: binary solvent mixtures and solvent localization. *J. Chromatogr.*, 214, 1981, 1–19.
33. Snyder, R. L., Mobile phase effects in liquid-solid chromatography: importance of adsorption-site geometry, adsorbate delocalization and hydrogen bonding. *J. Chromatogr.*, 255, 1983, 3–26.
34. Jaroniec, M., Klepacka, B., and Narkiewizc, J., Liquid adsorption chromatography with a two-component mobile phase. *J. Chromatogr.*, 170, 1979, 299–307.
35. Scott, R. P. and Kucera, P., Solute interactions with the mobile and stationary phases in liquid-solid chromatography. *J. Chromatogr.*, 112, 1975, 425–442.
36. Paanakker, J. E., Kraak, J. C., and Poppe, H., Some aspects of the mobile phase composition in normal-phase liquid-solid chromatography, with special attention to the role of water present in binary organic mixtures. *J. Chromatogr.*, 149, 1978, 111–126.
37. Schoenmakers, P. J., Billiet, H. A. H., and de Galan, L., Systemic study of ternary solvent behavior in reversed-phase chromatography. *J. Chromatogr.*, 218, 1981, 261–284.
38. Jandera, P., Colin, H., and Guiochon, G., Interaction indexes for prediction of retention in reversed-phase liquid chromatography. *Anal. Chem.*, 54, 1982, 435–441.
39. Horváth, C., Melander, W., and Molnár, I., Solvophobic interactions in liquid chromatography with nonpolar stationary phases. *J. Chromatogr.*, 125, 1976, 129–156.
40. Kováts, E., Gas-chromatographische characterisierung organischer Verbindungen. Teil 1: Retention indices aliphatischer Halogenide, Alkohole, Aldehyde und Ketone. *Helv. Chim. Acta*, 41, 1958, 1915–1918.
41. Novák, J., Vejrosta, J., Roth, M., and Janák, J., Correlation of gas chromatographic specific retention volumes of homologous compounds with temperature and methylene number. *J. Chromatogr.*, 199, 1980, 209–217.
42. Budahegyi, M. V., Lombosi, T. S., Mèszáros, S. Y., Nyiredy, Sz., Tarján, G., Timár, I., and Takács, J., Twenty-fifth anniversary of the retention index system in gas-liquid chromatography. *J. Chromatogr.*, 271, 1983, 213–229.
43. Schomburg, G., *Gas Chromatography. A Practical Course.* VCH, Weinheim, Germany, 1990.
44. Bruner, F., *Gas Chromatographic Environmental Analysis.* VCH, New York, 1993.
45. Fowlis, I., *Gas Chromatography.* 2nd Edition. John Wiley & Sons, New York, 1995.
46. Rohrschneider, L., Explanatory coefficients for stationary phases in gas chromatography from McReynolds phase constants. *Chromatographia*, 38, 1994, 679–688.
47. Rohrschneider, L., Eine Methode zur Characterisierung von gaschromatographischen Trennflüssigkeiten. *J. Chromatogr.*, 22, 1966, 6–22.
48. Abraham, M. H., Whiting, G. S., Doherty, R. M., and Shuely, W. J., Hydrogen bonding. XV. A new characterisation of the McReynolds 77–stationary phase set. *J. Chromatogr.*, 518, 1990, 329–348.
49. Poole, S. K., Kollie, T. O., and Poole, C. F., Influence of temperature on the mechanism by which compounds are retained in gas-liquid chromatography. *J. Chromatogr. A*, 664 (1994) 229–251.
50. Kersten, B. R. and Poole, C. F., Influence of concurrent retention mechanisms on the determination of stationary phase selectivity in gas chromatography. *J. Chromatogr.*, 399, 1987, 1–31.
51. Riedo, F. and Kováts, E. Sz., Effect of adsorption on solute retention in gas-liquid chromatography. *J. Chromatogr.*, 186, 1979, 47–62.

52. Fritz, D. F., Sahil, A., and Kováts, E. Sz., Study of the adsorption effects at the surface of poly(ethylene glycol)-coated column packings. *J. Chromatogr.*, 186, 1979, 63–80.
53. Liao, H.-L. and Martire, D. E., Concurrent solution and adsorption phenomena in gas-liquid chromatography — A comparative study. *Anal. Chem.*, 44, 1972, 498–502.
54. Nikolov, R. N., Identification and evaluation of retention mechanisms in gas-liquid chromatographic systems. *J. Chromatogr.*, 241, 1982, 237–256.
55. Karger, B. L. and Liao, H. S., The use of gas chromatography in the study of the adsorption and partition of nonelectrolytes with aqueous tetraalkylammonium bromide solution. *Chromatographia*, 7, 1974, 288–292.
56. Blumberg, L. M. and Berger, T. A., Molecular basis of peak width in capillary gas chromatography under high column pressure drop. *Anal. Chem.*, 65, 1993, 2686–2689.
57. Seybold, P. G. and Bertrand, J., A simple model for the chromatographic retentions of polyhalogenated biphenyls. *Anal. Chem.*, 65, 1993, 1631–1634.
58. Hasan, M. N. and Jurs, P. C., Computer-assisted prediction of gas chromatographic retention times of polychlorinated biphenyls. *Anal. Chem.*, 60, 1988, 978–982.
59. Hasan, M. N. and Jurs, P. C., Prediction of gas and liquid chromatographic retention indices of polyhalogenated biphenyls. *Anal. Chem.*, 62, 1990, 2318–2323.
60. Robbat A., Jr., Xyrafas, G., and Marshall, D., Prediction of gas chromatographic retention characteristic of polychlorinated biphenyls. *Anal. Chem.*, 60, 1988, 982–985.
61. Höfler, F., Melzer, H., Möckel, H. J., Obertson, L. W., and Anklam, E., Relationship between liquid and gas chromatographic retention behavior and calculated molecular surface area of selected polyhalogenated biphenyls. *J. Agric. Food Chem.*, 36, 1988, 961–965.
62. Havelec, P. and Sevcik, J. G. K., Concept of additivity for a nonpolar solute-solvent criterion log L^{16}. Nonaromatic compounds. *J. Chromatogr. A*, 677, 1994, 319–329.
63. McCoy, B. J., Partitioning and adsorption in gas-liquid-solid chromatography: models of temporal moments for retention and variance. *Chromatographia* 36, 1993, 234–240.
64. Surowiec, K. and Rayss, J., The structure and chromatographic properties of poly(ethyleneglycol) 20M layer on the surface of a diatomaceous earth type support. *Chromatographia*, 37, 1993, 444–450.
65. Surowiec, K. and Rayss, J., The properties of poly(ethylene glycol) 20M layer on the surface of a silanized support. *Chromatographia*, 27, 1989, 412–416.
66. Surowiec, K. and Rayss, J., The retention mechanism in poly(ethylene glycol) 20M layer on the surface of a silanized support. *Chromatographia*, 30, 1990, 630–634.
67. Yin, H. F., Zhu, Y., and Sun, Y. L., Variation of retention index of a solute due to the presence of a large neighboring peak. *Chromatographia*, 28, 1989, 502–504.
68. Berezkin, V. G., Adsorption in gas-liquid chromatography. *J. Chromatogr.*, 159, 1978, 359–396.
69. Berezkin, V. G. and Korolev, A. A., Investigation of the role of adsorption at the stationary phase interface in capillary columns prepared with cross-linked phases. *Chromatographia* 20, 1985, 482–486.
70. Hennig, P. and Engewald, W., Influence of adsorption effects on retention indices of selected C_{10}-hydroxy compounds at various temperatures. *Chromatographia*, 38, 1994, 93–97.
71. Kupchik, E. J., Structure-gas chromatographic retention time models of tetra-n-alkylsilanes and tetra-n-alkylgermanes using topological indexes. *J. Chromatogr.*, 630, 1993, 223–230.
72. Kupchik, E. J., Structure-gas chromatographic retention time models of tetra-n-alkanes and tetra-n-alkylgermanes using topological indexes. A correction. *J. Chromatogr.*, 645, 1993, 182–184.
73. Semlyen, J. A., Walker, G. R., Blofeld, R. E., and Pillips, C. S. G., The chromatography of gases and vapours. Part VIII. Some alkyl derivatives of silicon and germanium. *J. Chem. Soc. B31*, 1964, 4948–4953.
74. Kier, L. B., Murray, W. J., Randic, M., and Hall, L. H., Molecular connectivity V: connectivity series concept applied to density. *J. Pharm. Sci.*, 65, 1976, 1226–1230.
75. Kier, L. B. and Hall, L. H., A differential molecular connectivity index. *Quant. Struct. Act. Relat.*, 10, 1991, 134–140.
76. Hall, L. H. and Kier, L. B., Determination of topological equivalence in molecular graphs from the topological state. *Quant. Struct. Act. Relat.*, 9, 1990, 115–131.
77. Hall, L. H., Mohney, B., and Kier, L. B., The electrotopological state: an atom index for QSAR. *Quant. Struct. Act. Relat.*, 10, 1991, 43–51.
78. Schmidt, T. J., Merfort, I., and Willuhn, G., Gas chromatography-mass spectrometry of flavonoid aglycones. II. Structure-retention relationships and a possibility of differentiation between isomeric 6- and 8-methoxyflavones. *J. Chromatogr. A*, 669, 1994, 236–240.
79. Katritzky, A. R., Ignatchenko, E. S., Barcock, R. A., Lobanov, V. S., and Karelson, M., Prediction of gas chromatographic retention times and response factors using a general quantitative structure-property relationship treatment. *Anal. Chem.*, 66, 1994, 1799–1807.
80. Zhang,Y., Dallas, A. J., and Carr, P. W., Critical comparison of gas-hexadecane partition coefficients as measured with packed and open tubular capillary columns. *J. Chromatogr.*, 638, 1993, 43–56.

81. Abraham, M. H., Grellier, P. L., and McGill, R. A., Determination of olive oil-gas and hexadecane-gas partition coefficients, and calculation of the corresponding olive oil-water and hexadecane-water partition coefficients. *J. Chem. Soc. Perkin Trans.*, 2, 1987, 797–803.
82. Kollie, T. O., Poole, C. F., Abraham, M. H., and Whiting, G. S., Comparison of two free energy of solvation models for characterizing selectivity of stationary phases used in gas-liquid chromatography. *Anal. Chim. Acta*, 259, 1992, 1–13.
83. Poole, C. F., Kollie, T. O., and Poole, S. K., Recent advances in solvation models for stationary phase characterization and the prediction of retention in gas chromatography. *Chromatographia*, 34, 1992, 281–302.
84. Abraham, M. H., Andonian-Haftvan, J., Hamerton, I., Poole, C. F., and Kollie, T. O., Hydrogen bonding. XXVIII. Comparison of the solvation theories of Abraham and Poole, using a new acidic gas-liquid chromatography stationary phase. *J. Chromatogr.*, 646, 1993, 351–360.
85. Li, J. and Carr, P. W., Gas chromatographic study of solvation enthalpy by solvatochemically based linear solvation energy relationships. *J. Chromatogr. A*, 659, 1994, 367–380.
86. Li, J., and Carr, P. W., Extra-thermodynamic relationships in chromatography. Enthalpy-entropy compensation in gas chromatography. *J. Chromatogr. A*, 670, 1994, 105–116.
87. Li, J., Dallas, A. J., and Carr, P. W., Empirical scheme for the classification of gas chromatographic stationary phases based on solvatochromic linear solvation energy relationships. *J. Chromatogr.*, 517, 1990, 103–121.
88. Li, J., Zhang, Y., and Carr, P. W., Development of a gas chromatographic scale of solute hydrogen bond acceptor basicity and characterization of some hydrogen bond donor phases by use of linear solvation energy relationships. *Anal. Chem.*, 65, 1993, 1969–1979.
89. Li, J., Zhang, Y., Ouyang, H., and Carr, P. W., Gas chromatographic study of solute hydrogen bond basicity. *J. Am. Chem. Soc.*, 114, 1992, 9813–9828.
90. Guillaume, Y. and Giunchard, C., Effect of esther molecular structure and column temperature on the retention of eight esters in gas chromatography. *Chromatographia*, 39, 1994, 438–442.
91. Abraham, M. H., Whiting, G. S., Doherty, R. M., and Shuely, W. J., Hydrogen bonding. XV. A new characterisation of the McReynolds 77-stationary phase set. *J. Chromatogr.*, 518, 1990, 329–348.
92. Abraham, M. H., Whiting, G. S., Doherty, R. M., and Shuely, W. J., Hydrogen bonding. XVI. A new solute solvation parameter, π_2^H, from gas chromatographic data. *J. Chromatogr.*, 587, 1991, 213–228.
93. Li, J., Zhang, Y., and Carr, P. W., Novel triangle scheme for the classification of gas chromatographic phases based on solvatochromic linear solvation energy relationships. *Anal. Chem.*, 64, 1992, 210–218.
94. Ettre, L. S., The retention index system; its utilization for substance identification and liquid phase characterization. Part II: Correlation between retention index, structure and analytical characteristics. *Chromatographia*, 7, 1974, 39–46.
95. Haken, J. K., Retention indices in gas chromatography. *Adv. Chromatogr.*, 14, 1976, 376–407.
96. Tarján, G., Nyiredy, Sz., Györ, M., Lombosi, E. R., Lombosi, T. S., Budahegyi, M. V., Mèszáros, S. Y., and Takács, J. M., Thirtieth anniversary of the retention index according to Kováts in gas-liquid chromatography. *J. Chromatogr.*, 472, 1989, 1–92.
97. Evans, M. B. and Haken, J. K., Recent development in the gas chromatographic retention index scheme. *J. Chromatogr.*, 472, 1989, 93–127.
98. Saura-Calixto, F., Garcia-Raso, A., Cannelas, J., and Garcia-Raso, J., Correlations between GLC retention indices and boiling temperatures. II. Influence of the stationary phase polarity. *J. Chromatogr. Sci.*, 21, 1983, 267–271.
99. Chelghoun, C., Haddad, Y., Abdeddaim, K., and Guermouche, M. H., Indices de rétention et propriétés physico-chimiques de hydrocarbures en C_6. *Analusis*, 14, 1986, 49–51.
100. Bermejo, J., Canga, J. S., Gayol, O. M., and Guillen, M. D., Utilization of physico-chemical properties and structural parameters for calculating retention indices of alkylbenzenes. *J. Chromatogr. Sci.*, 22, 1984, 252–255.
101. Kaliszan, R. and Höltje, H. D., Gas chromatographic determination of molecular polarity and quantum chemical calculation of dipole moments in a group of substituted phenols. *J. Chromatogr.*, 324, 1982, 303–311.
102. Voelkel, A., Retention indices and thermodynamic functions of solution for model non-ionic surfactants in standard stationary phases determined by gas chromatography. *J. Chromatogr.*, 387, 1987, 95–104.
103. Calixto-Saura, F. and Garcia-Raso, A., Retention index, connectivity index and van der Waals volume of alkanes (GLC). *Chromatographia*, 15, 1987, 521–524.
104. Lamparczyk, H. and Radecki, A., The role of electric interactions in the retention index concept; implications in quantitative structure-retention studies. *Chromatographia*, 18, 1984, 615–618.
105. Garcia-Raso, A., Saura-Calixto, F., and Raso, M. A., Study of gas chromatographic behaviour of alkenes based on molecular orbital calculations. *J. Chromatogr.*, 302, 1984, 107–117.
106. Ong, V. S. and Hites, R. A., relationship between gas chromatographic retention indexes and computer-calculated physical properties of four compound classes. *Anal. Chem.*, 63, 1991, 2829–2834.
107. Burgan, P., Macak, J., Triska, J., Vodicka, L., Berlizov, Y. S., Omitrikov, V. R., and Navibach, V. M., Kováts retention indices of alkylquinolines on capillary columns. *J. Chromatogr.*, 391, 1987, 89–96.

108. Lamparczyk, H., Wilczinska, D., and Radecki, A., Relationship between the average molecular polarizabilities of polycyclic aromatic hydrocarbons and their retention indices determined on various stationary phases. *Chromatographia*, 17, 1984, 300–302.
109. Patte, F., Tcheto, M. E., and Laffort, P., Solubility factors for 240 solutes and 207 stationary phases in gas-liquid chromatography. *Anal. Chem.*, 54, 1982, 2239–2247.
110. Peng, C. T., Ding, S. F., Hua, R. L., and Yang, Z. C., Prediction of retention indexes. I. Structure-retention index relationship on apolar column. *J. Chromatogr.*, 436, 1988, 137–172.
111. Peng, C. T., Yang, Z. C., and Maltby, D., Prediction of retention indexes. III. Silylated derivatives of polar compounds. *J. Chromatogr.*, 586, 1991, 113–129.
112. Randic, M., On characterization of molecular branching. *J. Am. Chem. Soc.*, 97, 1975, 6609–6615.
113. Oszczapovic, J., Limitations of additivity of Kováts retention indices. *J. Chromatogr. A*, 668, 1994, 435–439.
114. Donnelly, J. R., Abdel-Hamid, M. S., Jeter, J. L., and Gurka, D. F., Application of gas chromatographic retention properties to the identification of environmental contaminants. *J. Chromatogr.*, 642, 1993, 409–411.
115. Abraham, M. H., Whiting, G. S., Doherty, R. M., and Shuely, W. J., Hydrogen bonding. XVII. The characterisation of 24 gas-liquid chromatographic stationary phases studied by Poole and co-workers, including molten salts, and evaluation of solute-stationary phase interactions. *J. Chromatogr.*, 587, 1991, 229–236.
116. Abraham, M. H., Whiting, G. S., Doherty, R. M., and Shuely, W. J., Hydrogen bonding. Part 13: A new method for the characterisation of GLC stationary phases — the Laffort data set. *J. Chem. Soc. Perkin Trans. 2*, 1990, 1451–1460.
117. Abraham, M. H., Whiting, G. S., Andonian-Haftvan, J., Steed, J. W., and Grate, J. W., Hydrogen bonding. XIX. The characterisation of two poly(methylphenylsiloxane)s. *J. Chromatogr.*, 588, 1991, 361–364.
118. Abraham, M. H. and Whiting, G. S., Hydrogen bonding. XXI. Solvation parameters for alkylaromatic hydrocarbons from gas-liquid chromatographic data. *J. Chromatogr.*, 594, 1992, 229–241.
119. Abraham, M. H., Hydrogen bonding. XXVII. Solvation parameters for functionally substituted aromatic compounds and heterocyclic compounds, from gas-liquid chromatographic data. *J. Chromatogr.*, 644, 1993, 95–139.
120. Zhang, H. and Hu, Z., Utilization of total solubility parameter for calculating retention indices of alkylbenzenes. *Chromatographia*, 33, 1992, 575–580.
121. Hu, Z. and Zhang, H., Prediction of gas chromatographic retention indices of alkenes from the total solubility parameters. *J. Chromatogr.*, 653, 1993, 275–282.
122. Hilal, S. H., Carreira, L. A., Karickhoff, S. W., and Melton, C. M., Estimation of gas-liquid chromatographic retention times from molecular structure. *J. Chromatogr. A*, 662, 1994, 269–280.
123. Rohrschneider, L., Prediction of retention data from stationary phase constants in gas chromatography. *Chromatographia*, 37, 1993, 250–258.
124. Heinzen, V. E. F. and Yunes, R. A., Correlation between gas chromatographic retention indices of linear alkylbenzene isomers and molecular connectivity indices. *J. Chromatogr.*, 654, 1993, 183–189.
125. Szász, G., Valkó, K., Papp, O., and Hermecz, I., Relationship between connectivity indexes of pyrido(1,2a)pyrimidin-4-ones and their liquid-liquid partition data obtained by gas-liquid chromatography. *J. Chromatogr.*, 243, 1982, 347–351.
126. Szász, G., Papp, O., Vámos, J., Hankó-Novák, K., and Kier, L. B., Relationship between molecular connectivity indices, partition coefficients and chromatographic parameters. *J. Chromatogr.*, 269, 1983, 91–95.
127. Arruda, A. C., Heinzen, V. E. F., and Yunes, R. A., Relationship between Kováts retention indices and molecular connectivity indices of tetralones, coumarins and structurally related compounds. *J. Chromatogr.*, 630, 1993, 251–256.
128. Jalali-Heravi, M. and Garkani-Nejad, Z., Prediction of gas chromatographic retention indices of some benzene derivatives. *J. Chromatogr.*, 648, 1993, 389–393.
129. Egolf, L. M. and Jurs, P. C., Quantitative structure-retention and structure-odor intensity relationships for a diverse group of odor-active compounds. *Anal. Chem.*, 65, 1994, 3119–3126.
130. Woloszyn, T. F. and Jurs, P. C., Prediction of gas chromatographic retention data for hydrocarbons from naphthas. *Anal. Chem.*, 65, 1993, 582–587.
131. Tóth, T., Use of capillary gas-chromatography in collecting retention and chemical information for the analysis of complex petrochemical mixtures. *J. Chromatogr.*, 279, 1983, 157–165.
132. Wehrli, A. and Kováts, E. Sz., Gas-chromatographische Characterisierung organischer Verbindungen. Teil 3 Berechnung der Retentionindices aliphatischer, alicyclischer und aromatischer Verbindungen. *Helv. Chim. Acta*, 42, 1959, 2709–2736.
133. Kováts, E. Sz., Gas chromatographic characterization of organic substances in the retention index system. *Adv. Chromatogr.*, 1, 1965, 229–247.
134. Bautz, D. E., Dolan, J. W., Raddatz, W. D., and Snyder, L. R., Computer simulation (based on a linear-elution-strength approximation) as an aid for optimizing separations by programmed-temperature gas chromatography. *Anal. Chem.* 62, 1990, 1560–1567.

135. Bautz, D. E., Dolan, J. W., and Snyder, L. R., Computer simulation as an aid in method development for gas chromatography. I. The accurate prediction of separation as a function of experimental conditions. *J. Chromatogr.*, 541, 1991, 1–19.
136. Dolan, J. W., Snyder, L. R., and Bautz, D. E., Computer simulation as an aid in method development for gas chromatography. II. Changes in band spacing as a function of temperature. *J. Chromatogr.*, 541, 1991, 21–34.
137. Snyder, L. R., Bautz, D. E., and Dolan, J. W., Computer simulation as an aid in method development for gas chromatography. III. Examples of its application. *J. Chromatogr.*, 541, 1991, 35–58.
138. Helaimia, F. and Messadi, D., Calcul des indices de rétention de quelques plastifiant en chromatographie en phase gazeuse avec programmation linéaire de température. *C.R. Acad. Sci. Paris*, 309(II), 1989, 37–42.
139. Messadi, D., Helaimia, F., and Ali-Mokhnache, S., Comparaison des indices de rétention calculés par interpolations linéaire et spline cubique en chromatographie gazeuse avec programmation de température. *C.R. Acad. Sci. Paris*, 310(I), 1990, 1447–1451.
140. Messadi, D., Helaimia, F., Ali-Mokhnache, S., and Boumahraz, H., Accurate determination of retention indices in programed temperature gas chromatography. *Chromatographia*, 29, 1990, 429–434.
141. Messadi, D. and Ali-Mokhnache, S., Calculation of retention indices and peak widths in temperature programmed gas-liquid chromatography. *Chromatographia*, 37, 1993, 264–270.
142. Kokko, M., Effect of variations in gas chromatographic conditions on the linear retention indices of selected chemical warfare agents. *J. Chromatogr.*, 630, 1993, 231–249.
143. Castello, G., Moretti, P., and Vezzani, S., Comparison of different methods for the prediction of retention times in programmed-temperature gas chromatography. *J. Chromatogr.*, 635, 1993, 103–111.
144. Habgood, H. W. and Harris, W. E., Retention temperature and column efficiency in programmed temperature gas chromatography. *Anal. Chem.*, 32, 1960, 450–453.
145. Bemgard, A., Colmsjö, A., and Wrangskog, K., Prediction of temperature-programmed retention indexes for polynuclear aromatic hydrocarbons in gas chromatography. *Anal. Chem.*, 66, 1994, 4288–4294.
146. Sun, Y., Zhang, R., Wang, Q., and Xu, B., Programmed-temperature gas chromatographic retention index. *J. Chromatogr. A*, 657, 1993, 1–15.
147. Takács, J. M., Contribution to the retention index system in gas-liquid chromatography. *J. Chromatogr. Sci.*, 29, 1991, 382–389.
148. Fernández Sáchez, E., Fernández Torres, Garcia Dominguez, J. A., and Santiuste, J. M., Kováts coefficients for predicting polarities in silicone stationary phases and their mixtures. *Chromatographia*, 31, 1991, 75–79.
149. Santiuste, J. M., Molecular structure coefficients of Takacs. Their dependence on column temperature, stationary phase polarity and solute chemical nature. *Chromatographia*, 38, 1994, 701–708.
150. Szentirmai, Z., Tarján, G., and Takács, J., Contribution to the theory of the retention index system. V. Characterization of polarity of stationary phases. McReynolds system. *J. Chromatogr.*, 73, 1972, 11–18.
151. Santiuste, J. M., Prediction of specific retention volumes in gas chromatography by using Kováts and molecular structural coefficients. *J. Chromatogr. A*, 690, 1995, 177–186.
152. Matisová, E., Rukriglová, M., Krupcik, J., Kovacicová, E., and Holotik, S., Identification of alkylbenzenes up to C_{12} by capillary gas chromatography and combined gas chromatography-mass spectrometry. I. OV-101 as stationary phase. *J. Chromatogr.*, 455, 1988, 301–309.
153. Matisová, E., Kovacicová, E., Ha, P. T., Kolek, E., and Engewald, W., Identification of alkylbenzenes up to C_{12} by capillary gas chromatography and combined gas chromatography-mass spectrometry. II. Retention indices on OV-101 columns and retention-molecular structure correlations. *J. Chromatogr.*, 475, 1989, 113–123.
154. Dimov, N. and Mekenyan, O., Quantitative relationships between the structure of alkylbenzenes and their gas chromatographic retention on stationary phases with different polarity. *J. Chromatogr.*, 471, 1989, 227–236.
155. Svob, V. and Deur-Siftar, D., Kováts retention indices in the identification of alkylbenzene degradation products. *J. Chromatogr.*, 91, 1974, 677–689.
156. Matisová, E. and Kurán, P., Reproducibility of retention indices measurement of alkylbenzenes on crosslinked methyl silicone phase fused silica capillaries. *Chromatographia* 30, 1990, 328–332.
157. Johansen, N. G. and Ettre, L. S., Retention index values of hydrocarbons on open-tubular columns coated with methylsilicone liquid phase. *Chromatographia*, 15, 1982, 625–630.
158. Lubeck, J. A. and Sutton, D. L., Kováts retention indices of selected hydrocarbons through C_{10} on bonded phase fused silica capillaries. *HRC&CC*, 6, 1983, 328–332.
159. Pacáková, V., Koch, K., and Smolková, E., The effect of instrumentation on the precision of retention indices. *Chromatographia*, 6, 1973, 320–324.
160. Dimov, N., Unified retention index of hydrocarbons separated on squalane. *J. Chromatogr.*, 347, 1985, 366–374.
161. Boneva, S. and Dimov, N., Unified retention index of hydrocarbons separated on dimethylsilicone OV-101. *Chromatographia*, 21, 1986, 697–700.

162. Skrbic, B. D. and Cvejanov, J. Dj., Unified retention indices of alkylbenzenes on OV-101 and SE-30. *Chromatographia*, 37, 1993, 215–217.
163. Soják, L., Krupcik, J., and Janák, J., Gas chromatography of all C_{15}-C_{18} linear alkenes on capillary columns with very high resolution power. *J. Chromatogr.*, 195, 1980, 43–64.
164. Bemgard, A. and Blomberg, L., Some factors influencing the precision in the determination of retention indices on polar capillary columns for gas chromatography. *J. Chromatogr.*, 473, 1989, 37–52.
165. Bemgard, A. and Blomberg, L., Influence of adsorption at the gas-liquid interface on the determination of gas chromatographic retention indices on open tubular columns coated with cyanosilicones. *J. Chromatogr.*, 502, 1990, 1–9.
166. Shibamoto, T., Harada, K., Yamaguchi, K., and Aitoku, A., Gas chromatographic retention characteristics of different coating thicknesses on a glass capillary column. *J. Chromatogr.*, 194, 1980, 277–284.
167. Soják, L., Berezkin, V. G., and Janák, J., Effect of adsorption on the reproducibility of retention indices of hydrocarbons in capillary gas-liquid chromatography. *J. Chromatogr.*, 209, 1981, 15–20.
168. Orav, A., Kuningas, K., and Rang, S., A comparison of different retention index systems with unsaturated and aromatic hydrocarbons in capillary gas chromatography on Peg 20M. *Chromatographia*, 37, 1993, 411–414.
169. De Frutos, M., Sanz, J., Martinez-Castro, I., and Jimenez, M. I., New parameters for the characterization of relationship between gas chromatographic retention and temperature. *Anal. Chem.*, 65, 1993, 2643–2649.
170. Heldt, V. and Köser, H. J. K., Different bases for the gas chromatographic retention index system. *J. Chromatogr.*, 192, 1980, 107–116.
171. Robbins, G. A., Wang, S., and Stuart, J. D., Using the static headspace method to determine Henry's law constants. *Anal. Chem.*, 65, 1993, 3113–3118.
172. Li, J., Dallas, A. J., Eikens, D. I., Carr, P. W., Bergmann, D. L., Hait, M. J., and Eckert, C. A., Measurement of large infinite dilution activity coefficients of nonelectrolytes in water by inert gas stripping and gas chromatography. *Anal. Chem.*, 65, 1993, 3212–3218.
173. Roth, M. and Novak, J., Utilization of the solution-of-groups concept in gas-liquid chromatography. *J. Chromatogr.*, 258, 1982, 23–33.
174. Petrovic, S. M., Lomic, S., and Sefer, I., Utilization of the functional group contribution concept in liquid chromatography on chemically bonded reversed phases. *J. Chromatogr.*, 348, 1985, 49–65.
175. Li, J. and Carr, P. W., Measurement of water-hexadecane partition coefficients by headspace gas chromatography and calculation of limiting activity coefficients in water. *Anal. Chem.*, 65, 1993, 1443–1450.
176. Abraham, M. H., Whiting, G. S., Fuchs, R., and Chambers, E., Thermodynamics of solute transfer from water to hexadecane. *J. Chem. Soc. Perkin Trans.*, 2, 1990, 291–300.
177. Khalfaoui, B. and Newsham, D. M. T., Determination of infinite dilution activity coefficients and second virial coefficients using gas-liquid chromatography. I. The dilute mixtures of water and unsaturated chlorinates hydrocarbons and of water and benzene. *J. Chromatogr. A*, 673, 1994, 85–92.
178. Khalfaoui, I. and Newsham, D. M. T., Determination of second cross viral coefficients from gas-liquid chromatographic data. II. Dilute mixtures of water and brominated hydrocarbons. *J. Chromatogr. A*, 688, 1994, 117–123.
179. Voelkel, A. and Janas, J., Inverse gas chromatographic determination of solubility parameter and binary parameters of (α,ω)-diamino oligoethers. *J. Chromatogr. A*, 669, 1994, 89–95.
180. DiPaola-Baranyi, G. and Guillet, J. E., Estimation of polymer solubility parameters by gas chromatography. *Macromolecules*, 11, 1978, 228–235.
181. Guillet, J. E., Molecular probes in the study of polymer structure. *J. Macromol. Sci. Chem.*, 4, 1970, 1669–1674.
182. Voelkel, A. and Jonas, J., Solubility parameters of broad and narrow distributed oxyethylates of fatty alcohols. *J. Chromatogr.*, 645, 1993, 141–151.
183. Spieksma, W., Luijk, R., and Govers, H. A. J., Determination of the liquid vapour pressure of low-volatility compounds from the Kováts retention index. *J. Chromatogr. A*, 672, 1994, 141–148.
184. Hamilton, D. J., Gas chromatographic measurement of volatility of herbicide esters. *J. Chromatogr.*, 195, 1980, 75–83.
185. Sherma, J. and Fried, B. (Eds.), *Handbook of Thin-Layer Chromatography*. Marcel Dekker, New York, 1991.
186. Poole, C. F. and Schouette, S. A., *Contemporary Practice of Chromatography*. Elsevier, Amsterdam, 1984, pp. 619–701.
187. Touchstone, J. C. and Dobbins, M. F., *Practice of Thin-Layer Chromatography*, Wiley, New York, 1983.
188. Gasparic, J., Chromatography on thin layers impregnated with organic stationary phases. In *Advances in Chromatography* (Eds. Giddings, J. C., Grushka, E., and Brown, P. R.). Marcel Dekker, New York, 1992, pp. 153–252.
189. Geiss, F., *Fundamentals of Thin-Layer Chromatography (Planar Chromatography)*, Dr. Alfred Huethig Verlag, Heidelberg, Germany, 1987.
190. Snyder, L. R. and Kirkland, J. J., *Introduction to Modern Liquid Chromatography*. 2nd Edition, John Wiley & Sons, New York, 1979.

191. Soczewinski, E., Solvent composition effects in thin-layer chromatography systems of the type silica gel-electron donor solvent. *Anal. Chem.*, 41, 1969, 179–182.
192. Snyder, L. R. and Glajch, J. L., Solvent strength and multicomponent mobile phases in liquid-solid chromatography. Binary-solvent mixtures and solvent localization. *J. Chromatogr.*, 214, 1981, 1–19.
193. Glajch, J. L. and Snyder, L. R., Solvent strength and multicomponent mobile phases in liquid-solid chromatography. Mixtures of three or more solvents. *J. Chromatogr.*, 214, 1981, 21–34.
194. Merkku, P., Yliruusi, J., Vuorela, H., and Hiltunen, R., Determination of specific heats and adsorption energies of ternary TLC eluents and silica gel. *J. Planar Chromatogr.*, 7, 1994, 305–308.
195. Nyiredy, Sz., Dallenbach-Tölke, K., and Sticher, O., Correlation and prediction of the k values for mobile phase optimization in HPLC. *J. Liquid Chromatogr.*, 12, 1989, 95–116.
196. Kowalska, T., Klama, B. and Sliwiok, J., On the mechanism of retention of different solutes in adsorption TLC with 2-propanol n-hexane eluents. *J. Planar Chromatogr.*, 5, 1992, 452–457.
197. Kowalska, T. and Klama, B., On the mechanism of retention in adsorption TLC with iso-propanol-n aliphatic hydrocarbon eluents. *J. Planar Chromatogr.*, 7, 1994, 63–69.
198. Kowalska, T. and Klama, B., On the mechanism of retention in adsorption TLC with aliphatic alcohol-n-hexane eluents. *J. Planar Chromatogr.*, 7, 1994, 147–152.
199. Rao, C. N. R., Dwivedi, P. C., Ratajczak, H., and Orville-Thomas, W. J., Relation between O-H stretching frequency and hydrogen bond energy: re-examination of the Badger-Bauer rule. *J. Chem. Soc. Faraday Trans.* II, 71, 1975, 955–966.
200. Ibbotson, D. A. and Moore, L. F., Association of n-alcohols in non-polar solvents. Part I. The dielectric polarisations, apparent dipole moments, and near infrared spectra of n-alcohols in carbon tetrachloride and cyclohexane. *J. Chem. Soc. B*, 1967, 76–83.
201. Arnett, E. M. and Small, L. E., Ionization of group 6 and 7 protonic acids in dimethyl sulfoxide. *J. Am. Chem. Soc.*, 99, 1977, 808–815.
202. Baranowska, I. and Swierczek, S., On the mechanism of retention of azines and diazines in adsorption TLC with n-propanol n-hexane eluents. *J. Planar Chromatogr.*, 7, 1994, 251–253.
203. Rozylo, J. K., Oscik-Mendik, B., and Janicka, M., Retention parameters of some organic substances and physico-chemical properties of chromatographic systems with two-component mobile phases. *J. Chromatogr.*, 395, 1987, 463–471.
204. Rozylo, J. K. and Janicka, M., Thermodynamic description of liquid-solid chromatography process in the optimization of separation conditions or organic compound mixture. *J. Liquid Chromatogr.*, 14, 1991, 3197–3212.
205. Janicka, M. and Rozylo, J. K., The use of TLC in sandwich chambers for theoretical prediction of mixture separation in HPLC. *J. Planar Chromatogr.*, 6, 1993, 362–367.
206. Soczewinski, E. and Golkiewicz, W., Application of the law of mass action to thin-layer adsorption chromatography systems of the type electron donor solvent-silica gel. *Chromatographia*, 4, 1971, 501–507.
207. Snyder, L. R., Role of the solvent in liquid-solid chromatography. A review. *Anal. Chem.*, 46, 1974, 1384–1393.
208. Schoenmakers, P. J., Billiet, H. A. H., and de Galan, L., Influence of organic modifiers on the retention behaviour in reversed-phase liquid chromatography and its consequences for gradient elution. *J. Chromatogr.*, 185, 1979, 179–195.
209. Schoenmakers, P. J., Billiet, H. A. H., and de Galan, L., Use of gradient elution for rapid selection of isocratic conditions in reversed-phase high-performance liquid chromatography. *J. Chromatogr.*, 205, 1981, 13–30.
210. Petrovic, S. M. and Loncar, E., Support-solvent effects in thin layer chromatography on silica. *J. Planar Chromatogr.*, 5, 1992, 359–363.
211. Petrovic, S. M., Lomic, S., and Stefer, I., Solute retention in the stationary phase of a liquid-solid chromatographic system. *Chromatographia*, 23, 1987, 915–924.
212. Petrovic, S. M. and Spika, M., Effect of the mobile phase on solute retention in liquid-solid chromatography. *J. Liquid Chromatogr.*, 14, 1991, 3061–3075.
213. Tesic, Z. L., Janjic, T. J., Tosic, R. M., and Celap, M. B., Effect of electronegativity of donor atoms on R_F values of tris(β-diketonato) complexes of cobalt(III), chromium(III) and ruthenium(III) by thin-layer chromatography on silica gel. *Chromatographia*, 37, 1993, 599–602.
214. Martinez, B., Orte, J. C., Miro, M., Crovetto, G., and Thomas, J., Thin-layer chromatography of N,N-disubstituted dithiocarbamates of nickel(II) and cobalt(II). *J. Chromatogr. A*, 655, 1993, 45–49.
215. Bazylak, G., Solute-solvent interactions of aliphatic aminoalcohols in underivatized silica-binary eluent HPTLC systems. *J. Planar Chromatogr.*, 5, 1992, 239–245.
216. Jaroniec, M., Rozylo, J. K., Jaroniec, J. A., and Oscik-Mendyk, M., Liquid adsorption chromatography with mixed mobile phases. IV. A new equation for the capacity ration involving the solute-solvent interactions. *J. Chromatogr.*, 188, 1980, 27–32.
217. Oscik, J., Jaroniec, M., and Malinowska, I., Thermodynamic approach to TLC with mixed mobile phase. Determination of parameters characterizing TLC systems. *J. Liquid Chromatogr.*, 6, 1983, 81–93.

218. Qin-Sun, W., Hua-Zheng, Y., and Bing-Wen, Y., The correlation between the molecular structures of O-ethyl-O-aryl-N-isopropyl phosphoroamidothioates and their retardation factors in HPTLC. *J. Planar Chromatogr.*, 6, 1993, 144–146.
219. Nurok, D., Habibi-Goudarzi, S., and Kleyle, R., Statistical approach to solvent selection as applied to two-dimensional thin-layer chromatography. *Anal. Chem.*, 59, 1987, 2424–2428.
220. Nurok, D., Julian, L. A., and Uhegbu, C. E., Parameter for predicting retention in planar chromatography. *Anal. Chem.*, 63, 1991, 1524–1529.
221. Nurok, D., Knotts, K. D., Kearns, M. L., Ruterbories, K. J., Uhegbu, C. E., and Alberti, P., A quantitative study of thin-layer chromatographic separation in a series of mobile phases. *J. Planar Chromatogr.*, 5, 1992, 350–358.
222. Knotts, K. D., Kearns, M. L., Julian, L. A., Lipkowitz, K. B., and Nurok, D., The relationship between separation quality and certain properties of the mobile phase. *J. Planar Chromatogr.*, 6, 1993, 105–111.
223. Jaroniec, M. and Martire, D., Models of solute and solvent distribution for describing retention in liquid chromatography. *J. Chromatogr.*, 387, 1987, 55–64.
224. Soczewinski, E. and Maciejewicz, W., The eluent strength of homologous series of ethers and ketones in normal phase thin-layer chromatography. *J. Planar Chromatogr.*, 7, 1994, 152–156.
225. Pyka, A., The topological indices and R_M values of benzoic acid derivatives: a structure — activity investigation. Part IV. *J. Planar Chromatogr.*, 7, 1994, 108–116.
226. Pyka, A., The application of topological indexes (I_t) for prediction of the R_M values of isomeric alcohols. Part V. *J. Planar Chromatogr.*, 7, 1994, 41–49.
227. Gutman, I., Ruscic, B., Trinajstic, N., and Wilcox, C. F., Graph theory and molecular orbitals. XII. Acyclic polyenes. *J. Chem. Phys.*, 62, 1975, 3399–3405.
228. Wiener, H., Structural determination of paraffin boiling points. *J. Am. Chem. Soc.*, 69, 1947, 17–20.
229. Balaban, A. T., Highly discriminating distance-based topological index. *Chem. Phys. Lett.*, 89, 1982, 399–402.
230. Pyka, A., A new stereoisomeric topological index (I_{STI}) for predicting the separation of stereoisomers in TLC. Part VII. *J. Planar Chromatogr.*, 7, 1994, 389–393.
231. Kowalska, T. and Podgorny, A., Physicochemical modeling of solute retention in reversed phase TLC with acetonitrile — water eluents. *J. Planar Chromatogr.*, 4, 1991, 313–315.
232. Kowalska, T., Podgorny, A. and Witkowska-Kita, B., Partial correlation coefficients in a study of retention mechanism in RPTLC and RPHPLC. *J. Planar Chromatogr.* 5, 1992, 192–196.
233. Kowalska, T. and Wikowska-Kita, B., A study of the retention of higher fatty alcohols and acids in RPTLC with bonded alkyl stationary phases. *J. Planar Chromatogr.*, 5, 1992, 424–430.
234. Kowalska, T., Witkowska-Kita, B., and Sajewicz, M., The effect of the coverage density of bonded octadecyl groups on the thermodynamics of retention in RPHPTLC systems. *J. Planar Chromatogr.*, 6, 1993, 149–152.
235. Baranowska, I. and Swierczek, S., A study on the retention of azines and diazines in RPTLC with bonded alkyl stationary phases. *J. Planar Chromatogr.*, 7, 1994, 399–405.
236. Podgorny, A. and Kowalska, T., Physico-chemical modeling of solute retention in reversed phase liquid chromatography with acetonitrile — phosphate buffer eluents. *J. Planar Chromatogr.*, 7, 1994, 291–293.
237. Cserháti, T. and Illés, Z., Influence of various salts on the reversed-phase retention of some dansylated amino acids in TLC. *Chromatographia*, 36, 1993, 302–306.
238. Cserháti, T., Lipophilicity determination of non-homologous series of nonionic surfactants by means of reversed-phase thin-layer chromatography. *J. Biochem. Biophys. Methods*, 27, 1993, 133–142.
239. Bazylak, G., Retention behavior and molecular structure parameters of diethanolamine isomers in some reversed phase HPTLC systems. *J. Planar Chromatogr.*, 5, 1992, 275–279.
240. Cserháti, T., Relationship between the physicochemical parameters of 3,5–dinitrobenzoic acid esters and their retention behaviour on β-cyclodextrin polymer support. *Anal. Chim. Acta*, 292, 1994, 17–22.
241. Cserháti, T. and Forgács, E., Retention of 4-cyanophenyl herbicides on water-insoluble β-cyclodextrin support. *J. Chromatogr. A*, 685, 1994, 295–302.
242. Ruane, R. J. and Wilson, I. D., Ion-pair reversed-phase thin-layer chromatography of basic drugs using sulphonic acids. *J. Chromatogr.*, 441, 1988, 355–360.
243. Bieganowska, M. L., Petruczinik, A., and Doraczynska-Szopa, A., Ion pair, reversed phase thin layer chromatography of some basic drugs and related pyridine derivatives. *J. Planar Chromatogr.*, 5, 1992, 184–191.
244. Körner, A., Ion pair reversed phase thin layer chromatography of sulfonated dyes: a theoretical approach towards optimization. *J. Planar Chromatogr.*, 6, 1993, 138–143.
245. Kossoy, A. D., Risley, D. S., Kleyle, R. M., and Nurok, D., Novel computation method for the determination of partition coefficients by planar chromatography. *Anal. Chem.*, 64, 1992, 1345–1349.
246. Biagi, G. L., Guerra, M. C., Barbaro, A. M., Barbieri, S., Recanatini, M., Borea, P. A., and Pietrogrande, M. C., Study of the lipophilic character of xanthine and adenosine derivatives. *J. Chromatogr.*, 498, 1990, 179–190.

247. Biagi, G. L., Guerra, M. C., Barbaro, A. M., Barbieri, S., Recanatini, M., and Borea, P. A., Study of the lipophilic character of xanthine and adenosine derivatives. II. Relationships between log k', R_M and log P values. *J. Liquid Chromatogr.*, 13, 1990, 913–927.
248. Dross, K. P., Mannhold, R., and Rekker, R. F., Drug lipophilicity in QSAR practice: II. Aspects of R_M-determinations; critics of R_M corrections; interrelations with partition coefficients. *Quant. Struct. Act. Relat.*, 11, 1992, 36–44.
249. Reichardt C., Empirical parameters of solvent polarity and chemical reactivity. *Molec. Interact.*, 3, 1982, 241–283.
250. Rohrscheider L., Solvent characterization by gas-liquid partition coefficients of selected solutes. *Anal. Chem.*, 45, 1973, 1241–1247.
251. Snyder L. R., Classification of the solvent properties of common liquids. *J. Chromatogr.*, 92, 1974, 223–230.
252. Karch, K., Sebastian, I., and Halász, I., Preparation and properties of reversed phases. *J. Chromatogr.*, 122, 1976, 3–16.
253. Wells M. J. M. and Clark C. R., Liquid chromatographic elution characteristics of some solutes used to measure column void volume on C_{18} bonded phase. *Anal. Chem.*, 53, 1981, 1341–1345.
254. Berendsen, G. E., Schoenmarkers, P. J., Galan, G., Vigh, G. Z., Varga Puchonyi, Z., and Inczédy, I., On the determination of hold-up time in reversed-phase liquid chromatography. *J. Liquid Chromatogr.*, 3, 1980, 1669–1682.
255. Knox, J. H. and Kaliszan, R., Theory of solvent disturbance peaks and experimental determination of thermodynamic dead-volume in column liquid chromatography. *J. Chromatogr.*, 349, 1985, 211–219.
256. Melander, W. R., Campbell, D. E., and Horváth, Cs., Enthalpy-entropy compensation in reversed-phase chromatography. *J. Chromatogr.*, 167, 1978, 215–225.
257. Unger, K. K., *Porous Silica, Journal of Chromatography Library, Vol. 16.* Elsevier, Amsterdam, 1979.
258. Snyder, L. R., *Principles of Adsorption Chromatography,* Marcel Dekker, New York, 1968.
259. Berthod, A., Silica: backbone material of liquid chromatographic column packings. *J. Chromatogr.*, 549, 1991, 1–28.
260. Nondek, L. and Vyskocil, V., Determination of surface hydroxyl groups on silanized silica gels for reversed-phase liquid-chromatography. *J. Chromatogr.*, 206, 1981, 581–585.
261. Poumeliotis, P. and Unger, K. K., Structure and properties of n-alkyldimethylsilyl bonded silica reversed-phase packings. *Chromatography*, 149, 1978, 211–224.
262. Köhler J. J., Chase, D. B., Farlee, R. D., Vega, A. J., and Kirkland, J. J., Comprehensive characterization of some silica-based stationary phases for high-performance liquid chromatography. *J. Chromatogr.*, 352, 1986, 275–305.
263. Hair, M. L., in Corey, E. R., Corey, J. P., and Gaspar, P. P. (Eds.), *Silicon Chemistry,* Ellis Horwood, Chichester, 1988, 481–489.
264. Ilier, R.K., *The Chemistry of Silica.* Wiley, New York, 1979.
265. Engelhard, H. and Müller, H., Chromatographic characterization of silica surfaces. *J. Chromatogr.*, 218, 1981, 395–407.
266. Hair M. L. and Hertl, W., Acidity of surface hydroxyl groups. *J. Phys. Chem.*, 74, 1970, 91–94.
267. Unger K. K., Janzen, R., Jilge, G., Lork, K. D., and Anspach, B., in Horváth Cs. (Ed.), *High Performance Liquid Chromatography: Advances and Perspectives, Vol. 5.* Academic Press, London, 1988, pp. 1–93.
268. Kiselev, A.V. and Lygin, V. J. *Infrared Spectra of Surface Compounds.* Wiley-Interscience, New York, 1975.
269. Brunaue, S., Emmeth, P. H., and Teller, E., Adsorption of gases in multimolecular layers. *J. Am. Chem. Soc.*, 60, 1938, 309–312.
270. Gregg, S. J. and Sing, K. S. W., *Adsorption, Surface Area and Porosity.* Academic Press, New York, 1982.
271. Everett, D. H., in Unger, K. K., Rouquerol, J., Sing, K. S. W., and Kral, H. (Eds.), *Characterization of Porous Solids.* Elsevier, Amsterdam, 1988, pp 1–21.
272. Scott, R. P. and Kucera, P., Some aspects of the chromatographic properties of thermally modified silica gel. *J. Chromatogr., Sci.*, 13, 1975, 337–346.
273. El Rassi, Z., Gonnet, C., and Rocca, J. L., Chromatographic studies of the influence of water and thermal treatment on the activity of silica gel. *J. Chromatogr.*, 125, 1976, 179–201.
274. Unger, K. K., Regnier, F. E., and Majors, R. E. (Eds.), Liquid chromatography packings. *J. Chromatogr.*, 544, 1991.
275. Miller, N. T. and Di Bussolo, J. M., Studies on the stability of n-alkyl-bonded silica gels under basic pH conditions. *J. Chromatogr.*, 499, 1990, 317–332.
276. Stadalius, M. A., Gold, H. S., and Snynder, L. R., Optimization model for the gradient elution separation of peptide mixtures by reversed-phase high-performance liquid chromatography. Verification of band width relationships for acetonitrile-water mobile phases. *J. Chromatogr.*, 327, 1985, 27–35.
277. Stancher, B. and Zonta, F., Comparison between straight and reversed-phase in the high-performance liquid chromatographic fractions of retinol isomers. *J. Chromatogr.*, 234, 1982, 244–248.
278. Snyder, L. R., Glajch, J. L., and Kirkland, J. J., *Practical HPLC Method Development,* Wiley, New York, 1988, Ch.3.

279. Köhler, J. and Kirkland, J. J., Improved silica-based column packings for high-performance liquid chromatography. *J. Chromatogr.*, 385, 1987, 125–133.
280. Kirkland, J. J., Dilks, C. H., Jr., and DeStefano, J. J., Normal phase high-performance liquid chromatography with highly purified silica microspheres. *J. Chromatogr.*, 635, 1993, 19–30.
281. Nahum A. and Horváth Cs., Surface silanols in silica-bonded hydrocarbonaceous stationary phases, *J. Chromatogr.*, 203, 1981, 53–64.
282. Hearn, M. T. W. and Grego, B., High performance liquid chromatography of amino acids, peptides and proteins XL. Further studies on the role of the organic modifier in the reversed-phase high-performance liquid chromatography of polypeptides. *J. Chromatogr.*, 255, 1983, 125–136
283. Ohtsu, Y., Shiojima, Y., Ohumura, T., Koyama, J. I., Nakamura, K., Nakata, O., Kimata, K., and Tanaka N., Performance of polymer-coated silica C_{18} packing materials prepared from high-purity silica gel. *J. Chromatogr.*, 481, 1989, 147–157.
284. Jane, I., The separation of wide range of drugs of abuse by high-performance liquid chromatography. *J. Chromatogr.*, 111, 1975, 227–233.
285. Sugden, K., Cox, G. B., and Loscombe, C. R., Chromatographic behaviour of basic amino compounds on silica and ODS-silica using aqueous methanol mobile phases. *J. Chromatogr.*, 149 1978, 377–390.
286. Stou, R. W., Cox, J. B., and Odiorne, T. J., Surface treatment and porosity control of porous silica microspheres. *Chromatographia*, 24, 1987, 602–612.
287. Derek, D. and Novák, I., Silica gel and carbon column packings for use in high-performance liquid chromatography. *Chromatographia*, 30, 1990, 582–590.
288. Buszewski, B., Derek, D., Garaj, J., Novák, I., and Suprynowicz, Z., Influence of the porous silica gel structure on the coverage density of a chemically bonded C_{18} phase for high-performance liquid chromatography. *J. Chromatogr.*, 446, 1988, 191–201.
289. Gilpin, R. K., The bonded phase: stucture and dynamics. *J. Chromatogr. Sci.*, 22, 1984, 371–377.
290. Buszewski, B., Chracterization of C_{18} chemically modified silica-gel and porous glass phases by high resolution solid state NMR spectroscopy and other physico-chemical methods. *Chromatographia*, 28, 1990, 233–242.
291. Pleiderer, B. and Bajer, E., Silanediol groups of the silica gel nucleosil: active sites involved in the chromatographic behaviour of bases. *J. Chromatogr.*, 486, 1989, 67–71.
292. Buszewski, B., Influence of silica gel surface purity on the coverage density of chemically bonded phases, *Chromatographia*, 34, 1992, 573–579.
293. Kimata, K., Tanaka, N., and Araki, T., Suppression of the effect of metal impurities in alkylsilylated silica packing materials. *J. Chromatogr.*, 594, 1992, 87–96.
294. Heidrich, D., Volkmann, D., and Zurawski, B., The place of SiOH groups in the absolute acidity scale from ab initio calculations. *Chem. Phys. Lett.*, 80, 1981, 60–63.
295. Sadek, P. C., Koester, C. J., and Bowers, L. D., The investigation of the influence of the trace metals in chromatographic retention processes. *J. Chromatogr. Sci.*, 25, 1987, 489–493.
296. Nawrocki, J., Silica surface controversies, strong adsorption sites, their blockage and removal. Part I. *Chromatographia*, 31, 1991, 177–192.
297. Bij, K. E., Horváth, Cs., Melander, W., and Nahum, A., Surface silanols in silica bonded hydrocarbonaceous stationary phases. Irregular retention behaviour and effect of silanol masking. *J. Chromatogr.*, 203, 1981, 65–84.
298. Cox, G. B. and Stout, R. W., Study of the retention mechanism for basic compounds on silica under "pseudo-reversed-phase" conditions. *J. Chromatogr.*, 384, 1987, 315–336.
299. Gilpin, R. K. and Burke, M. F., Role of tri- and dimethylsilanes in tailoring chromatographic adsorbents. *Anal. Chem.*, 45, 1973, 1383–1389.
300. Kong, T. M. and Simpson, C. F., Novel method for the preparation of chemically bonded phases of their chromatographic performance. *Chromatographia*, 24, 1987, 385–394.
301. Akopo, S. O., Scott, R. P., and Simpson, C. F., The production of stable uniform reverse-phases for liquid chromatography by oligomeric synthesis. *J. Liquid Chromatogr.*, 14, 1991, 217–236.
302. Scott R.P., Control of retention by molecular interactions in RP-HPLC chromatography. *J. Chromatogr.*, 656, 1993, 51–68.
303. Lochmüller, C. H. and Wilder, D. R., The sorption behaviour of alkyl bonded phases in reverse-phase, high performance liquid chromatography. *J. Chromatogr. Sci.*, 17, 1979, 574–579.
304. Laub, R. J. and Purnell, J. H., Criteria for the use of mixed solvents in gas liquid chromatography. *J. Chromatogr.*, 112, 1975, 71–79.
305. Laub, R. J., in Kuwana, P. (Ed.), *Physical Methods in Modern Chromatographic Analysis.* Academic Press, New York, 1983, Ch. 4.
306. Katz, D. E., Ogan, K., and Scott, R. P., Distribution of a solute between two-phases. The basic theory and its application to the prediction of chromatographic retention. *J. Chromatogr.*, 352, 1986, 67–90.
307. Murakami, F., Retention behaviour of benzene derivatives on bonded reversed-phase columns. *J. Chromatogr.*, 178, 1979, 393–399.

308. Geng, X. and Regnier, E., Stoichiometric displacement of solvent by non-polar solutes in reversed-phase liquid chromatography. *J. Chromatogr.*, 332, 1985, 147–168
309. Hafkenscheid, T. L. and Tomlinson, E., Estimation of physicochemical properties of organic solutes using HPLC retention parameters. *Adv. Chromatogr.*, 25, 1986, 1–55.
310. Tanford, C., *The Hydrophobic Effect: Formation of Micelles and Biological Membranes*. Wiley-Interscience, New York, 1980. Ch. 1 and 2.
311. Tanaka, N. Tanigawa, T., Kimata, K., Hosoya, K., and Araki, T., Selectivity of carbon packing materials in comparison with octadecylsyl- and pyrenylethylsilylsilica gels in reversed-phase liquid chromatography. *J. Chromatogr.*, 549, 1991, 29–41.
312. Horváth, Cs., Melander, W., and Molnár, I., Solvophobic interactions in liquid chromatography with non-polar stationary phases. *J. Chromatogr.*, 125, 1976, 129–156.
313. Tanaka, N., Sakagami, K., and Araki, N., Effect of alkyl chain length of the stationary phase on retention and selectivity in reversed-phase liquid chromatography. Participation of solvent molecules in the stationary phase. *J. Chromatogr.*, 199, 1980, 327–337.
314. Wise, C. A. and May, W. E., Effect of C_{18} surface coverage on selectivity in reversed-phase liquid chromatography of polycyclic aromatic hydrocarbons. *Anal. Chem.*, 55, 1983, 1479–1485
315. Sander, L. C. and Wise, S. A., Investigations of selectivity in RPLC of polycyclic aromatic hydrocarbons. *Adv. Chromatogr.*, 25, 1986, 139–155.
316. Jinno, K., Nagoshi T., Tanaka, N., Okomoto, M., Fetzer, J. C., and Biggs, W. R., Elution behaviour of planar and non-planar polyciclic aromatic hydrocarbons on various chemically bonded stationary phases in liquid chromatography. *J. Chromatogr.*, 392, 1987, 75–82.
317. Jinno, K., Ibuki, T., Tanaka, N., Okomoto, M., Fetzer, J. C., Biggs, W. R., Griffiths, P. R., and Olinger, J. M., Retention behaviour of large polycyclic aromatic hydrocarbons in reversed-phase liquid chromatography on a poymeric octadecylsilica stationary phase. *J. Chromatogr.*, 461, 1989, 209–227.
318. Bakalyar, S. R., McIlwrick, R., and Roggendorf, E., Solvent selectivity in reversed-phase high-pressure liquid chromatography. *J. Chromatogr.*, 142, 1977, 353–365.
319. Tanaka, N., Goodell, H. and Karger, B. L., The role of organic modifiers on polar group selectivity in reversed-phase liquid chromatography. *J. Chromatogr.*, 158, 1978, 233–248.
320. Tanaka, N., Kimata, K., Hosoya, K., Miyanishi, H., and Araki, T., Stationary phase effects in reversed-phase liquid chromatography. *J. Chromatogr.*, 656, 1993, 265–287.
321. Lork, K. D., Unger, K. K., Bruckner, H., and Hearn, M. T. W., Retention behaviour of paracelsin peptides on reversed-phase silicas with varying n-alkylchain length and ligand density. *J. Chromatogr.*, 476, 1989, 135–145.
322. Tanaka, N. and Araki, M., Polymer-based packing materials for reversed-phase liquid chromatography. *Adv. Chromatogr.*, 30, 1989, 81–102.
323. Tanaka, N., Hashizume, K. and Araki, M., Comparison of polymer-based stationary phases with silica-based stationary phases in reversed-phase liquid chromatgraphy. Selective binding of rigid, compact molecules by alkylated polymer gels. *J. Chromatogr.*, 400, 1987, 33–45.
324. Nevejans, F. and Verzele, M., On the structure and chromatographic behaviour of polystyrene phases. *J. Chromatogr.*, 406, 1987, 325–342.
325. Hosoya, K., Maruya, S., Kimata, K., Kinoshita, H., Araki, T., and Tanaka, N., Polymer-based packing materials for reversed-phase liquid chromatography. Steric selectivity of polymer gels provided by diluents and cross-linking agents in suspension polymerization. *J. Chromatogr.*, 625, 1992, 121–129.
326. Tanaka, N., Ebata, T., Hashizume, K. and Araki, M., Polymer-based packing materials with alkyl backbones for reversed-phase liquid chromatography. Performance and selectivity. *J. Chromatogr.*, 475, 1989, 195–208.
327. Tanaka, N., Kimata, K., Mikawa, Y., Hosoya, K., Araki, T., Ohtsu, Y., Shiojima, Y., Tsuboi, R., and Tsuchiva, H., Performance of wide-pore silica and polymer-based packing materials in polypeptide separation: effect of pore size and alkyl chain length. *J. Chromatogr.*, 535, 1990, 13–31.
328. Uchida, T., Ohtani, T., Kasai, M., Yanagihara, Y., Noguchi, K., Izu, H., and Hara, S., Chemically behaviour and analysis on a chemically stable C_{18}-bonded vinyl alcohol copolymer column with alkaline and acidic eluents. *J. Chromatogr.*, 506, 1990, 327–334.
329. Afeyan, N. B., Fulton, S. P., and Regnier, F. E., Perfusion chromatography packing material for peptides and proteins. *J. Chromatogr.*, 544, 1991, 267–279.
330. Colin, H., Eon, C., and Guichon, G., Reversed-phase liquid-solid chromatography on modified carbon black, *J. Chromatogr.*, 122 1976, 223–242.
331. Colin H. and Guichon G., Comparison of some packings for reversed-phase high-performance liquid-solid chromatography. *J. Chromatogr.*, 158, 1978, 183–205.
332. Scott, R. P., The role of molecular interactions. *J. Chromatogr.*, 122, 1976, 35–53.
333. Katz, H. D., Ogan, K., and Scott, R. P., Distribution of a solute between two phases. The basic theory and its application to the prediction of chromatographic retention. *J. Chromatogr.*, 352, 1986, 67–90.

334. Jandera, P. and Churácek, J., Gradient elution in liquid chromatography. The influence of the composition of the mobile phase on the capacity ratio (retention volume, band width, and resolution) in isocratic elution — theoretical considerations. *J. Chromatogr.*, 91, 1974, 207–221.
335. Schoenmakers, P., Billiet, H. A. H., Tijjssen, R., and de Galan, L., Gradient elution in reversed-phase liquid chromatography. *J. Chromatogr.*, 149, 1978, 519–537.
336. Karger, B. L., Gant, J. R., Hartkopf, A., and Weiner, P. H., Hydrophobic effects in reversed-phase liquid chromatography. *J. Chromatogr.*, 128, 1976, 65–78.
337. Horváth, Cs. and Melander, W., Liquid chromatography with hydrocarbonaceous bonded phases; theory and practice of reversed phase chromatography. *J. Chromatogr.*, Sci., 15, 1977, 393–404.
338. McCormick, R. M. and Karger, B. L., Distribution phenomena of mobile-phase components and determination of dead volume in reversed-phase liquid chromatography. *Anal. Chem.*, 52, 1980, 2249–2257
339. Slaats, H. E., Markowski, W., Fekete, J., and Poppe, H., Distribution equilibra of solvent components in reversed-phase liquid chromatographic columns and relationship with the mobile phase volume. *J. Chromatogr.*, 207, 1981, 299–323.
340. Jandera, P., Colin, H., and Guichon, G., Interaction indexes for prediction of retention in reversed-phase liquid chromatography. *Anal. Chem.*, 54, 1982, 435–441.
341. Jandera, P., Method for characterization of selectivity in reversed-phase liquid chromatography. 1. Derivation of the method and verification of the assumptions. *J. Chromatogr.*, 352, 1986, 91–110.
342. Snyder, L. R. and Kirland, J. J., *Introduction to Modern Liquid Chromatography*. Wiley-Interscience, New York, 1979.
343. Jandera, P., Correlation of retention and selectivity in reversed-phase high performance liquid chromatography with interaction indices and with lipophilic and polar structural indices. *J. Chromatogr.*, 656, 1993, 437–467.
344. Kaliszan, R., *Quantative Structure-Chromatographic Retention Relationships*, Wiley, New York, 1987.
345. Reichardt, C., *Solvent Effects in Organic Chemistry*. Verlag Chemie, Weinheim, 1979.
346. Hansch, C. and Fujita, T., p-σ-π Analysis. A method for the correlation of biological activity and chemical structure. *J. Am. Chem. Soc.*, 86, 1964, 1616–1626.
347. Free, S. M., Jr. and Wilson, J. F., A mathematical contribution to structure-activity studies. *J. Med. Chem.*, 7, 1964, 395–399.
348. Kaliszan R., in Brown, P. R. and Hartwick, R. A. (Eds.), *High Performance Liquid Chromatography*, Wiley, New York, 1989, Ch. 4.
349. Tijssen, R., Billiet, H. A. H., and Schoenmarkers, P. J., Use of the solubility parameter for predicting selectivity and retention in chromatography. *J. Chromatogr.*, 122, 1976, 185–203.
350. Martine, D. E. and Boehmn, R. E., Molecular theory of liquid adsorption chromatography. *J. Liquid Chromatogr.*, 3, 1980, 753–774
351. Carr, P. W., Doherty, R. M., Kamlet, M. J., Taft, R. W., Melander, W., and Horváth, Cs., Study of temperature and mobile-phase effects in reversed-phase high-performance liquid chromatography by the use of the solvatochromic comparison method. *Anal. Chem.*, 58, 1986, 2674–2689
352. Synder, L. R., Carr, P. W., and Rutan, S. C., Solvatochromically based solvent-selectivity triangle. *J. Chromatogr.*, 656, 1993, 537–547.
353. Dill, K. A., The mechanism of solute retention in reversed-phase liquid chromatography. *J. Phys. Chem.*, 91, 1987, 1980–1988.
354. Ying, P. T., Dorsey, J. G., and Dill, K. A., Retention mechanism of reversed-phase liquid chromatography: determination of solute-solvent interaction free energies. *Anal. Chem.*, 61, 1989, 2540–2546.
355. Sentell, K. B. and Dorsey, J. G., Retention mechanism in reversed-phase liquid chromatography. Stationary-phase bonding density and partitioning. *Anal. Chem.*, 61, 1989, 930–934.
356. Johnson, B. P., Khaledi, M. G., and Dorsey, J. G., Solvatochromic solvent polarity measurements and retention in reversed-phase liquid chromatography. *Anal. Chem.*, 58, 1986, 2354–2385.
357. Valkó, K., Snyder, L. R., and Glajch, J. L., Retention in reversed-phase liquid chromatography as a function of mobile-phase composition. *J. Chromatogr.*, 656, 1993, 501–520.
358. Fredeslund, A., Jones, R. L., and Prausnitz, J. M., Group-contribution estimation of activity coefficients in nonideal liquid mixtures. *AIChE J.*, 21, 1975, 1086–1099.
359. Smith, R. M. and Burr, C. M., Retention prediction of analytes in reversed-phase high-performance liquid chromatography based on molecular structure. *J. Chromatogr.*, 550, 1991, 335–356.
360. Tsantili-Kakoulidou, A., El Tayar, N., van de Waterbeemdand, H., and Testa, B., Sructural effects in the lipohilicity of di- and polysubstituted benzenes as measured by reversed-phase high-performance liquid chromatography. *J. Chromatogr.*, 389, 1987, 33–45.
361. Boyce, C. B. C. and Millborrow, B. V., A simple assessment of partition data for correlating structure and bilogical activity using thin-layer chromatography. *Nature*, 208, 1965, 537–539.
362. Soczewinski, E. and Wachtmeister, C. A., The correlation between the composition of certain ternary two-phase solvent system and R_M values. *J. Chromatogr.*, 7, 1962, 311–320.

363. Snyder, L. R., Dolan, J. W., and Gant, J. R., Gradient elution in high-performance liquid chromatography. Theoretical bases for reversed-phase systems. *J. Chromatogr.*, 165, 1979, 3–30.
364. Sherblom, P. M. and Eganhouse, R. P., Correlation between octanol-water partition coefficients and reversed-phase high-performance liquid chromatography capacity factors. *J. Chromatogr.*, 454, 1988, 37–50.
365. Braumann, T. and Jastorff, B., Physico-chemical characterization of cyclic nucleotides by reversed-phase high-performance liquid chromatography. II. Quantitative determination of hydrophobicity. *J. Chromatogr.*, 350, 1985, 105–118.
366. Michels, J. J. and Dorsey, J. D., Estimation the reversed-phase liquid chromatographic lipophilicity parameter log k'_w using ET-30 solvatochromism. *J. Chromatogr.*, 499, 1990, 435–451.
367. Braumann, T., Genieser, H. G., Lullmann, C., and Jastorff, B., Determination of hydrophobic parameters by reversed-phase high-performance liquid chromatography. Effect of stationary phase. *Chromatographia*, 24, 1987, 777–782.
368. Clark, C. R., Barksdale, J. M., Mayfield, C. A., Ravis, W. R., and DeRuiter, J., Liquid chromatographic measurement of hydrophobicity constants for n-arylsulfonylglycine aldose reductase inhibitors. *J. Chromatogr. Sci.*, 28, 1990, 83–92.
369. Unger, S. H. and Chiang, G. H., Octanol-physiological buffer distribution coefficients of lipophilic amines by reversed-phase high-performance liquid chromatography and their correlation with biological activity. *J. Med. Chem.*, 24, 1981, 262–270.
370. Pietrogrande, M. C., Dondi, F., Blo, G., Borea, P. A., and Bighi, C., Octadecyl, phenyl and cyano phases comparison for the RP-HPLC prediction of octanol-water partition coefficient. *J. Liquid Chromatogr.*, 10, 1987, 1065–1075.
371. Biagi, G. L., Recanatini, M., Barbaro, A. M., Guerra, M. C., Sapone, A., Borea, P. A., and Pietrogrande, M. C., in Silipo, C. and Vittoria, A. (Eds.), *QSAR: Rational Approaches to the Design of Bioactive Compounds.* Elsevier, Amsterdam, 1991, p. 83.
372. Demotes-Mainard, F., Jarry, C., Thomas, J. and Dallet, P., RPHPLC retention data of new 2–amino-2–oxazolines an approach of their lipophilic properties. *J. Liquid Chromatogr.*, 14, 1991, 795–805.
373. Jinno, K. and Yokoyama, Y., Retention prediction for polymer additives in reversed-phase liqiuid chromatography. *J. Chromatogr.*, 550, 1991, 325–334.
374. Petrauskas, A. A. and Svedas, V. K., Hydrophobicity of ß-lactam antibiotics. Explanation and prediction of their behaviour in various partitioning solvent systems and reversed-phase chromatography. *J. Chromatogr.*, 585, 1991, 3–34.
375. Yamagami, C. and Takao, N., Hydrophobicity parameters determined by reversed-phase liquid chromatography. II. Dependence of retention behaviour of pyrazines on mobile phase composition. *Chem. Pharm. Bull.*, 39, 1991, 1217–1221.
376. Pereira, A. L., Barreiro, E. J. L., Freitas, A. C. C, Correa, C. J. C., and Gomes, L. N. L. F., Investigation of the lipophilicity of antiphlogistic pyrazole derivatives: relationships between log k_w and log P values of 5-arylamino and arylhydrazono-3-metyl-4-nitro-1-phenylpyrazoles. *J. Liquid Chromatogr.*, 14, 1991, 1161–1171.
377. Patel, H. B., King, D. N., and Jefferies, T. M., Application of solute and mobile phase partition coefficient to describe solute retention in reversed-phase high-performance liquid chromatography. *J. Chromatogr.*, 555, 1991, 21–31.
378. Tchapla, A., Colin, A., and Guiochon, G., Linearity of homologous series retention plots in reversed-phase liquid chromatography. *Anal. Chem.*, 56, 1984, 621–625.
379. Timerbaev, A. R., Semenova, O. P., Tsoi, I. G., and Petrukhin, O. M., Correlation analysis in liquid chromatography of metal chelates. Multi-dimensional models in reversed-phase liquid chromatography. *J. Chromatogr.*, 648, 1993, 307–314.
380. Snyder, L. R., *Principles of Adsorption Chromatography.* Marcel Dekker, New York, 1968, Ch. 8.
381. Snyder, L. R., Role of the solvent in liquid-solid chromatography. *Anal. Chem.*, 46, 1974, 1378–1393.
382. Soczewinski, E., Solvent composition effects in thin-layer chromatography systems of the type silica gel-electron donor solvent. *Anal. Chem.*, 41, 1969, 179–182.
383. Scott, R. P. W. and Kucera, P., Solute interactions with the mobile and stationary phases in liquid-solid chromatography. *J. Chromatogr.*, 112, 1975, 425–442
384. Scott, R. P. W., The role of molecular interactions in chromatography. *J. Chromatogr.*, 122, 1976, 35–53.
385. Soczewinski, E. and Wachtmeister, C. A., The relation between the composition of certain ternary two-phase solvent systems and R_M values. *J. Chromatogr.*, 7, 1962, 311–320.
386. Soczewinski, E. and Matysik, G., Two types of R_M-composition relationships in liquid-liquid partition chromatography. *J. Chromatogr.*, 32, 1968, 458–471.
387. Schoenmakers, P. J., Billiet, H. A., and de Galan, L., Systematic study of ternary solvent behaviour in reversed-phase liquid chromatography. *J. Chromatogr.*, 218, 1981, 261–287.
388. Schoenmakers, P. J., Billier, H. A., and de Galan, L., Use of gradient clution for rapid selection of isocratic conditions in reversed-phase liquid chromatography. *J. Chromatogr.*, 205, 1981, 13–30.

389. Hasan, M. H. and Jurs, P. C., Prediction of gas and liquid chromatographic retention indices of polyhalogenated biphenyl. *Anal. Chem.*, 62, 1990, 2318–2323.
390. Mokrosz, J. and Duszynska, B., Topological indices in correlation analysis. Molecular shape indices in modelling of the chromatographic retention parameters. *Quant. Struct. Act. Relat.*, 9, 1990, 33–71.
391. Vuorela, H., Lehtonen, P., and Hiltunen, R., Optimization of the high-performance liquid chromatography of coumarins in *Angelica archangelica* with reference to molecular structure. *J. Chromatogr.*, 507, 1990, 367–380.
392. Osmialowski, K. and Kaliszan, R., Studies of performance of graph theoretical indices in QSAR analysis. *Quant. Struct. Act. Relat.*, 10, 1991, 125–134.
393. Valko, K. and Slegel, P., Chromatographic separation and molecular modelling of triazines with respect to their inhibition of the growth of L1210/R71 cells. *J. Chromatogr.*, 592, 1992, 59–63.
394. Jinno, K., Retention behaviour of large polycyclic aromatic hydrocarbons in reversed-phase liquid chromatography. *Adv. Chromatogr.*, 30, 1989, 123–165.
395. Wise, S. A., Sander, L. C., Lapoyade, R., and Garrigues, P., Anomalous behaviour of selected methyl-substituted polycyclic aromatic hydrocarbons in reversed-phase liquid chromatography. *J. Chromatogr.*, 514, 1990, 111–122.
396. Kaliszan, R., Osmialowski, K., Tomellini, S. A., Hsu, S. H., Fazio, S. D., and Hartwick, R. A.,Non-empirical descriptors of sub-molecular polarity and dispersive interactions in reversed-phase HPLC. *Chromatographia* 20, 1985, 705–708.
397. Kaliszan, K., Osmialowski, K., Tomellini, S. A., Hsu, S. H., Fazio, S. D., and Hartwick, R. A., Quantative retention relationships as a function of mobile and C_{18} stationary phase composition for non-cogeneric solutes. *J. Chromatogr.*, 352, 1986, 141–155.
398. Kaliszan, R. and Osmialowski, K., Correlation between chemical structure of non-congeneric solutes and their retention on polybutadiene coated alumina. *J. Chromatogr.*, 506, 1990, 3–16.
399. Vogel, A. I., *Textbook of Practical Organic Chemistry.* Chaucer, London, 1977, p. 1034.
400. Arenas, R. V. and Folley, J. P., Solvent strength, selectivity and retention mechanism studies on polybutadiene-coated alumina columns in reversed-phase liquid chromatography. *Anal. Chem. Acta*, 246 1991, 113–130.
401. Lu, P., Zou, H. F., and Zhang, Y., Effects of molecular structure parameters a, b, and c in the fundamental retention equation. *J. Chromatogr.*, 509, 1990, 171–187.
402. Zou, H. F., Wang, Q. S., Gao, R. Y., Yang, H. Z., Yang, B. W., Zhang, Y. Z., and Lu, P. C., Prediction of retention of O-ethyl, O-aryl and N-isopropyl phosphoroamidothiates in RP-HPLC from molecular structure parameters. *Chromatographia*, 31 ,1991, 143–146.
403. Kaliszan, R., Kaliszan, A., Noctor, T. G., Purcell, W. P., and Wainer, I. W., Mechanism of retention of benzodiazepines in affinity, reversed-phase and adsorption high-performance liquid chromatography in view of quantative structure retention relationships. *J. Chromatogr.*, 609, 1992, 69–81.
404. Taylor, P. J., in Hansch, C., Sammes, P. G., and Taylor, J. B. (Eds.), *Comprehensive Medical Chemistry*, Vol. 4. Pergamon Press, Oxford, 1990, p. 241.
405. Kaliszan, R., Quantative structure-retention relationships applied to high-performance liquid chromatography. *J. Chromatogr.*, 656, 1993, 417–435.
406. Chen, N., Zhang, Y., and Lu, P., The S index in the retention equation in reversed-phase high-performance liquid chromatography. *J. Chromatogr.*, 603, 1992, 35–42.
407. Smith, R. M. and Burr, C. M., Retention prediction of analytes in reversed-phase high-performance liquid chromatography based on molecular structure. *J. Chromatogr.*, 373, 1986 191–204.
408. El Tayar, N., van de Waterbeemd, N., and Testa, B., Lipophilicity measurements of protonated basic compounds by reversed-phase high-performance liquid chromatography. I. Relationship between capacity factors and methanol concentration in methanol-water eluents. *J. Chromatogr.*, 320, 1985, 293–304.
409. Dolan, J. W., Grant, J. R., and Snyder, L. R., Gradient elution in high-performance liquid chromatography. II. Practical application to reversed-phase systems. *J. Chromatogr.*, 165, 1979, 31–58.
410. Snyder, L. R., Quarry, M. A., and Glajch, J. L., Solvent-strength selectivity in reversed-phase HPLC. *Chromatographia*, 24, 1987, 82–96.
411. Cooper, H. A. and Hurtubise, R. J., Prediction of retention of hydroxyl aromatics in reversed-phase liquid chromatography with slope-intercept relations. *J. Chromatogr.*, 360, 1986, 327–341.
412. Stadalius, M. A., Gold, H. S., and Snyder, L. R., Optimization model for gradient elution separation of peptide mixtures by reversed-phase high-performance liquid chromatography. Verification of retention relationships. *J. Chromatogr.*, 296, 1984, 31–59.
413. Chen, N., Zhang, Y., and Lu, P., Effects of molecular structure on the S index in the retention equation in reversed-phase high-performance liquid chromatography. *J. Chromatogr.*, 606 1992, 1–8.
414. Chen, N., Zhang, Y., and Lu, P., Effects of molecular structure on log k'_w index and linear S- log k'_w index correlation in reversed-phase high-performance liquid chromatography. *J. Chromatogr.*, 633, 1993, 31–41.

415. Park, J. H., Jang, M. D., and Shin, M. J., Solvatochromic hydrogen bond donor acidity of cyclodextrins and reversed-phase liquid chromatographic retention of small molecules on β-cyclodextrin-bonded silica stationary phase. *J. Chromatogr.*, 595, 1992, 45–52.
416. Zou, H., Zhang, Y., and Lu, P., Quantative correlation of parameters log k'_w -S in retention equation in reversed-phase high-performance liquid chromatographic and solvatochromic parameters. *J. Chromatogr.*, 522, 1990, 49–55
417. Kammlet, M. J., Doherty, R. M., Abraham, M. H., Marcus, Y., and Taft, R. W., Linear solvation energy relationships. 46. An improved equation for correlation and prediction of octanol/water partition coefficients of organic nonelectrolytes (including strong hydrogen bond donor solutes. *J. Phys. Chem.*, 92, 1988, 5244–5255.
418. Riley, C. M., Tomlinson, E., and Jeffries, T. M., Functional group behaviour in ion-pair reversed-phase high-performance liquid chromatography using surface-active pairing ions. *J. Chromatogr.*, 185, 1979, 197–224.
419. Hafkenscheid, T. and Tomlinson, E., Estimation of aqueous solubilities of organic non-electrolytes using liquid chromatographic retention data. *J. Chromatogr.*, 218, 1981, 409–425.
420. Unger, S. H., Cook, J. R., and Hollenberg, J. S., Simple procedure for determining octanol-aqueous partition, distribution, and ionization coefficients by reversed-phase high-pressure liquid chromatography. *J. Pharm. Sci.*, 67, 1978, 1364–1367.
421. Altomare, C., Carotti, A., Cellamare, S., and Ferappi, M., Lipophilic measurements of benzenesulfonamide inhibitors of carbonic anhydrase by reversed-phase HPLC. *Int. J. Pharm.*, 56, 1989, 273–281.
422. Gago, F., Alvarez-Builla, J., and Elguero, J., Group contributions to hydrophobicity and elution behaviour of pyrine derivatives in reversed-phase high-performance liquid chromatography. *J. Chromatogr.*, 449, 1988, 95–101.
423. Lurie, L. S. and Allen, A. C., Reversed-phase high-performance liquid chromatographic separation of fentanyl homologues and analogues. Variables affecting hydrophobic group contribution. *J. Chromatogr.*, 292, 1984, 283–294.
424. Smith, R. M., Functional group contributions to the retention of analytes in reversed-phase high-performance liquid chromatography. *J. Chromatogr.*, 656, 1993, 381–415.
425. Smith, R. M. and Burr, C. M., Retention prediction of analytes in reversed-phase high-performance liquid chromatography based on molecular structure. I. Monosubstituted aromatic compounds. *J. Chromatogr.*, 481, 1989, 71–74.
426. Valkó, K. and Slégel, P., New chromatographic hydrophobicity index φ based on the slope and the intercept of the log k' versus organic phase concentration plot. *J. Chromatogr.*, 631, 1993, 49–61.
427. Buszewski, B., Jaroniec, M., and Gilpin, R. K., Influence of eluent composition on retention and selectivity of alkylamide phases under reversed-phase conditions. *J. Chromatogr.*, 668, 1994, 293–299.
428. Gilpin, R. K., Jaroniec, M., and Lin, S., Dependence of the methyl selectivity on the composition of hydroorganic eluents for reversed-phase liquid chromatographic systems with alkyl bonded phases. *Chromatographia*, 30, 1990, 393–399.
429. Gilpin, R. K., Jaroniec, M., and Lin, S., Studies of the surface composition of phenyl and cyanopropyl bonded phases under reversed-phase liquid chromatographic conditions using alkanoate and perfluoroalkanate esters. *Anal. Chem.*, 63, 1991, 2849–2852.
430. Kowalska, T., Physico-chemical modelling of solute retention in reversed-phase HPLC with methanol-water mobile phase. *Chromatographia*, 27, 1989, 628–630.
431. Dimova, N., Kowalska, T., and Dimov, N., Physico-chemical modelling of solute retention in reversed-phase HPLC with phosphate buffer mobile phase. *Chromatographia*, 31, 1991, 600–602.
432. Hammers, W. E., Meurs, G. J., and de Ligny, C. L., Correlation between liquid chromatographic capacity ratio data on Lichrosorb RP-18 and partition coefficients in the octanol-water system. *J. Chromatogr.*, 247, 1982, 1–13.
433. Wells, M. J. M. and Clark, C. R., Investigated of N-alkylbenzamides by reversed-phase liquid chromatography. *J. Chromatogr.*, 235, 1982, 31–41.
434. Schoenmarkers, P. J., Biliet, H. A., and de Galan, L., Influence of organic modifiers on the retention behaviour in reversed-phase liquid chromatography and its consequences for gradient elution. *J. Chromatogr.*, 185, 1979, 179–195.
435. Wells, M. J. M. and Clark, C. R., Study of the relationship between dynamic and static equilibrium methods for the measurement of hydrophobicity. Comparison of capacity factors and partition coefficients for some 5,5-disubstituted barbituric acids. *J. Chromatogr.*, 284, 1984, 319–335.
436. Yamagami, C., Takao, N. and Fujita, T., Hydrophobicity parameters of diazines (1) analysis and prediction of partition coefficients of monosubstituted diazines. *Quant. Stuct. Act. Relat.*, 9, 1990, 313–320.
437. Yamagami, C., Yokota, M., and Takao, N., Hydrogen-bond effects of ester and amide groups in heteroaromatic compounds on the relationship between the capacity factor and the octanol-water partition coefficient. *J. Chromatogr.*, 662, 1994, 49–60.

438. Korakas, D., Valkó, K., Wood, I., Gibbons, W. A., and Toth, I., Structure-retention relationships of diastereomeric mixtures of lipidic amino acid conjugates on reversed-phase stationary phases. *J. Chromatogr.*, 659, 1994, 307–315.
439. Cupid, B. C., Nicholson, J. K., Davis, P., Ruane, R. J., Wilson Glen, R. C., Rose, V. S., Beddell, C. R., and Lindon, J. C., Quantitative structure chromatography relationships in reversed-phase high-performance liquid chromatography: prediction of retention behaviour using theoretically derived molecular properties. *Chromatography*, 37, 1993, 241–249.
440. Park, J. H., Jang, M. D., and Shin, M. J., Solvatochromic hydrogen bond acidity of cyclodextrins and reversed-phase liquid chromatographic retention of small molecules on a β-cyclodextrin-bonded silica stationary phase. *J. Chromatogr.*, 595, 1992, 45–52.
441. Park, J. H., Carr, P. W., Abraham, M. H., Taft, R. W., Doherty, R. M., and Kamlet, M. J., Some observations regarding different retention properties of HPLC stationary phases. *Chromatographia*, 25, 1988, 373–381.
442. Forgács, E. and Cserháti, T., Retention behaviour of some barbituric acid derivatives on polyethylene-coated silica column. *J. Chromatogr.*, 656, 1994, 233–238.
443. Cserháti, T., Use of principal component and cluster analysis for the comparison of reversed-phase HPLC column. *Anal. Lett.*, 27, 1994, 2615–2637.
444. Jandera P., A method for characterization and optimization of reversed-phase liquid chromatographic separation based on retention behaviour in homologous series. *Chromatographia*, 19, 1984, 101–112.
445. Breuer, M. R. P., Claessens, H. A., and Cramers, C. A., Statistical evaluation of the validity of a method for characterizing stationary phases for reversed-phase liquid chromatography based on retention of homologous series. *Chromatographia* 38, 1994, 127–146.
446. Karger, B. L., Snyder, L. R., and Eon, C., Expanded solubility parameter treatment for classification use chromatographic solvents and adsorbents. *Anal. Chem.*, 50, 1978, 2126–2145.
447. Knox, J. H., Kaur, B., and Millward, J. R., Structure and performance of porous graphitic carbon in liquid chromatography. *J. Chromatogr.*, 352, 1986, 3–25.
448. Bassler, B. J. and Hartwick, R. A., The appication of porous graphitic carbon as an HPLC stationary phase. *J. Chromatogr. Sci.*, 27, 1989, 162–174.
449. Gu, G. and Lim, C. K., Separation of anionic and cationic compounds of biomedical interest by high-performance liquid chromatography on porous graphitic carbon. *J. Chromatogr.*, 515, 1990, 183–192.
450. Kiselev, A.V., Adsorbents in gas chromatography. *Adv. Chromatogr.*, 4, 1967, 1113–206.
451. Engelhard, W., Pörschmann, J., and Welsch, T., Graphitized carbon black as a shape-selective stationary phase in GC. *Chromatographia*, 30, 1990, 537–543.
452. Unger, K., Roumeletios, P., Müller, H., and Goetz, H., Novel porous carbon packings in reversed-phase high-performance liquid chromatography. *J. Chromatogr.*, 202, 1980, 3–14.
453. Gilbert, M. T., Knox, J. H., and Kaur, B., Porous glassy carbon, a new columns packing material for gas chromatography and high-perforformance liquid chromatography. *Chromatographia*, 16, 1982, 138–146.
454. Knox, J. H., Unger, K., and Müller, H., Prospects for carbon as packing material in high-performance liquid chromatography. *J. Liquid Chromatogr.*, 6, 1983, 1–32.
455. Kaur, B., The use of porous graphitic carbon in high-performance liquid chromatography. *LC-GC Int.*, 3, 1990, 41–48.
456. Knox, J. H. and Kaur, B., in Brown, P. R. and Hartwick, R. A. (Eds.), *High-Performance Liquid Chromatography*. John Wiley and Sons, New York, 1989, pp. 189–222.
457. Kaliszan, R., Osmialowsky, K., Bassler, B., and Hartwick, R., Mechanism of retention in high-performance liquid chromatography on porous graphitic carbon as revealed by principal component analysis of structural descriptors of solutes. *J. Chromatogr.*, 499, 1990, 333–344.
458. Bassler, B., Kaliszan, R., and Hartwick, R., Retention mechanism on metallic stationary phases. *J. Chromatogr.*, 461, 1989, 139–147.
459. Forgács, E., Valkó, K., and Cserháti, T., Influence of physicochemical parameters of some ring-substituted phenol derivatives on their retention on a porous graphitized carbon column. *J. Liquid Chromatogr.*, 14, 1991, 3457–3473.
460. Forgács, E. and Cserháti, T., High-performance liquid chromatographic retention behaviour of ring-substituted aniline derivatives on a porous graphitized carbon column. *J. Chromatogr.*, 600, 1992, 43–49.
461. Forgács, E. and Cserháti, T., Dependence of the retention of some barbituric acid derivatives on a porous graphitized carbon column on their physicochemical parameters. *J. Pharm. Biomed. Anal.*, 10, 1992, 861–865.
462. Forgács, E., Cserháti, T., and Bordás, B., Application of multivariate mathematical-statistical methods for the comparison of the retention behaviour of porous graphitized carbon and octadecylsilica column. *Anal. Chim. Acta*, 279 1993, 115–122.
463. Forgács, E., Cserháti, T. and Bordás, B., Comparison of the retention behaviour of phenol derivatives on porous graphitized and octadecylsilica column. *Chromatographia*, 36, 1993, 19–26.
464. Berridge, J. C., Analysis tioconazole using high-performance liquid chromatography with porous graphitic carbon column. *J. Chromatogr.*, 449, 1988, 317–321.

465. Forgács, E., Valkó, K., and Cserháti, T., Porous graphitized carbon and octadecylsilica columns in the separation of some monoamine oxidase inhibitory drugs. *J. Chromatogr.*, 631, 1993, 207–213.
466. Cserháti, T. and Forgács, E., Separation of some chlorophenoxyacetic acid congeners on porous graphitized carbon column. *J. Chromatogr.*, 643, 1993, 331–336.
467. Forgács, E. and Cserháti, T., Retention behaviour of some commercial pesticides on a porous graphitized carbon column. *Analyst*, 120, 1995, 1941–1944.
468. Nawrocki, J., Rigney, M. P., McCormick, A., and Carr, P. W., Chemistry of zirconia and its use in chromatography. *J. Chromatogr.*, 657, 1993, 229–282.
469. Mecera, P. D. L., van Ommen, J. G., Doesburg, E. B. M., Burggaaf, A. J., and Ross, J. H. R., Zirconia as a support for catalysists. *Appl. Catal.*, 57, 1990, 127–148.
470. Trüdinger, U., Müller, G., and Unger, K. K., Porous zirconia and titania as packing materials for high-performance liquid chromatography. *J. Chromatogr.*, 535, 1990, 111–125.
471. Homes, H. F., Fuller, E. L., and Gammage, R. B., Heats of immersion in the zirconium oxide-water system. *J. Phys. Chem.*, 76, 1972, 1497–1510.
472. Rigney, M. P., Weber, T. P., and Carr, P. W., Preparation and evaluation of a polymer-coated zirconia reversed-phase chromatographic support. *J. Chromatogr.*, 484, 1989, 273–291.
473. van Graff, M. A. C. G. and Burggraaf, A. J., *Adv. Ceramics*, 12, 1983, 744.
474. Arshady, R., Beaded polymer supports and gels. I. Manufacturing techniques. *J. Chromatogr.*, 586, 1991, 181–197.
475. Kawahara, M., Nakamura, H., and Nakajima, T., Titania and zirconia: possible new ceramic microperticulates for high performance liquid chromatography. *J. Chromatogr.*, 515, 1990, 149–158.
476. Rigney, M. P., Funkenbusch, E. F., and Carr, P. W., Physical and chemical characterization of microporous zirconia. *J. Chromatogr.*, 499, 1990, 291–304.
477. Blackwell, J. A. and Carr, P. W., The role of Lewis acid base processes in ligand-exchange chromatography of benzoic acid derivatives on zirconium oxide. *Anal. Chem.*, 64, 1992, 853–862.
478. Blackwell, J. A., A chromatographic comparison of relative Lewis acidities of alumina and zirconia. *Chromatographia*, 35, 1993, 133–138.
479. Blackwell, J. A. and Carr, P. W., Ion- and ligand-exchange chromatography of proteins using porous zirconium oxide supports in organic and inorganic Lewis base eluents. *J. Chromatogr.*, 596, 1992, 27–41.
480. Blackwell, J. A. and Carr, P. W., Ligand exchange chromatography of free amino acids and proteins on porous microparticulate zirconium oxide. *J. Liquid Chromatogr.*, 15, 1992, 1487–1506.
481. Yu, J. and El Rassi, Z., Reversed-phase liquid chromatography with microsperical octadecyl-zirconia bonded stationary phases. *J. Chromatogr.*, 631, 1993, 91–106.
482. Kolla, P., Köhler, J., and Schomburg, G., Polymer-coated cation exchange stationary phases on the basis of silica. *Chromatographia*, 23, 1987, 465–472.
483. Bien-Vogelsag, U., Deege, A., Figge, H., Köhler, J., and Schomburg, G., Synthesis of stationary phases for reversed-phase LC using silanization and polymer coating. *Chromatographia*, 19, 1984 170–179.
484. Forgács, E. and Cserháti, T., in press.
485. Cserháti, T. and Forgács, E., in press.
486. Funkenbush, E. F., Carr, P. W., Hanggi, D. A., and Weber, T. P., U.S. Patent No. 5108597, 1992.
487. Weber, T. P. and Carr, P. W., Comparison of isomer separation on carbon-clad microporous zirconia and on conventional reversed-phase high-performance liquid chromatography supports. *Anal. Chem.*, 62, 1990, 2620–2625.
488. Weber, T. P., Carr, P. W., and Funkenbusch, E. F., Evaluation of a zirconia-based carbon-polymer composite reversed-phase chromatographic support. *J. Chromatogr.*, 519, 1990, 31–52.
489. Cserháti, T., Retention characteristics of an aluminium oxide HPLC column. *Chromatographia*, 29, 1990, 593–596.
490. Poisson, R., Brunelle, J. P., and Nortier, P., in: Stiles, A. B. (Ed.), *Catalytic Supports, Supported Catalysis*. Butterworth, Boston, 1987, p. 11.
491. Forgács, E. and Cserháti, T., Use of alumina support for the separation of ethoxylated oligomer surfactants according to the length of the ethyleneoxide chain. *Anal. Lett.*, 29(2), 1996, 321–340.
492. Forgács, E. and Cserháti, T., Retention behaviour of tributylphenilethylene oxide oligomers on an alumina high-performance liquid chromatographic column. *J. Chromatogr.*, 661, 1994, 239–243.
493. Haky, J. E., Vemulapalli, S., and Wieserman, L. F., Comparison of octadecyl-bonded alumina and silica for reversed-phase high-performance liquid chromatography. *J. Chromatogr.*, 505, 1990, 307–318.
494. Buchberger, W. and Winsauer, K., Alumina as stationary phase for ion chromatography and column-coupling techniques. *J. Chromatogr.*, 482, 1982, 401–406.
495. Blackwell, J. A., A Chromatographic comparison of the relative Lewis acidities of alumina and zirconia. *Chromatographia*, 35, 1993, 133–138.
496. Nondek, L., Liquid chromatography on chemically bonded electron donors and acceptors. *J. Chromatogr.*, 373, 1986, 61–80.

497. Kimata, K., Hosoya, K., Tanaka, N., Araki, T., and Patterson, D. G., Jr., Preparation of nitrophenylethylsilylated silica gel and its chromatographic properties in the separation of polychlorinated dibenzo-p-dioxins. *J. Chromatogr.*, 595, 1992, 77–88.
498. Antle, P. E., Goldberg, A. P., and Synder, L. P., Characterization of silica-based reversed-phase columns with respect to retention selectivity. Solvophobic effects. *J. Chromatogr.*, 321, 1985, 1–32.
499. De Biasi, V., Lough, W. J., and Evans, M. B., Study of the lipophilicity of organic bases by reversed-phase liquid chromatography with alkaline eluents. *J. Chromatogr.*, 353, 1986, 279–284.
500. Bitteur, S. and Rosset, R., Comparison of octadecyl-bonded silica and styrene-divinylbenzene copolymer sorbents for trace enrichment purposes. *J. Chromatogr.*, 394, 1987, 279–293.
501. Bechalany, A., Rothlisberger, T., El Tayar, N., and Testa, B., Comparison of various nonpolar stationary phases used for assessing lipophilicity. *J. Chromatogr.*, 473, 1989, 115–124.
502. Dawkins, J. V., Gaggot, N., Lloyd, L. L., McConville, J. A., and Warner, F. P., Reversed-phase high-performance liquid chromatography with a C_{18} polyacrylamide-based packing. *J. Chromatogr.*, 452, 1988, 145–156.
503. Yamaguchi, J. and Hanai, T., Selectivity of alkyl-bonded phases in reversed-phase liquid chromatography of hydrophobic compounds. *Chromatographia*, 27, 1989, 371–377.
504. Sentell, K. B., Bornes, K. W., and Dorsey, J. G., Ultrasound driven synthesis of reversed-phase stationary phaes for liquid chromatography using 4-dimethylaminopyridine as acid-acceptor. *J. Chromatogr.*, 455, 1988, 95–104.
505. Ascah, T. L. and Feibush, B., Novel, highly deactivated reversed-phase for basic compounds. *J. Chromatogr.*, 506, 1990, 357–369.
506. Gami-Yilinkou, R. and Kaliszan, R., Determination of hydrophobicity of organic bases on poly(butadiene)-coated alumina and on octadecylsilica columns. *Chromatographia*, 30, 1990, 277–282 .
507. Kaliszan, R., Blain, W. R., and Hartwick, R. A., A new HPLC method of hydrophobicity evaluation employing poly(butadiene)-coated alumina columns. *Chromatographia*, 25, 1988, 5–11.
508. Bien-Vogelsang, U., Deege, A., Figge, H., Kohler, J., and Schomburg, G., Synthesis of stationary phases for reversed-phase LC using silanization and polymer coating. *Chromatographia*, 19, 1984, 170–179.
509. Knox, J. H. and Kaur, B., in Brown, P. B. and Hartwick, R. A. (Eds.), *High-Performance Liquid Chromatography*, Wiley, New York, 1989.
510. Tanaka, N., Tanigawa, T., Kimata, K., Hosoya, K., and Araki, T., Selectivity of carbon packing materials in comparison with octadecylsilyl- and pyrenylethylsilylsilica gels in reversed phase liquid chromatography. *J. Chromatogr.*, 549, 1991, 29–41.
511. Miyake, K., Kitaura, F., Mizuno, N., and Terada, H., Phosphatidylcholine-coated silica as a useful stationary phase for high-performance liquid chromatographic determination of partition coefficients between octanol and water. *J. Chromatogr.*, 389, 1987, 47–56.
512. Pidgeon, C. and Venkatarum, U. V., Immobilized artifical membrane chromatography: supports composed of membrane lipids. *Anal. Biochem.*, 176, 1989, 36–47.
513. Armstrong, D. W., Stalcup, A. W. M., Hilton, M. L., Duncan, J. D., Faulkner, J. R., Jr., and Chang, S. C., Derivatized cyclodextrin for normal-phase liquid chromatographic separation of enantiomers. *Anal. Chem.*, 62, 1990, 1610–1615.
514. Forgács, E. and Cserháti, T., Retention behaviour of barbituric acid derivatives on β-cyclodextrin polymer-coated silica column. *J. Chromatogr.*, 668, 1994, 395–402.
515. Forgács E., Use of principal component analysis for the evaluation of the retention behaviour of monoamine oxidase inhibitory drugs on β-cyclodextrin column. *J. Pharm. Biomed. Anal.*, 13, 1993, 525–532.
516. Cserháti, T. and Forgács, E., Use of multivariate mathematical methods for the evaluation of retention data matrices. *Adv. Chromatogr.*, 36, 1996, 1–63.
517. Kaliszan, R., Quantitative structure-retention relationships, *Anal. Chem.*, 64, 1992, 619A-631A.
518. Mager, H., *Moderne Regressionsanalyse*. Salle, Sauerlander, Frankfurt am Main, 1982, p. 135.
519. Orloci, L., Rao, C. R., and Stitiler, W. M., *Multivariate Methods in Ecological Work*. International Cooperative Publishing House, Fairland, MD, 1979.
520. Benzeri, J. P., *l'Analyse des Donnees, Vol. 2*. Dunod, Paris, 1973.
521. Malinowski, E. R. and Howery, D. C., *Factor Analysis in Chemistry*. Wiley, New York, 1980.
522. Fucart, T., *Analyse Factorielle*. 2nd ed., Masson, Paris, 1985.
523. Morrison, D. F., *Multivariate Statistical Methods*. 2nd ed. McGraw-Hill, New York, 1976.
524. Mardia, K. V., Kent, J. T., and Bibby, J. M., *Multivariate Analysis*. Academic Press, London, 1979.
525. Llinas, J. R. and Ruiz, J. M., in Vernin, G. and Chanon, M. (Eds.), *Computer Aids to Chemistry*. Ellis Horwood, Chichester 1985, p. 200.
526. Schaper, K. J. and Kaliszan, R., in Mutschler, E. and Winterfeldt, E. (Eds.), *Trends in Medicinal Chemistry*. VCH, Weinheim, Berlin, 1987, p. 200.
527. Sammon, J. W., Jr., A nonlinear mapping for data structure analysis. *IEEE Trans. Comput.*, C18, 1969, 401–407.
528. Willett, P., *Similarity and Clustering in Chemical Information System*. Research Studies Press, New York, 1987.

INDEX

A

Acceptor strength, 153
Acetic acid, 44, 45
Acetone, 44, 45
Acetonitrile
    enthalpy of the retention process, 44
    Gibbs free energy, 45
    RPHPLC, 118
Acid-base interactions, 148
Acidity
    of hydrogen bonds, 17, 18, 41
    of silanol groups, 122–123, 131
Acids, higher fatty, 96–100, 107–109
Activity coefficients, 84, 86, 89
    adsorptive thin-layer chromatography, 96, 99
    RPHPLC distribution coefficients and, 149
Adsorption, 3–13, 91
    competitive
        of analytes, 7–10
        of eluent components, 5–7
    HPLC, fundamentals, 116
    interfacial, 20
    molecular basis for separation, thin-layer chromatography, 94–106
    monolayer, 7, 155
    side effects of, and retention behavior, 24–34
    of water in silica hyroxyl groups, 123
Adsorption capacity, 15
Adsorption coeffients, 29, 31
Adsorption-desorption isotherms, 125
Adsorption energy, 96
Adsorptive thin-layer chromatography, 92, 94–106
Aggregates, formed by analyte, 91
Alcohols
    amino, 111–112
    higher fatty, retention indices, 96–100, 107–109
Aldehydes, 48, 49
n-Alkanes, 71, 73, 74
    chemical warfare agents with, 65–66
    retention, 52–54, 62, 63
    retention indices of tetradecynes with, 77
Alkanes, steric selectivity and, 182–183
Alkenes, retention, 50–52
Alkylbenzenes, 194
    linear, retention and chemical structure, 54–55
    unified retention indices, 73
n-Alkylbis(trifluoromethyl)phospine sulfide, 65–66
Alkyl-bonded silica phase, 137–140
n-Alkylcarboxylic acids, 47, 48
n-Alkyl esters, 48, 49
Alkylgermanes, 32, 34–36
Alkylsilanes, 32–36
n-Alkyl-substituted benzenes, 48, 49
Alkyl-substituted phenols, 47, 48
n-Alkynes, retention indices of tetradecynes with, 77
Alumina, 116
    octadecyl-bonded, 202–203
    polybutadiene-coated, 203
    polyethylene-coated, 176
Aluminum oxide-based supports, 195–204
Aniline
    enthalpy of the retention process, 44
    Gibbs free energy, 45
    retention of alumina and, 195
Asahipak ODP-50 polymer gel, 141–143
Azides, 98, 101

B

Basicity, of hydrogen bonds, 17, 18, 41, 49
Benzaldehyde, 44, 45
Benzene
    1,4-disubstituted, 164
    enthalpy of the retention process, 44
    Gibbs free energy, 45
    molecular structural coefficients, 72
    physicochemical parameters, 19
    position of substituents on, 49, 51
Benzene derivatives, 56
Benzenesulfonamides, 164
Benzodiazepine, 159
Benzoic acid esters, 46
Benzoic acids, 164
Benzonitrile, 44, 45
Benzyl alcohol, 44, 45
Binary eluent mixtures, partition coefficient, 8

Boiling point, 47, 50, 59
Bonded phases, 132–133
Bonding density, carbon content and, 126–127
Branching, 55
Brush phase, 132
Bulk phase, 133
n-Butanol, 72
2-Butanone, 44, 45
Butylamine, 44, 45
Butyronitrile, 72

C

Canonical correlation analysis, 185–186, 212
Capacity factors, 7
   alkylsilanes and alkylgermanes, 35–36
   correlation with molecular parameters, 37–46
   ethoxylated nonylphenol derivatives, 198
   NPHPLC, 128
   in thin-layer chromatography, 92–93, 104
Capacity ratio, 78–79
   HPLC, 117
   RPHPLC, 118–119
Carbon
   bonding density and content of, 126–127
   number in molecule, retention and, 49–50, 62
   packing materials, 182–188
Carbon-coated zirconia, 194–195
Carbon tetrachloride, 44, 45
Carrier gas, 15
   flow rate, 28, 65, 66
   molecular parameters and peak width, 20–21
   pressure of, 84–86, 89–90
   viscosity of, 61, 78
Carvacrol, 32, 34
Charge-transfer stationary phases, 204
Chemical methods, hydroxyl group determination, 121
Chemical potential, 5, 6, 95–96
Chemical warfare agents, 64–68
Chemometrics. *See* Canonical correlation analysis; Principal component analysis; Stepwise regression analysis
α-Chloroacetophenone, 65–68
2-Chlorobenzalmalononitrile, 65–68
Chloroform, 30, 95
Chromatographic hydrophobicity index, 166–168
Chromatography. *See also* High-performance liquid chromatography
   adsorption phenomena, 5–13
   liquid, 91–116
   molecular forces in, 10–12
   pH range for, 127
   temperature programmed gas, 59–69
   thin-layer, 92–116
Citronellol, 32, 34
Cluster dendogram, 181, 197, 200, 201
Column liquid chromatography, 91
Column pressure, retention volume and, 84
Columns
   historical background, 120

NPHPLC, 127–128
   parameters of, 15
   prolonged use of, and the S-index, 159
   RPHPLC, 118
Column temperature, 15, 20, 28–29
   capacity ratio and, 78
   S-index and, 160
   in temperature programmed gas chromatography, 59–64
Compressibility factor, 78
Connectivity indices, 54, 55, 68. *See also* Molecular connectivity
m-Cresol, 44, 45
Cyanoalkyl-bonded phases, 142
4-Cyanophenyl herbicides, 113, 115
Cyclic alkenes, 48, 49
Cycloalkanes, steric selectivity and, 182–183
β-Cyclodextrin, 112, 113, 206–210
Cyclohexane
   enthalpy of the retention process, 44
   Gibbs free energy, 45
   physicochemical parameters, 19
Cytochrome-C, 188

D

Dansylated amino acids, 108, 110, 111, 193, 194
Dead time, 39, 78
   chromatographic index and, 166
   RPHPLC, 119
Dead volume, 119
Decane
   enthalpy of the retention process, 44
   Gibbs free energy, 45
   physicochemical parameters, 19
Dendograms, cluster, 181, 187, 200, 201
Density, of solvent, 104
Desorption, of water in silica hyroxyl groups, 123
Develosil, 130
Devosil, 131
α,ω-Diamino oligo-ethers, 87–88
Diazines, 98, 101
Dibenz[b,f]-1,4-oxazepin, 65–68
Dibutyl ether, 44, 45
Dichlorosilane, 132–133
Dielectric constant, 10
Dielectric interactions, 116
Diethyl benzene, 19
Diethyl ether, 44, 45
Diffusion coefficient, 25, 26
Dilution coefficient, 79, 81, 82, 84
Dimethylacetamide, 44, 45
Dimethyl aniline, 72
N,N-Dimethylaniline, 44, 45
Dimethylformamide, 44, 45
Dimethylsufoxide, 44, 45
3,5-Dinitrobenzoic acid esters, 112–115
Dioxan, 19
1,4-Dioxane, 72
Dipalmitoyl-phosphatidylcholine, 205
Dipolarity, 43

Dipolarity/polarizability parameter, 41, 49, 104, 153
Dipole-dipole interactions, 58, 59, 160
Dipole forces, 116
Dipole-induced dipole interactions, 160
Dipole moment, 10, 47, 56–58, 158
Dipropyl ether, 44, 45
Direct (normal) phase high-performance liquid chromatography, 120–132
Dispersion forces/interactions, 11–12, 43, 116, 148, 160
Displacement model, 155
Distribution coefficients, 135, 136, 153
1,4-Disubstituted benzenes, 164
N,N-Disubstituted dithiocarbamates, 102
Dodecane
   molecular structural coefficients, 71, 72
   physicochemical parameters, 19
   retention polarities, 74
   specific retention volume, 53–54, 74
Dyes, sulfonated, 114, 116

E

Effective charge, 153
Electronegativity, 153
Electron polarizability, 102
Electron withdrawing power, 58, 61
Electrostatic forces, 9
Eluotropic strength, of organic solvents, 184
End-capping, 130
Enthalpy, 46, 62
   relationship with entropy, RPHPLC, 119–120
   of solution, 43, 44. *See also* Solvation enthalpy
Entropy, 46, 61, 62, 119–120
Environmental pollutants
   environmental fate of xenobiotics, 79
   herbicides, 113, 115
   organochlorines, 62–64
   pesticides, 188
   retention indices, 48
Equipment, 15. *See also* Columns; Packing materials
Ethanol, 44, 45
Ethers, 44, 45, 87–88, 96–100
Ethyl acetate
   enthalpy of the retention process, 44
   Gibbs free energy, 45
   heat capacity, 94–95
   molecular structural coefficients, 72
Ethylamine, 44, 45
Ethylbenzene, 44, 45
15 O-Ethyl-O-aryl-N-isopropyl phosphoroamideothioate derivatives, 103–104
(O-Ethyl S-2-diisopropylamino)ethyl methylphosphonothioate, 65–68

F

Factor analysis, 212–213. *See also* Principal component analysis
F factors, 73, 74

Film thickness, 31
Flavonoid aglycones, 36
Fragmental contribution, 149, 160
Fragrance compounds, 57
Free energy of association, 9, 43. *See also* Gibbs free energy; Linear free energy relationships
Functional group contribution, 164–166

G

Gas. *See* Carrier gas
Gas chromatography, 15–90
Gas flow, 28, 65, 66
Gas-liquid chromatography (GLC), 15
   adsorption phenomena, 3–5
   adsorption side effects and retention behavior, 24–34
   models of retention and separation, 17–23
Gas-solid chromatography (GSC), 15
GCB (graphitized carbon black), 184
Geometrical descriptors, 61
Geraniol, 32, 34
Gibbs free energy, 5, 43, 45, 78. *See also* Free energy of association
Glass support, 183–188
GLC. *See* Gas-liquid chromatography
Graphitized carbon black (GCB), 184
Graphitized carbon packing materials, 182–188
GSC (gas-solid chromatography), 15

H

Halogens
   position on benzene ring, 49, 51
   in retention of polyhalogenated biphenyls, 22–24
Hammett's constant, 113, 185, 196
Hansch-Fujita substituent constant, 112, 185, 196
Headspace method, 79–81
Heat capacities, 94–95
Heat of vaporization, 89
Helium gas, 15
Henry's law constant, 25, 27, 79–81, 86
Herbicides, 113, 115
Heterocyclic amines, 108
n-Hexadecane, partition coefficients, 40
Hexafluoroisopropanol, 44, 45
Hexane, 44, 45
1-Hexene, 44, 45
Higher fatty alcohols, 96–100, 107–109
High-performance liquid chromatography (HPLC), 116–210
   direct (normal)-phase, 120–132
   equations, 6
   fundamentals, 116–120
   reversed-phase, 118, 128, 132–181
Hildebrand-Schatchard interaction parameter, 87
Hildebrand solubility parameter theory, 143
cis-Hydrindane, 72
Hydrocarbonaceous silica, 205

Hydrocarbons. *See also* Polycyclic aromatic hydrocarbons
  brominated, 86
  enthalpy of the retention process, 44
  Gibbs free energy, 45
  interfacial adsorption, 20
  retention, 59, 61
  unified retention indices, 73, 75
  unsaturated, 73, 77
Hydrogen bond acidity, 17, 18, 41
Hydrogen bond basicity, 17, 18, 41, 49
Hydrogen bond energy, 158
Hydrogen bonds, 11, 43, 88
  between carbonyl and hydroxyl groups, 96–100
  HPLC and, 116
  indicator variables, 151, 171
  retention time and, 37
  retention with two-component mobile phase and, 95
  role in retention strength and retention selectivity, 102
  RPHPLC and, 169
  RPTLC and, 106, 107, 108
  in silica hyroxyl groups, 123
  in total solubility factor, 50
Hydrogen gas, 15
Hydrophilic forces, 84, 102
Hydrophobic interactions, 13, 50
  to reduce the hydrophobic surface area, 137
  RPHPLC, 170
Hydrophobicity, 39, 56, 82, 92, 103
  of carbon packing materials, 182
  concept of, in chromatography, 150–151
  of 4-cyanophenyl herbicides, 113, 115
  determination of, with new RPHPLC phases, 204–206
  with molecular structure parameters, retention and, 110–112
  in RPTLC, 106
  S-index and, 162
Hydrophobicity index, 166–168
Hydrophobic ligand, 108
Hydroxyl groups, 96–100, 121
Hypothetical retardation coefficients, 170

I

Immobilized artificial membranes, 205–206
Indicator variables, 151, 171
Induced dipole-induced dipole interactions, 11–13
Infinite dilution coefficients, 79, 81, 82, 84
Infrared spectroscopy, 121
Injectors, 15
Interaction energy, 5–6
Interaction indices, RPHPLC, correlation with retention and selection, 143–146, 178
Inverse gas chromatography, 28, 87
1-Iodobutane, 72
Ion exchange distribution coefficient, 129
Ionic interactions, 129–131
Ion-induced dipole interaction, 10–11
Ion-ion interactions, 10
Ion-pair formation, 10
Ion-pair reversed-phase thin-layer chromatography, 114, 116
Ion-permanent dipole interactions, 10
Isobutylacetate, 19
Isocratic separation mode, 47–59
Isotopic exchange methods, hydroxyl group determination, 121

K

Ketones
  retention indices, 96–100
  solvent strength, 104
Kováts index, 51, 69
  correlation with molecular parameters, 47–68
  correlation with vapor pressure, 89, 90
  isocratic separation mode, 47–59
  n-decane, 74
  relationship with characteristics of analyte, 54
  temperature programming and, 59–69
Kromasil, 130, 131

L

Length squared per time, 21
Lennard-Jones potential, 12
Leo hydrophobic constants, 144
LiChrospher Zorbax SIL, 130
Ligands
  alkyl chain, on silica surface, 130
  hydrophobic, 108
Linear free energy relationships, 146
Linear solvation energy relationships (LSER), 149, 162
Lipophilic indices, RPHPLC, correlation with retention and selection, 143–146
Lipophilicity, 82, 92
  of nonionic surfactants, 112
  RPTLC, 106
Liquid chromatography, 5–13, 91–116. *See also* High-performance liquid chromatography
London dispersion forces, 11–12
LSER (linear solvation energy relationships), 149, 162

M

Mass-transfer coefficient, 26
Mechanical strength, 127
Menthol, 105–106
Metal chelates, correlation analysis in LC of, 152–155
Metal increment, 153
Metal ions
  effect on silica, 130–131
  in zirconia, 191
Metal-oxide based supports, 188–204
Methanol
  enthalpy of the retention process, 44
  Gibbs free energy, 45

NPHPLC, 128
   polar group selectivity and, 139
   RPHPLC, 118
N-Methylaniline, 44, 45
2-Methyl pentanol-1, 19
2-Methyl-2-pentanol, 72
3-Methyl pentanol-3, 19
2-Methyl-2-propanol, 44, 45
Migration distance, of solvent in thin-layer chromatography, 93
Migration velocity, 3
Miscibility, 117
Mobile phase, 3, 91
   effects on solute retention, 119
   effects on steric selectivity, 142
   in solvophobic theory, 144
   two-component, 95
Modeling
   adsorption side effects and retention behavior, 24–27
   capacity factor and retention time, 39, 40
   distribution of a solute, 143
   GLC retention and separation, 17–23
   liquid chromatography, 91
   multiparametric retention, 152, 154
   partition-displacement, 169
   retention, 60
      multiparameteric, 152, 154
      validation of, 155
   retention capacity in RPTLC, 106–107
   retention indices, 50–51
   separation, 17–23, 51, 91
   thermodynamic, 98–99
Molar refraction, 41, 102, 112, 113, 185
Molar volume, 50, 104, 153
Molecular connectivity, 32, 143, 153. *See also* Connectivity indices
Molecular orbital methods, 173–174
Molecular refractivity, 158
Molecular structural coefficients
   analytes on OV-105 and QF-1, 72
   dodecane, 69, 71
Monoamine oxidase inhibitory drugs, 207–209, 210
Monolayer adsorption, 7
Multiparametric retention models, 152, 154
Multivariate regression analysis, 52, 55
Mustard, 65–68
Myoglobin, 188

N

Neighboring peak effect, 29
Net retention volume, 3–5, 16
   in GLC, 19–20
   n-alkane, 52
   PEG 20M and, 30
   surface inhomogeneity and, 6–7
Nitroaromatic stationary phase, 204
Nitroethane, 44, 45
Nitrogen gas, 15
Nitromethane, 44, 45

Nitropropane, 44, 45
1-Nitropropane, 72
Nonylphenyl ethylene oxide oligomers, 196, 197
Normal-phase high-performance liquid chromatography (NPHPLC), 120–132
Nuclear magnetic resonance (NMR), 121, 130
Nucleosil, 130

O

Octadecyl-bonded alumina, 202–203
Octadecyl-bonded silica, 203, 204
Octadecyl-coated zirconia, 194
Octadecyl-modified polystyrene divinylbenzene, 203
Octane, 44, 45
n-Octane, 30
n-Octanol, 72
Octanol-water partition system, 150
   correlation with RPHPLC retention parameters, 151–152
   log P values of, 170
2-Octyne, 72
Oligomeric phase, 132–133
Oligosaccharides, 193, 194
Organochlorines, 62–64
Orientation interactions, 148
Ostwald absorption coefficient, 17
Ostwald dilution law, 169
Ovoalbumin, 188
Oxazoline derivatives, 151

P

Packing materials. *See also* Alumina; Silica
   aluminum oxide-based supports, 195–204
   graphitized carbon, 182–188
   metal-oxide based, 188–204
   polymer-based, 140–143
Particle diameter, 93, 118
   silica, 123–124
   zirconia, 190–191
Partition coefficient, 7–8, 10, 29
   capacity ratio and column temperature, 78
   environmental fate and, 79
   gas-liquid, on hexadecane, 40, 41, 49
   gas-water, of nonelectrolytes, 82–84, 86
   PEG 20M and, 31
   in two-component eluent, 102
   water-hexadecane, 82, 84, 85
   zero pressure, 85
Partition-displacement model, 169
Partitioning, 3
   adsorptive thin-layer chromatography, 94–95
   GLC and, 16, 19, 20, 37–39
Peak shape, 27. *See also* Neighboring peak effect
Peak width, 20–22, 64
   in thin-layer chromatography, 93
   with unmodified silica columns and totally organic solvents, 127

PEG 20M, 27–28
  partition coefficient and, 31
  retention volume and, 30
  with tetradecenes and tetradecynes, 74, 76–77
Pentadecane, 44, 45
Pentane, 44, 45
Pentanone-2, 19
2-Pentanone
  enthalpy of the retention process, 44
  Gibbs free energy, 45
  molecular structural coefficients, 72
Peptides, 193, 194
Permanent dipole-induced dipole interactions, 11, 13
Permanent dipole-permanent dipole interactions, 11, 13
Permeability constant, 93
Pesticides, retention behavior, 188
PGC (porous graphitized carbon support), 183–188
pH
  range for chromatography, 127
  retention and, 129
PHB (polyhalogenated biphenyls), 22–24
Phenols, 44, 45, 47, 48
Phenylalcohol homologous series, 194
Phenylazapurines, 164
Phenyltriazines, 164
Physicochemical parameters
  aniline derivatives, 195–196
  cluster dendogram of, 181, 197, 200, 201
  meaning of QSRR equations and, 155–158
  relationship with isocratic separation mode, 47–54
  relationship with retention indices, 41–43, 58, 60
    applications, 79–90, 166–177
  thin-layer chromatography, 98, 101, 103
trans-Pinocarveol, 32, 34
Planar liquid chromatography, 91
PLRP-S 300 polymer gel, 142, 143
Polar group selectivity, 139, 142–143
Polar interactions, 94
Polarity, 56, 117
  silanol groups and, 125–126, 138
  submolecular, 157
Polarizability, 10–12, 37
  dipolarity/polarizability parameter, 41, 49, 104, 153
  electron, 102
  molecular, 47
Polar structural indices, RPHPLC, 143–146
Polybutadiene-coated alumina, 203
Polybutadiene-coated zirconia, 193
Polycyclic aromatic hydrocarbons (PAH), 47, 68
  partition coefficient and, 102
  retention indices, 96–100
  retention in RPTLC, 107–108
  retention on stationary phase, 138
  separation of, with zirconia, 193, 194
Polyethylene-coated alumina, 176
Polyethylene-coated silica, 176
Polyethylene-coated zirconia, 193–194
Poly(ethylene glycol). See PEG 20M
Polyhalogenated biphenyls (PHB), 22–24
Polymer-based packing materials, 140–143

Polymer-coated carbon-clad zirconia, 194
Polymeric bonded phase, 133
Polynuclear aromatic hydrocarbons. See Polycyclic aromatic hydrocarbons
Polyoxyethylene sorbitol strearate, 188
Polystyrene-coated zirconia, 193
Poly(styrene-dininylbenzene), 205
Pore diameter, surface area and, 126
Pore size, packing and, 127
Pore volume, 126, 190
Porosity
  effect of polymer biporous structure on chromatographic properties, 142
  of particles, 26
  relationship with surface area, 125
  silica, 124–125
Porous glass, 116
Porous graphitized carbon support (PGC), 183–188
Potential energy, 147–148
Primary alkenes, 48, 49
Principal component analysis (PCA), 176, 179, 197, 199, 209, 210, 212–213
PRISMA model, 94
Propanol, 44, 45, 128
Propionaldehyde, 44, 45
Propionitrile, 44, 45, 72
Propyl acetate, 44, 45
Propylamine, 44, 45
Propylbenzene, 44, 45
Proteins, separation of, 188, 192–194
Pyrazine derivatives, 151
2-(1-Pyrenyl)ethylsilated silica, 182, 183
Pyridines
  3- and 4-substituted, 164
  molecular structural coefficients, 72
Pyridol[1,2-α]pyrimidin derivatives, 55
Pyridol[1,2-α]pyrimidin-4-one, 55

Q

Quantitative structure activity relationships (QSAR), 147, 185
Quantitative structure-retention relationships (QSRR), 146–151
  physicochemical meaning of, 155–158
  statistical theory and, 149
Quantum chemical descriptors, 157
Quinoline derivatives, 101, 102

R

References, 215–233
Refraction, 49
Refractive index, 47
Refractivity, molecular, 158
Regression analysis, 103–104. See also Multivariate regression analysis; Stepwise regression analysis

Remoxipride, 187
Resolution, in thin-layer chromatography, 94
Response factor, 37
Retardation coefficients, 170
Retention
    determination of, 118–120
    ion exchange character of, 129–131
    metal chelates, 152–155
    pesticides, 188
    RPHPLC, 118, 143–146, 149–150
        correlation with 1-octanol-water partition, 151–152
        functional group contribution, 164–166
        porous graphitized carbon column, 185–186
        quantitative structure-retention relationships, 146–150
        S-index in, 158–163
    RPLC, 134–135
    RPTLC, 106–116
    silicas, 128
    with two-component mobile phase, 95–96
Retention indices, 16–18, 49, 60. *See also* Kováts index
    acids, 96–100
    adsorptive thin-layer chromatography, 94
    n-alkylcarboxylic acids, 48
    alkyl-substituted phenols, 48
    chemical warfare agents, 66
    correlation with molecular parameters, 32–79
    environmental pollutants, 48
    ethers, 96–100
    higher fatty alcohols, 96–100
    ketones, 96–100
    n-alkanes, 62, 63
    organochlorines, 63, 64
    physicochemical parameters and, 41–43, 58, 60, 79–90
    polycyclic aromatic hydrocarbons, 96–100
    prediction of, SPARC program, 50–51
    Takács, 68–73
    tetradecenes, 76, 77
    tetradecynes, 77
    unified, 73–79
Retention of analytes. *See also* Net retention volume; Retention indices
    adsorption and, 3
    calculated parameters, 19
    influence of adsorption side effects on, 24–34
    interfacial adsorption and, 20
    modeling, 17–23
    nonpolar, 18
Retention of solutes, in TLC and RPTLC, 92
Retention polarities, 73, 74
Retention selectivity, 102
Retention strength, 102, 176, 178, 183–184
Retention time, 27, 28, 60–62
    capacity factor and, 38–39, 119
    chemical warfare agents, 68
    chromatographic index and, 166
    correlation with molecular parameters, 32–37, 173
    flavonoid aglycones, 36

Retention volume. *See also* Net retention volume; Specific retention volume
    bonded phases and, 133
    inlet column pressure and, 84
    n-alkanes, 71, 73
Reversed-phase high-performance liquid chromatography (RPHPLC), 118, 128, 132–181
Reversed-phase thin-layer chromatography (RPTLC), 92
    ion-pair, 114, 116
    molecular basis of separation, 106–116

S

Salt-bridge, 10
Salting-out effect, 108–111
Sarin, 65–68
Secondary ion mass spectrometry (SIMS), 130
Selectivity. *See also* Steric selectivity
    of polar groups, effect of polymer structure on, 139, 142–143
    of polymer-based packing materials, 140
    of retention, 102
    shape, 142
    solvent, in HPLC, 117
    between two stationary phases, 137
Semiempirical molecular orbital methods, 173–174
Separation
    isocratic separation mode, 47–59
    modeling, 17–23, 51, 91
    of proteins, 188, 192, 193
    RPHPLC, 143–146, 188
    RPTLC, molecular basis for, 106–116
    TLC, molecular basis for, 94–106
Separation distance, 104
Separation efficiency, 27, 28
Separation number, 93
Shodex DE-613 polymer gel, 140–141, 143
Silanization, 193
Silanol groups
    acidity, 122–123, 131
    polarity and, 125–126
    reduction of the effects of, 129
    structure, 120–121
Silanophilic interaction, 128–129
Silica, 116, 118
    alkyl-bonded, 137–140, 176, 177
    β-CD-bonded, 174–175
    β-cyclodextrin-coated, 206–210
    chemical properties, 122–132
    hydrocarbonaceous, 205
    octadecyl-bonded, 203, 204
    packing *vs.* polymer gels, 143
    polyethylene-coated, 176
    problems with, in NPHPLC, 127–128
    purity of, 129–131
    2-(1-pyrenyl)ethylsilated, 182, 183
    retention capacity, 176–177
    reversed-phase character of, 132

supports, 120–121
   types, 128
Silicic acid, 122
Siloxane groups, 132
SIMS (secondary ion mass spectrometry), 130
Snyder's polarity indices, 144
Soczewinski-Wachtmeister relationship, 150
Solubility, 87
   parameters of, 148, 155
   of silica, 123
   of sparingly soluble components, 79
   total solubility parameter, 49–50, 52
Solute polarity parameter, 117
Solvation enthalpy, 41. *See also* Enthalpy of solution
Solvatochromic parameters, 162, 163
Solvent miscibility, 117
Solvent polarity, 117, 154
Solvents
   in HPLC, fundamentals, 116–117
   organic, eluotropic strength of, 184
   in RPLC, stationary phase interactions, 133–137
Solvent selectivity, HPLC, 117
Solvent strength, 104, 117, 153
Solvophobic theory, 137, 143, 144
Soman, 65–68
SPARC computer program, 50–51
Specific retention volume, 17, 69, 72, 73
   dodecane, 53–54, 74
   n-alkanes, 74
   vapor pressure and, 89
Spectral mapping technique, 176
Spectroscopic methods, hydroxyl group determination, 121
Square of the dipole moment, 158
Stability, zirconia, 191–192
Stability constant, 153
Static headspace method, 79–81
Stationary phase, 3, 20
   effects in RPLC, 135–137
   HPLC, 116
   mechanical strength of, 127
   new materials, RPHPLC, 204–210
   OV-1, 78
   OV-17, 55, 56
   OV-101, 52, 57–58, 60
   OV-105, 69, 71
   SE-30, 90
   selectivity between two, 137
   solvent interactions in RPLC, 133–137
Statistical thermodynamic theory, 155
Statistics. *See also* Principal component analysis; Stepwise regression analysis
   canonical correlation analysis, 185–186, 212
   of QSRR, 149
Stepwise regression analysis, 32, 56, 211–212
   β-cyclodextrin-coated silica column, 206–208
   retention of alkenes, 52
   retention of aniline derivatives and physicochemical parameters, 195–196
   retention of surfactants, 112
   salting-out effect, 111

Sterical characteristics, 105, 112, 138–140
Steric selectivity
   of alkyl-bonded phase, 138–140
   of carbon packing materials, 182–183
   effect of mobile phase on, 142
   of polymer gels, 140–142
Structure-retention relationships, quantitative, 146–150, 152
Submolecular polarity parameter, 157
Sulfonated dyes, 114, 116
Surface area, 3, 15, 59, 104. *See also* Adsorption phenomena
   pore diameter and, 126
   pore volume and, 126
   relationship with porosity, 125
   silica, 124
   zirconia, 190
Surface tension, 4
Surfactants
   cluster dendogram, 181
   nonionic, retention, 110, 112, 176–179, 198, 202
Swain-Lupton electronic parameter, 112–113, 185, 196

T

Tabun, 65–68
Taft's constant, 185
Takács retention index, correlation with molecular parameters, 68–73
Temperature. *See* Column temperature
Temperature programmed gas chromatography, 59–69
Terachloromethane, 19
Tetrabutylammonium bromide, 114, 116
Tetradecane, 44, 45
n-Tetradecenes, 74, 76, 77
n-Tetradecynes, 74, 77
Tetrahydrofuran
   enthalpy of the retention process, 44
   Gibbs free energy, 45
   heat capacity, 94–95
   physicochemical parameters, 19
   polar group selectivity and, 139
Tetrahydrofuran-water solvent system, 174
Theoretical plates, 93
Thermodynamic parameters, 62, 63, 159
Thin-layer chromatography (TLC), 92–116
Thujol derivatives, 105–106
Thymol, 32, 34
Thyroglobulin, 188
TLC (thin-layer chromatography), 92–116
Toluene
   enthalpy of the retention process, 44
   Gibbs free energy, 45
   molecular structural coefficients, 72
Topological indices
   adsorptive thin-layer chromatography, 104–106
   hydrocarbons, 57, 61
   retention behavior and, 32, 35, 36, 57
Total solubility parameter, 49–50, 52

Total topological indexes (TTV), 32, 35, 36
Transferrin, 188
Tributylphenol ethyleneoxide oligomer surfactants, 198–199, 202, 203
Trichlorosilane, 133
Triethylamine, 44, 45
Trifluoroethanol, 44, 45
Trimethylsilyl group, 129, 130
Trioxepane, 19
TSK C_{15}-4PW polymer gel, 142, 143
TTV (total topological indexes), 32, 35, 36
Tween 80, 188
Two-dimensional nonlinear map, 176, 180, 197, 200, 201, 209, 210

U

Undecane, 44, 45
Unified retention indices, correlation with molecular parameters, 73–79
Unsaturation, 55

V

Valeraldehyde, 19
Valeronitrile, 72
Validation, of retention models, 155
van der Waals interactions, 9, 11–13, 160
van der Waals volume, 47, 56–58, 158
Vapor pressure, 88–90
cis-Verbenol, 32, 34
Virial coefficients, 85–86
Viscosity of carrier gas, 61, 78

X

Xenobiotics, 79
p-Xylene, 44, 45

Z

Zirconia, chemistry, 188–191
Zorbax RX SIL, 128, 130